# The Origins of Language Revisited

Nobuo Masataka
Editor

# The Origins of Language Revisited

## Differentiation from Music and the Emergence of Neurodiversity and Autism

Springer

*Editor*
Nobuo Masataka
Primate Research Institute
Kyoto University
Inuyama, Aichi, Japan

ISBN 978-981-15-4249-7 ISBN 978-981-15-4250-3 (eBook)
https://doi.org/10.1007/978-981-15-4250-3



This Springer imprint is published by the registered company Springer Nature Singapore Pte Ltd.
The registered company address is: 152 Beach Road, #21-01/04 Gateway East, Singapore 189721, Singapore

# Preface

This book summarizes the advancement of the research and the consideration about the process of evolution and differentiation of language on the basis of empirical evidence after the publication of *The Origins of Language* published by Springer in 2008.

There is a general consensus that language gives the human mind the culturally evolved means to differentiate reality in great detail. The chapters about nonhuman primate communication included in the present book together reveal the fact that the evolution of language required the neural rewiring of circuits that controlled vocalization. The vocal tract muscles in nonhuman primates are controlled from an old emotional center. Nevertheless, the emergence of voluntary control over vocalization is recognized. Based upon this capability, humans are enabled to possess a remarkable degree of voluntary control over their voice. Correspondingly, conceptual and emotional systems, though limited, are differentiated in nonhuman primates as in humans. It is noteworthy that to accomplish such function, the rudimentary form of syntax (regularity of call sequences) has emerged in nonhuman primates.

In humans, readers will be known that emotional evaluations have totally separated from concept representation and from behavior (e.g., when sitting around a table and discussing snakes, humans do not jump on the table uncontrollably in fear, every time "snakes" are mentioned). Differentiation of psyche states with a significant degree of voluntary control over each part gradually evolved along with language and brain rewiring. Consequently, language contributed not only to differentiation of our conceptual ability but also to differentiation of psychic functions of concepts, emotion, and behavior. This differentiation destroyed the primordial synthesis of the psyche. With the evolution of language, the human psyche started to lose its cohesiveness, its entity. While every piece of conceptual knowledge is inextricably associated with the emotional evaluation of a situation as well as with the appropriate behavior for satisfying instinctual needs in nonhuman primates, this is not for the case for humans. Most of the knowledge that exists in culture and is expressed in language is not associated emotionally with human instinctual needs.

This is tremendously advantageous for the development of conceptual culture for science and technology.

Loss of synthesis, on the other hand, leads to internal crises and may cause clinical depression that would require something that enables humans to somehow restore the synthesis. That is the function of human music, which is some aspect of the rudimentary form of language that has been discarded during its further evolution. This book also reveals the process by which music has become differentiated from language, though the prelinguistic system found in nonhuman primates, which could probably be referred to as "prosodic protolanguage," provided a precursor for both modern language and music.

Inuyama, Aichi, Japan Nobuo Masataka

# Contents

# About the Editor

**Nobuo Masataka** has been a Professor at Kyoto University's Primate Research Institute since 2003. His doctoral dissertation was a study on vocal communication in New World primates. Particularly, he specialized in playback experiments, using synthesized sounds in order to investigate vocal perception in monkeys. Thereafter, he extended his research interests to include human infants and arrived at several intriguing findings on preverbal infants' vocal interactions with their mothers. These findings are summarized in his book Onset of Language, published by Cambridge University Press in 2003. In 2008, he edited The Origins of Language, Unravelling Evolutionary Forces, published by Springer. The present work is essentially the second edition of that book.

# Chapter 1
# Empirical Evidence for the Claim That the Vocal Theory of Language Origins and the Gestural Theory of Language Origins Are Not Incompatible with One Another

**Nobuo Masataka**

**Abstract** The author argues about the implications of the evolution of motherese for the emergence of language in the human history, and it occurred in both the vocal mode and the manual mode, the fact indicating that the gestural theory of and the vocal theory of language origins are not incompatible with one another. It is a commonplace observation that hearing adults tend to modify their speech in an unusual and characteristic fashion when they address infants and young children. The available data indicate that motherese or infant-directed speech is a prevalent form of language input to hearing infants and that its salience for preverbal infants results both from the infant's attentional responsiveness to certain sounds more readily than others and from the infant's affective responsiveness to certain attributes of the auditory signal. In the signing behavior of deaf mothers when communicating with their deaf infants, a phenomenon quite analogous to motherese in maternal speech is observed. Concerning the aspect of linguistic input, moreover, there is evidence for the presence of predispositional preparedness in human infants to detect motherese characteristics equally in the manual mode and in the vocal mode. Such cognitive preparedness, in fact, serves as a basis on which sign language learning proceeds in deaf infants. One can seek its evolutionary origins in the rudimentary form of teaching behavior that occurs in the adult–infant interaction in nonhuman primates as well as in humans, by which the cross-generational transmission of parenting is made possible, including that in the deaf community.

**Keywords** Motherese · Signed language · Language learning · Infant-directed signing · Infant-directed speech

---

N. Masataka (✉)
Primate Research Institute, Kyoto University, Inuyama, Aichi, Japan



N. Masataka (ed.), *The Origins of Language Revisited*,
https://doi.org/10.1007/978-981-15-4250-3_1

## 1.1 Importance of Motherese when Considering Language Evolution

It is well known that there have been controversies about the origin of language and that some scientists, including Charles Darwin, sought it in animal vocalizations whereas others including Wilhelm Wundt considered manual gestures as the origin. Here. The author argues from the ontogenetic perspective the implications of the evolution of motherese for the emergence of language in the human history, and that it occurred in both the vocal mode and the manual mode, the fact indicating that the gestural theory of and the vocal theory of language origins are not incompatible with each other.

Developmentally, the earliest non-cry sounds produced by human infants do not transmit meanings, unlike words uttered by older individuals, but rather reveal fitness or express states. On the one hand, we make no assumption that very young infants intend to communicate with such sounds. On the other hand, adults who receive the sounds are provided with broad reception/interpretation capabilities that allow them to provide differentiated feedback to signals, as noted by Masataka (2003). This assumption is in accordance with results in animal communication indicating nonhuman primates are far more flexible in recognition/reaction to signals than in signal production or functional usage. Flexible responsivity of receivers is a necessary condition for their being able to apply selection pressure on volubility and flexibility of infant vocal tendencies (Oller et al. 2016). Such reasoning could lead us to notice the importance of considering how selection pressures might engender increase in the tendency of an infant to produce spontaneous non-cry vocalization for arguing language evolution, and that in fact, evolution of increased infant spontaneous vocalizations began with developmental steps in individual infants under the selective pressure of their own caregivers.

Indeed, recent researches have revealed the fact that during modern human development, infant vocal capabilities emerge at least partly in response to social interaction, where caregivers react to vocal capabilities of infants in accordance with a scaffolding principle requiring parental discernment and intuitive parenting to reinforce vocal exploration and learning. Both endogenous inclination of infants to explore the vocal space and interactive feedback from caregivers thus foster growth in vocal capability.

If this assumption about selection pressures for spontaneous vocalizations is valid, it follows that hominin parents would also have been selected to become aware of the fitness reflected in infant vocalizations and capable of responding to those indicators with selective care and reinforcement of vocalizations. And the author believes this to have been illustrated so far in the phenomenon that is called "motherese."

## 1.2 Implications of Infant-Directed Speech or Motherese in the Vocal Mode for Spoken Language Learning

Although the fact that hearing adults tend to speak to infants in an odd and characteristic fashion has been a commonplace observation, it was Ferguson (1964) who first offered a coherent description of the linguistic and the paralinguistic features of child-directed speech. In the languages investigated by Ferguson, which included English, Spanish, Arabic, Comanche, Giyak, and Marathi, the use of elevated pitch and exaggerated pitch excursions were the most prominent characteristics observed across cultures. Since then, this sort of speech style has been the focus of considerable research, and such speech is commonly referred to as "motherese" or infant-directed speech. Most studies included the claim that hearing adults speak to children in a high-pitched, even "squeaky" voice (Fig. 1.1).

The data that are available to date all indicate that prosodic properties characterizing infant-directed speech and infant-directed singing are a key component of language input to preverbal infants and that they serve as important social and attentional features in early development (Saint-Georges et al. 2013). This fact has led several researchers to explore the possibility that the effectiveness of exaggerated

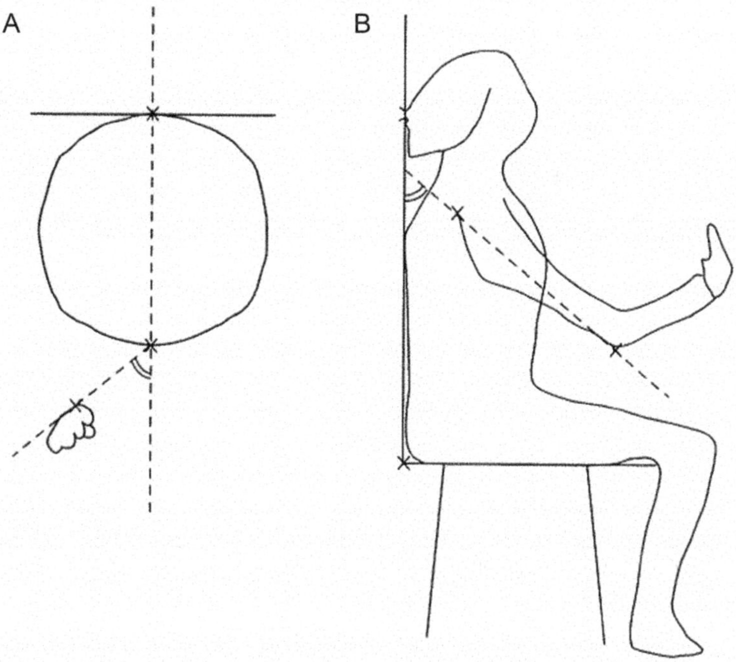


**Fig. 1.1** Schematic representation of the planar view (**a**) and the side view (**b**) of the mothers as recorded on film. The points marked by a light pen on the digitizer are indicated by an X. (Cited from Masataka 1992)

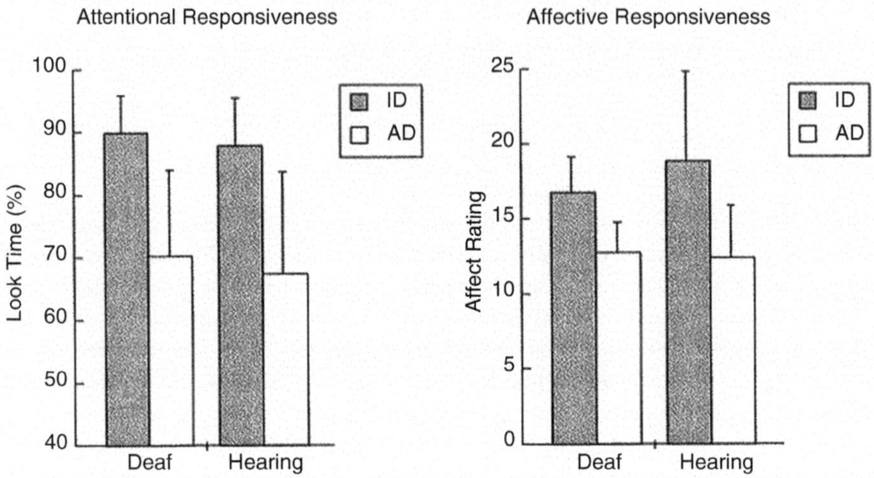


**Fig. 1.2** Comparison of responsiveness to infant-directed (ID) signing versus adult-directed (AD) signing between hearing infants and deaf infants. Error bars represent standard deviations. (Cited from Masataka 1998)

properties in infant-directed speech for modulating infant attentional and affective responsiveness results from innate predispositions to respond selectively to those properties. Cooper and Aslin (1990) examined behavioral preferences for infant-directed speech over adult-directed speech in two groups of infants, one made up of 12 one-month-old infants and the other consisting of 16 two-day-olds. Infants of both groups were tested according to the same visual-fixation-based auditory-preference procedure. The results showed that both one-month-old infants and newborns preferred infant-directed speech over adult-directed speech. Although the absolute magnitude of the infant-directed speech preference was significantly greater in the older infants, who showed longer looking durations than the younger infants, subsequent analyses showed no significant difference in the relative magnitude of this effect, indicating that infants' preference for the exaggerated prosodic features of infant-directed speech is present from birth and may not depend on any specific postnatal experience. From these results, it appears that postnatal experience with language does not have to be extensive for the infants to learn about the prosodic features of their native language. However, it has been shown that newborns prefer the intonational contour and temporal patterning of a prenatally experienced melody (Panneton 1985). Thus, the possibility remained that both prenatal and postnatal auditory experience could affect the relative salience of prosodic cues for young infants (Fig. 1.2).

In a subsequent study, 4- and 9-month-old English-learning infants were reported to show a robust attentional and affective preference for infant-directed speech over adult-directed speech in Cantonese, though this language was completely "foreign" to them (Werker et al. 1994). Similarly, 4- to 7-month-old English-learning infants were reported to show as strong attentional and affective responsiveness to infant-directed singing in foreign languages as to that in English (Trainor 1996). However, the infants included in these experiments had heard speech or songs before. In fact, all hearing infants of hearing parents are exposed to some form of speech and song as early as the prenatal period (DeCasper et al. 1994). Masataka (2003) investigated the attentional responsiveness to infant-directed speech and infant-directed song in two-day-old healthy infants, who were hearing and were born to congenitally deaf parents. A total of 15 infants participated in the experiment. All of the infants lived in typical nuclear families. The parents had acquired Japanese sign language as a first language and communicated with one another by the signed language exclusively. They lived in typical deaf communities. Since auditory experience for infants in utero mostly comes from their parents with regard to speech and song, the findings obtained from these infants were expected to provide a convincing answer to the question of whether preference for infant-directed speech and infant-directed song is predispositional or not.

For the stimuli presented, infant-directed and adult-directed speech, and infant-directed and adult-directed songs were prepared. Both speech samples and song samples were made up of a Japanese version and an English version, respectively. Speech samples and song samples were recorded from 10 adult females: five native speakers of Japanese and five of English were instructed to read or sing identical short scripts and play songs either to an infant or to an adult in Japanese and English, respectively. Acoustical comparisons of the speech samples and the song samples revealed that when directed to the infant, both speech and song showed such modification as reported to be typical of infant-directed speech or song; i.e., the pitch was elevated, the pitch contour was exaggerated, and the tempo became slower. When these infant-directed and adult-directed speech or song stimuli were presented, the infants were found to look longer at the infant-directed version as opposed to the adult-directed version whether the stimuli were those of speech samples or singing samples. Since intrauterine recordings taken near term have revealed that the maternal voice and heartbeat are audible in utero, but nonmaternal voices are rarely audible because of attenuation by maternal tissue and/or masking by intrauterine sounds (Querleu and Renard 1981; Querleu et al. 1981), these results strongly indicate a predispositional perceptual preference by human infants for exaggerated prosodic properties, whether they are in speech or song.

In addition to the fact that linguistic input to infants is modified prosodically, evidence has been presented that indicates the possibility of phonetic modification in motherse in a way that would enhance language learning. Kuhl et al. (1997) audiotaped 10 native-speaking women when speaking to their 2- to 5-month-old infants and when speaking to their adult friend in each of the US, Russia, and Sweden. Native-language words containing the vowels /i/, /a/, and /u/ were preselected for analysis in the three languages, and, with regard to each of the three vowels in each of the languages, the vowel triangle was compared between the speech sample recorded in the infant-directed condition and that recorded in the adult-directed condition.

The results revealed that when interacting with the infants, phonetic modifications occurred in the maternal speech as compared with when communicating with the adults. Across all the three languages, mothers were found to produce acoustically more extreme vowels when addressing their infants, resulting in an expansion of the vowel triangle in the speech sample collected in the infant-directed condition. They did not simply raise all formant frequencies when speaking to their infants, as they might have done if they were mimicking child speech. Rather, formant frequencies were selectively modified to achieve an expansion of the acoustic space encompassing the vowel triangle. When vowel triangle areas were compared between the infant-directed sample and the adult-directed sample for each mother, the results were highly consistent across individuals. For each of the 30 mothers, the area of the vowel triangle was greater in the infant-directed condition than in the adult-directed condition. On average, they expanded the vowel triangle by 92% when addressing their infants (English, 91%, Russian, 94%, Swedish, 90%). The extent of the expansion did not differ across languages, suggesting that the phonetic modification took place essentially in the similar manner regardless of the difference of the languages.

An expanded vowel triangle is assumed to increase the acoustic distance between vowels, which eventually enables the infants to make distinctions among them more easily. Moreover, the possibility arises that by stretching the triangle, vowels would come to be produced that go beyond those produced in typical adult conversation. From both an acoustic and articulatory perspective, these vowels are "hyperarticulated," and hyperarticulated vowels are perceived by adults as "better instances" of vowel categories. Indeed, a laboratory test showed that when listening to good instances of phonetic categories, infants show greater phonetic categorization ability (Iverson and Kuhl 1995). When prosodic properties of maternal speech are exaggerated, the infant listening to that is enabled to imitate the prosodic pattern more easily. Given the fact that the ability of vocal imitation of infants proceeds simultaneously in terms of prosody and in terms of phonetics during their first 6 months of life (Masataka 2003), it seems reasonable to hypothesize that the same would be the case for maternal speech with exaggerated phonetic properties: when the vowel triangle is expanded in maternal speech, the imitation of the vowel heard by the infant would be facilitated.

## 1.3 Characteristics of Signs of Deaf Mothers When Interacting with Their Deaf Infants

Results of the first systematic attempt to analyze parental signing to infants were reported by Erting, Prezioso, and O'Grady Hynes (1990). Since 1985, they had been collecting videotapes of deaf mothers interacting with their deaf babies. Their preliminary observations supported previous researchers' findings that the deaf mothers varied their signing regarding the type of movement, location on the infant's body, the intensity and speed of movement, and rhythmic patterning so that they could get and maintain the infant's attention effectively. These findings appeared to be consistent with the claim that motherese in spoken languages is mainly concerned with prosodic modification. Based on the findings, they focused on the American sign for MOTHER produced by two deaf mothers who had acquired ASL as their first language during two different types of interaction: interaction with their deaf infants when the infants were between 5 and 23 weeks of age and interaction with their adult friends. When a total of 27 MOTHER signs directed to the infants were compared with the same number of the signs directed to the adults, the mothers were found to (1) place the sign closer to the infant, perhaps the optimal signing distance for visual processing, and to (2) orient the hand so that the full handshape was visible to the infants. Moreover, (3) the mothers' face was fully visible for the infants to see, (4) eye gaze was directed at the infants, and (5) the sign was lengthened by repeating the same movement. The results appear to support the claim that parents use special articulatory features when communicating with infants, including parents from a visual culture whose primary means of communication is visual-gestural instead of auditory-vocal.

Following that study, Masataka (1992, 1996, 1998) conducted a series of experiments on the sign motherese phenomenon in deaf infants and their deaf parents who acquired Japanese sign l(JSL) as their first language. When starting the series of experiments concerning sign motherese, Masataka (1992) attempted to replicate the finding with a larger sample of participants who lived in a different cultural environment from those studied by Erting et al. (1990) and with the use of more exhaustive methodology for the analysis of signs. In most signed languages that have been investigated so far, facial expressions are known to play a multifunctional role for transmitting the meaning embedded in each signing movement. As with spoken language, they are used to express affect. However, unlike in spoken languages, specific facial behaviors in the signed languages also constitute the required grammatical morphology for numerous linguistic structures (for instance, relative clauses, questions, and conditional sentences in the case of ASL). Such characteristics would pose tremendous difficulty in the quantitative analysis of signing behavior. In this regard, JSL has a practical advantage in that it relies on cues produced by head movement exclusively. Consequently, it appears relatively easy to separate linguistic dimensions from paralinguistic dimensions in sign production.

In all, 13 mothers participated in the recordings. Eight of them were observed when freely interacting with their deaf infants and when interacting with their deaf adult friends (Masataka 1992). The remaining five were instructed to recite seven prepared sentences either toward their infants or toward their adult friends. Recordings were made with each mother and her infant or friend seated in a chair in a face-to-face position. The height of each chair was adjusted such that the eyes of the mother, her infant, and her friend were about 95 cm above the floor. The infant's body was fastened to the seat by a seat belt. The mother's behavior was monitored with two movie cameras. One of the two provided the frontal plane view and the other provided the right-side view. At the beginning of each recording session, she was instructed to interact with her infant or friend as she normally might when they were alone together. She was also told not to move her head during the session.

For each recognized sign recorded on this tape, the following four measurements were performed: (a) duration (the number of frames); (b) average angle subtended by the right hand with respect to the sagittal plane of the mother; (c) average angle subtended by the right elbow with respect to the body axis of the mother; and (d) whether the mother repeated the same sign consecutively or not. For measuring the positions of the hand and the elbow, each of the frames was projected by a movie projector onto a digitizer, which was connected to a minicomputer. By plotting the position of the hand and the head or the position of the elbow and the body axis on the digitizer with a light pen, the computer measured the angle between them with an accuracy of 0.5 degrees. Subsequently, the measured values were averaged for each sign by the mother. These measurements were performed to analyze the degree of exaggeration of signing by the mother. As signed languages are processed by deaf infants in the visual mode whereas spoken languages are processed by hearing infants mainly in the auditory mode, it was hypothesized that exaggeration of signing behavior should appear in the pattern of movements of hands and arms that were making the signing gestures.

When the pattern of such movement was compared when the mother was interacting with her infant and when she was interacting with another adult, striking differences were found with regard to all of the four parameters measured. All of the differences were statistically significant. For the analysis of the duration of signs, values were averaged across participants for each condition. The duration was longer in the case of signs directed to a mother's infant than in the case of signs directed to her adult friend.

When the angle of the hand and elbow subtended to the sagittal plane or body axis was calculated across participants for each condition, the same tendency was apparent. Mean scores for maximum values of angles for each sign directed to infants significantly exceeded those for signs directed to adults with respect to both the hand and the elbow. Similarly, mean scores of averaged values of angles for each sign directed to infants significantly exceeded those for signs directed to adults. With regard to all three of these parameters, post-hoc comparisons revealed that each participant demonstrated a significant increase in these scores when she interacted with her infant. Comparison of the rates of repetition of signs for signs directed to her infant and for signs directed to her adult friend also showed that the rates for the 13 mothers when they interacted with their infant exceeded the rates when they interacted with their adult friend, and that each mother demonstrated a significant increase in repetitions when she interacted with her infant.

Overall, the results indicate the presence of striking differences between the signed language used by Japanese deaf mothers when they interact with their infants and when they interact with their adult friends. When interacting with their infants, they use signs at a relatively slower tempo and are more likely to repeat the same sign, and movements used to make the signs are exaggerated. The outcome of this experiment revealed a phenomenon that is analogous to motherese in maternal speech. At the time of that study, it was well known that signed languages are organized in an identical fashion to spoken languages with regard to most linguistic aspects (Klima and Bellugi 1979). When adults produce manual actions that are part of the phonetic inventory of signed languages in communicating with young deaf children, who have only a rudimentary knowledge of language, one would predict that the social interaction should have as unusual a quality as when adults speak to young hearing children. That is, if the interaction is to be established and maintained, adults must be led to utilize special constraints in their activities and to produce signals, in either the signed or spoken modalities, in a somewhat modified manner according to the infant's level of attention and comprehension. This notion is strongly supported by the results of the experiments described above.

In particular, it is important to note that the mothers manipulated the duration, scope (angle), and word repetition rate of their sign production when signing to their infants as contrasted to their adult friends. This "sign motherese" is considered to be parallel to "speech motherese" in manipulating or varying the prosodic patterns of the signal because duration, scope, and repetition rate are all dimensions of prosody in sign, roughly analogous to duration, pitch, and repetition in speech. The fact that the mothers involved in the experiments clearly manipulated or changed these dimensions of sign production indicates that they were doing something in addition to and presumably apart from their manipulation of affective dimensions of sign production. In the case of speech motherese, actually, it is known to be frequently accompanied by pronounced modifications in facial expression (i.e., more frequent and more exaggerated smiling, arched eye-brows, and rhythmical head movement; Gusella et al. 1986; Sullivan and Horowitz 1983). Nevertheless, these components can all be regarded as paralinguistic in speech, and also in JSL, their occurrences, if they do occur, do not influence mother–infant communication from a grammatical perspective, as described earlier. Thus, it can be concluded that the phenomenon of motherese is a property of an amodal language capacity, at least with regard to prosodic dimensions.

## 1.4 Perception of Sign Motherese

A phenomenon quite analogous to speech motherese was thus considered to have been identified and designated sign motherese. However, although in the spoken modality the available data indicate motherese is a prevalent form of language input to infants, it is still unclear whether the same is the case for sign motherese. Certainly in the studies introduced above the experimenters were able to tell the difference between the form of infant-directed signing and that of adult-directed signing on behavioral and physical grounds. However, the question of whether deaf infants can perceive the difference remains unanswered. Continually monitoring the infants' degree of attention and understanding, mothers might modify various features of their signing so as to maintain the degree of the infants' responsiveness at an optimal level. Does it enhance the infants' acquisition of the basic units of signed language? In order to address this question, Masataka (1996, 1998) undertook the following study of the perception of sign motherese.

The experiment was made up of two parts; i.e., one involved congenitally deaf infants of deaf mothers while hearing infants of hearing mothers participated in the other. In the first experiment, seven deaf infants at 6 months of age were presented with a stimulus videotape that comprised excerpts of the infant- and adult-directed signs produced by the five deaf mothers described above. All of the deaf mothers of the infants, whose husbands were also deaf, were signers of Japanese sign language as their first language. In the stimulus tape, the following seven sentences were signed toward the infants of the mothers or toward their adult friends: "Good morning." "How are you today?" "Get up." "Come on now." "If you get up right away, we have a whole hour to go for a walk." "What do you want to do?" "Let's go for a walk."

Obviously, the question being asked in this study is whether deaf infants will attend to and prefer infant-directed signing over adult-directed signing. Therefore, the infant's reactions to the stimulus presentations were videotaped and later scored in terms of attentional and affective responsiveness to different video presentations. The stimulus presentation and recording took place in a black booth. Throughout the experiment, each of the seven infants was presented the stimulus tape only once. During each session, a mother stood with her back to the video monitor and held the infant over her shoulder so as to allow the infant to face the video display at a distance of approximately 60 cm. The session was conducted only when the infant was quiet and alert. During the session, the mother was told to wear a music-delivering headphone and to direct her gaze to a picture on the wall, 90 cm to her right. Although deaf, the mother was asked to wear the headphone so that hearing mother–infant dyads with no exposure to a signed language could be comparatively investigated, which will be described below.

The infant's reactions were evaluated by four raters, two for attentional responsiveness and the other two for affective responsiveness. As an indicator of attentional responsiveness, the percentage of the total video display time spent looking at the video screen, averaged between the two raters, was measured. For the measurement, the two watched the infant on screen independently and pushed a button whenever the infant fixated on the video display. These button presses were counted and timed by a computer. In order to measure affective responsiveness, each of the other two observers was independently told to attend to facial features as well as vocalizations of the infant and to rate the infant with a set of 9-point scales on the three dimensions that were originally used in the previous study of the perception of speech motherese (Werker and McLeod 1989). The three dimensions were treated as indexes of a single underlying factor, and one cumulative score of affective responsiveness was created for each infant by summing the ratings on the three dimensions by each rater and by averaging between the two raters. Thus, the maximum and the minimum cumulative scores would be 27 and 3, respectively, for each infant for each stimulus condition. The higher the scores received by the infants on the dimensions, the more positive emotions they were judged to be experiencing.

When the actual amount of time each infant fixated on the videotape during depictions of infant-directed signing versus adult-directed signing was analyzed statistically, a significant difference was found. The overall mean proportion of time each infant fixated on the videotape was 90.4% (SD = 6.3) for infant-directed signing and 69.8% (SD = 14.7) for adult-directed signing. When the videotape segment of infant-directed signing was presented, the infants apparently looked at it longer than when the segment of adult-directed signing was shown. Concerning affective responsiveness, the response scores to the segment of infant-directed signing exceeded those to the segment of adult-directed signing in every infant participant. Overall mean scores were 16.9 (SD = 2.5) for infant-directed signing and 12.8 (SD = 2.1) for adult-directed signing. Thus, the infants were affectively more responsive to infant-directed signing than to adult-directed signing.

The results of this experiment revealed that sign motherese evoked more robust responsiveness than adult-directed signing did from the deaf infants. This offers the long-term prospect of identifying features of motherese that are not specific to a particular language modality. Certainly it is important to take account of the possibility that the attractiveness of sign motherese is due to something in how deaf mothers manipulated their affective behavior rather than characteristics of signs themselves. In this regard, it should be noted that the head movements of the subject mothers were limited in the initial videotaping. As already noted, this restriction does not influence mother–infant communication from a grammatical perspective. From the paralinguistic perspective, however, the communicative sample obtained from the mothers under those circumstances should be affected by this restriction. Yet, the infants showed greater responsiveness to infant-directed signing than to adult-directed signing, suggesting that human infants may have an equal capacity to attend to motherese characteristics in speech or sign. They could be predisposed to attend to appropriately modified input, regardless of the modality of the input.

No doubt, in order to answer the question of whether the particular modal capacities must be triggered by some amount of experience or not, one must examine the response to these patterns in infants who lack substantive experience (linguistic exposure) in the modality. Indeed, concerning speech motherese, several researchers have treated this issue. Among others, Cooper and Aslin (1990) showed that the infant's preference for exaggerated prosodic features is present when tested at the age of 2 days. From the results, it appears that postnatal experience with language does not have to be extensive for the infant to learn about the prosodic features of their native language. However, it has also been shown that newborns still show a preference for the intonational contour and/or temporal patterning of a prenatally experienced melody (Panneton 1985). Thus, both prenatal and postnatal auditory experiences affect the relative salience of prosodic cues for young infants. In a subsequent study, four- and nine-month-old English-learning infants were found to show a robust attentional and affective preference for infant-directed speech over adult directed speech in Cantonese, though the language was completely "foreign" for them (Werker et al. 1994), but these infants had heard SPEECH before. No convincing evidence has ever been presented to prove the hypothesis. Rather, it can be said that as long as hearing infants are investigated, it is extremely difficult, if not possible, to address the question of whether exposure to any language is necessary for the infant to be able to react to the motherese form of the language.

Therefore, a subsequent study (Masataka 1998) was conducted to determine whether hearing infants with no exposure to signed language also prefer sign motherese. If such a preference exists, this could be a demonstration that human infants lock onto particular kinds of patterned input in a language modality completely independent of prior experience. The participants were 45 sets of hearing mothers and their first-born, full-term. Hearing infants with no exposure to a signed language (21 boys and 24 girls). All of the infants were 6 months old. Their mothers were all monolingual women who spoke only Japanese, were between 22 and 29 years old, and were middle class. The stimulus tape shown to each of the 45 infants was the same video tape used in the experiment on deaf infants. The protocol of stimulus presentations and that of scoring the infant's reactions to the stimulus presentations were also the same as in the previous experiment.

The analysis of the data obtained from the 45 infants revealed that there was a significant main effect for stimulus type with respect to their attentional responsiveness. They looked at the videotape filming infant-directed signing longer that at the tape filming adult-directed signing. The overall mean proportion of time that each of them fixated on the stimulus tape was 87.5% (SD = 8.2) for infant-directed signing and 66.0% (SD = 17.3) for adult-directed signing. Moreover, when comparing responsiveness for infant-directed signing versus adult-directed signing of the hearing infants with that of deaf infants involved in the above experiment, no significant difference was found in the scores between the two groups. Concerning affective responsiveness, too, a similar tendency was found. For the 45 hearing infants, overall mean scores were 18.9 (SD = 6.1) for infant-directed signing and 12.6 (SD = 3.4) for adult-directed signing. A statistically significant difference was found between the two stimulus conditions. Comparing these scores with those of the deaf infants, there was again not a significant difference between the two participant groups.

Clearly, hearing infants who have had no exposure to any form of signed language are attracted to motherese in JSL in a manner strikingly similar to that of deaf infants who have been exposed to the language from birth. Infants are attracted to sign motherese, regardless of whether they have ever seen sign language. Space and movement are known to be the means for transmitting morphological and syntactic information in signed languages. The continuous, analogue properties of space and movement are used in systematic, rule-governed ways in almost all signed languages that have been investigated so far (Armstrong et al. 1995; Newport 1981). JSL is not an exception for that. The abstract spatial and movement units are analogous in function to discrete morphemes found in spoken languages. Among those properties in spoken languages, it is hypothesized to be the grand sweeps in changes in fundamental frequency (dynamic peak-to-peak changes) that attract hearing infants to spoken motherese (Fernald 1985; Fernald and Simon 1984). As a visual analogue to such features, it is likely to be the sweeps of peak visual movement in sign motherese that attract deaf infants. These special properties evident in infant-directed communication may have universal attentional and affective significance, and human infants are predisposed to attend to the properties even if they have had no exposure to sign motherese before.

## 1.5 Emergence of Manual Babbling

The above findings would lead us to consider how the presence or absence of such linguistic input affects the linguistic behavior infants acquire. Concerning the issue of how the ontogenetic process of vocal behavior in deaf infants differs from that in hearing infants within the first year of their lives, little is known so far, partly because babbling has been anecdotally assumed to appear under maturational control. However, recent observations revealed that vocal ontogeny proceeds essentially similarly in deaf and hearing infants until the end of the marginal babbling stage (Masataka 2003). Thereafter, hearing infants come to produce canonical babbling whereas deaf infants are unable to do so. This was considered to be due to the necessity of auditory feedback for the acquisition of articulatory ability, an ability that makes the pronunciation of reduplicated consonant-vowel syllables possible.

Interestingly, the onset period of canonical babbling in hearing infants exactly coincides with the onset period of manual babbling which has been proposed to occur in deaf infants as well as in hearing infants by Petitto and Marentette. In the course of conducting research on signing infants' transition from pre-linguistic gesturing to first signs, they had been performing close analyses of the "articulatory" variables (analogous to "phonetic" level) of all manual activity produced by ASL deaf infants of ASL deaf parents from the age of 6 months. Actually, the presence of manual babbling had been noted by some researchers prior to the report of Petitto and Marentette. Prinz and Prinz (1979), on the basis of their observations between the age of 7 months and 21 months of a hearing infant born to a deaf mother, reported that the infant exhibited a group of manual behaviors that appeared to be produced as a result of imitation of signing of the mother. In another dyad of a hearing infant and its deaf mother, the infant was observed to clap and rub the hands in a circular motion by Griffith (1985). That behavior was interpreted as a prelexical form of a sign included in ASL. However, neither study had conducted such a systematic and quantitative analysis as Petitto and Marentette did. During the investigation, they noticed the presence of a class of manual activity that was unlike anything else that had been reported so far in previous literature.

These manual behaviors included linguistically relevant units but were produced in entirely meaningless ways, and they were wholly distinct in their pattern from all other manual activities, i.e., general motor activity, communicative gestures, and signs. They occurred between 9 and 12 months of age, the period corresponding to the canonical babbling stage for hearing infants acquiring speech. Indeed, subsequent analyses revealed that this class of manual activity was characterized by the identical timing, patterning, structure, and use of vocal behavior in hearing infants that is universally identified as babbling. As a result, it was termed "manual babbling." Further comparison of the ontogeny of manual action between hearing infants who were acquiring spoken language with no exposure to a signed language and deaf infants who were acquiring ASL as their first language revealed that both of them produced roughly equal proportions of communicative gestures (e.g., arms raised to request being picked up) at approximately 8 to 14 months of age.

Nevertheless, the deaf infants produced far more manual babbling forms (e.g., handshape-movement combinations exhibiting the phonological structure of formal signed language) than did the hearing infants, and manual babbling accounted for approximately 40% of manual activity in deaf infants and less than 10% of the manual activity of hearing infants.

Takei (2001) attempted to replicate the above findings in deaf infants growing up with exposure to JSL. He observed one male and one female infants, whose parents both acquired JSL as their first language. He made at least 1-h observations of each of the infants at their home when interacting with the mother freely with prearranged toys with the use of a video-recorder. Subsequently the video-recordings were transcribed in the laboratory. When the recorded manual activities were transcribed according to the same classification categories as employed by Pettito and Marentette (1991), manual babbling appeared first at the age of 6 months, and occurred most often when the infants were 10 months old. Immediately after that period, the first recognizable signs that were meaningful as JSL signs were recorded. After the onset of these first signs, the frequency of manual babbling tended to decline. Basically, the peak of manual babbling was found to be a reliable predictor of the onset of meaningful signs in the infant.

Taken together, these studies revealed that deaf infants with exposure to signed language showed manual babbling with much higher frequency than hearing infants. In addition, the phenomenon was most robust just prior to the onset of meaningful vocabulary in signed language in each of the deaf infants. In spite of this quantitative difference, however, even hearing infants, with no exposure to a signed language, were found to produce manual babbling. Given the fact that manual babbling precedes the onset of the first recognizable signs as vocal babbling precedes the onset of the first meaningful words, language capacity was taken by the authors of the article to be able to manifest itself equally as sign or speech in human infants; i.e., in hearing infants, the ability of vocal and gestural language can develop simultaneously and in a similar manner, at least during the preverbal period, though the degree to which they develop can vary, because some infants come to produce almost equal numbers of first words both in speech and sign around the identical period (Goodwyn and Acredolo 1993). In the case of deaf infants, the progression of acquiring spoken language is hindered earlier. This is to some extent due to the lack of auditory feedback that is crucial for the acquisition of articulatory ability in speech. As a consequence, they were considered to be obliged to exclusively rely on the manual mode to realize their linguistic capacity. This fact could partly account for a richer repertoire of manual babbling by the sign-exposed infants. However, it is not the only variable responsible for the difference.

Concerning other responsible variables, Masataka (2003) observed the progression of manual activity of 3 preverbal deaf infants of hearing parents with no exposure to a signed language and found that although the infants did perform manual babbling, the amount of the behavior was greater than that performed by hearing infants but much smaller than that performed by deaf infants with exposure to JSL. In that study, a comparison was made of the pattern of the development of manual activity between the age of 8 months and 12 months in three groups of infants, i.e., three deaf infants of deaf parents with exposure to JSL, three deaf infants of hearing parents with no exposure to any signed language, and two hearing infants of hearing parents with no exposure to any signed language. When all of the infants' manual activities were transcribed in an identical manner, they were found to be classified into two types in every infant: syllabic manual babbling and gestures, and the infants of the three groups produced similar types and quantities of gestures during the study period. However, they differed in their production of manual babbling. Manual babbling behavior accounted for 25 to 56% of manual activity in deaf infants with exposure to JSL, but a mere 2 to 6% of the manual activity of hearing infants. Manual babbling as a percent of manual activity [manual babbling/(manual babbling + gesture)] showed a significant increase as the deaf infants developed, but did not exhibit such a change in the hearing infants. On the other hand, the value was intermediate between the values of these two groups for deaf infants with no exposure to any signed language (between 10% and 15%). Moreover, the value was greatest at the age of 8 months in all three groups of infants and showed a significant decrease as they developed.

Thus, the phenomenon of manual babbling does not necessarily imply that babbling is purely an expression of a "brain-based language capacity" that is amodal, or that babbling is not a phenomenon of motor development but is strongly associated with the abstract linguistic structure of language. Even vocal babbling arises with the aid of properties of motor organization that are shared by the motor system controlling the manual articulators, and thus the parallel progression of manual babbling and spoken babbling during the time course of their development per se should not be a very astonishing fact.

## 1.6 Antecedents of Sign Motherese

In all, sign motherese apparently acts as a behavioral adjustment to help children to learn languages using the manual mode. This fact should lead us to question what is predispositionally evolved in humans as the antecedents of such capability. In fact, Csibra and Gergely (2006) proposed that humans are provided with an evolutionary adaptation such that when presenting new information in general, adults create and children respond to a pedagogical context. Apparently, the enhancement of sign language learning that occurs in children caused by adults' sign motherese can fall into this context.

According to this account, children are sensitive to cues from adults that highlight information if the information is new and important for their learning of novel action that is important for their living in general. Namely, children are more likely to imitate acts that are marked as intentional and demonstrated for themselves. For example, if infants simply witness an adult bend over and touch a lamp with the hand in order to illuminate it, rather than having the adult specifically show them after making eye contact, infants are less likely to imitate it (Csibra and Gergely 2006). Given the fact that sign motherese is important for language learning in the pedagogical context, one can expect similar levels of action learning in much broader contexts in infant-directed demonstrations relative to adult-directed demonstrations as long as both are presented in an intentional pedagogical fashion. Also, human children are extremely well known as voracious learners while adults are ubiquitous teachers!

In this regard, the findings reported by Brand, Baldwin, and Washburn (2002) should be noteworthy. In that study, teaching behavior about the usage of novel objects was video-recorded and subsequently analyzed in adults. When their performance was compared when it was directed toward 6- to 13-month-old infants and when it was directed toward another adult, a particular suite of embellished behavior was found to emerge when it was directed toward infants, which has been referred to as "motionese" or infant-directed action. Given young children's fledging attentional control, the motionese system of behaviors incorporates some previously studied features, such as eye gaze and emotional expressiveness (Chong et al. 2003), but also involves a variety of other modifications. These include closer proximity to a child versus an adult partner, greater enthusiasm, a large range of motion, simplified action sequences, greater repetitiveness, and higher interactiveness, including more and longer gazes to infants' faces and more turn-taking. Extensions of this work have also found evidence of longer pauses in infant-directed action compared with adult-directed action and a unique coordination of speech and action in demonstrations for children (Rohlfing et al. 2006).

On the basis of these findings, a question concerning the effectiveness of motionese as a teaching behavior would arise. Although the possibility remains that this medley of cues could be distracting or frustrating relative to the straightforward adult-directed action style, it should be more likely that these cues function to provide a richer learning experience than a standard demonstration. If so, then motionese would fit within a suite of behavioral adjustments for communication with infants and children, including the motherese in the vocal mode and in the manual mode that have been documented so far.

In fact, suggestive evidence for this possibility has been presented by Brand and Shallcross (2008), who reported that motionese preferentially attracted children's attention, namely, that 6- to 13-month-old infants looked longer at infant-directed versus adult-directed demonstrations when both were available to view simultaneously. This was the case when the faces were digitally blurred, to obscure eye gaze and expressive information, suggesting a role for the entire suite of behaviors and not simply facial cues. Increased scrutiny or interest in the behaviors could lead children to encode and remember the actions better for subsequent imitation.

The results of the study also indicated the possibility that motionese may have functions beyond attracting children's attention or preparing them for new information: motionese modifications may also help to make demonstrated acts easier to parse (e.g., by highlighting boundary points with repetition and eye gaze), stress which body parts and subtle physical motions are necessary (e.g., a horizontal twist before vertically pulling off a cup), and highlight the intentions behind the acts (e.g., by exaggerating facial expressions of surprise and satisfaction). If motionese functions in this manner, children may learn more easily, and imitate more faithfully, when the suite of cues is shown throughout the demonstration.

Therefore, in order to directly test whether motionese influences young children's observational learning of novel acts, whether the special infant-directed action modifications parent use when teaching their children really improved 2-year-olds' imitation was investigated in a subsequent study (Williamson and Brand 2014). In that study, a total of 48 children saw an adult perform a series of acts on four novel objects using either an infant-directed style (e.g., larger range of motion and enhanced boundary marking with more repetition) or an adult-directed style. After each demonstration, the children received a test period for that object during which they were allowed to play for 25 s. When the videos recorded during that period were coded, children who saw any demonstration (either infant-directed or adult-directed) were found to show imitation in that they were more likely to produce the target acts than were children in a non-demonstration baseline group. Moreover, children who saw demonstrations augmented with motionese exhibited even higher levels of imitation than did children who saw adult-directed demonstrations, indicating that infant-directed demonstrations were particularly effective teaching behaviors.

In fact, a rudimentary form of motionese is known in nonhuman animals. Importantly, such behavior is assumed to serve as a basis on which social learning is enhanced in the animals.

The population-level instrumental use of objects present in the external world has been regularly reported among nonhuman animals. Perhaps the most famous case of such an instance is that of Imo, a Japanese macaque (*Macaca fuscata*) that learned that washing sand-covered sweet potatoes made them more palatable (Kawai 1963). Imo's exercise spread through her community and became standard practice within approximately 10 years. It is not clear, however, just how the transmission occurred, but it seems likely that observational learning played a part.

More recently, Masataka et al. (2009) saw evidence of the social transmission of flossing in a troop of free-ranging long-tail macaques (*Macaca fascicularis*) in Thailand. The researchers first observed nine adult monkeys routinely pull the hair from the head of women tourists and use the hair to floss their teeth. Flossing has since become widespread in the troop. Although as with the potato washing, there is no clear demonstration of how the behavior spread, Masataka et al. (2009) found evidence that the mother monkeys attempted to teach flossing to their offspring through modeling. When the behavior of seven female macaques was videotaped when they were flossing and their behavior was compared when their infant was nearby and when it was not, they paused more often, repeated flossing more frequently, and flossed for a longer time when their infants were present than when their youngsters were not around. Though there is no direct evidence that the infants imitated their mothers' behavior, the motionese-like behavior observed suggests that modeling played a part in the spread of flossing in the troop and that the spread was enhanced by such behavior modifications.

## 1.7 Role of Sign Motherese as a Part of Multimodal Motherese in the Development of Social Understanding in Infants

Taken together, many findings show that the so-called motherese phenomenon occurs either in the vocal mode (infant-directed speech, or speech motherese) or in the manual mode (infant-directed signing, or sign motherese). Motherese, whether in the vocal mode or in the manual mode, highlights mother–child adjustments during interactions. This is because these behavioral characteristics are a product that has evolved as the historical extension of motionese as a device of enhancing social learning that has merged during the evolution of nonhuman primates.

Mothers are able to adjust their motherese to children's age, cognitive abilities, and linguistic level. Therefore, motherese may arouse children's attention by signaling mothers' linguistic behavior that is addressed to the children. Mothers also adapt their motherese to children's reactivity and preferences. Mothers' continuous adjustments to their children result in the facilitation of exchanges and interactions, with positive consequences for sharing emotions and for learning and language acquisition. Children's reactivity is also important given that their presence increases motherese, and children's positive, contingent feedback makes them more affective, which in turn increases the quality of the motherese.

In all, motherese mediates and reflects an interactive loop between the child and the caregiver, such that each person's response may increase the initial stimulation of the other partner. At the behavioral level, the interactive loop is underpinned by the emotional charge of the affective level and the construction of intersubjective tools, such as joint attention and communicative skills. Based upon such findings, Gogate, Bahrick, and Watson (2000) claim that infant-directed communication, even if it occurs between hearing infants and their caregives, is as seldom unimodal as it is in face-to-face communication between hearing adults, and that it is more complex and broader topic than has been assumed. They argue that the entire sensory system perceives information about the dialogue partner. Communication is so much more than speech, and manual and gestural movements. It encompasses more than simply focusing on visual and audible signals: variations in intonation, pitch, intensity, and gestures, facial expression, and eye gaze are recognizable. As reasoned by Ugur et al. (2015), humans need additional information to interpret ambiguous information in a dialogue, whose partners need to establish common ground in order to recognize each other's intentions and theory of mind. Infants may, however, have an advantage because child-directed communication is simpler, prosodic, and visual features are more exaggerated, which makes certain sensory input more salient, and intentions are therefore easier to recognize and patterns in the information stream can be more easily identified.

While there was a clear separation in child-directed research between that of different modality, mainly auditory and visual, the above reasoning has led researchers to orient toward the interplay of amodal input. Bahrick and Lickliter (2000) theorize that intersensory redundancy makes the given more salient in a constant stream of arbitrary information, stating that young hearing infants prefer temporal synchronous and amodal stimuli over asynchronous stimuli. For example, when hearing somebody speak in the same rhythm, they are more likely to see her/his movements of lips in the same rhythm than to see her/his movements of lips in the different rhythm.

Temporal synchronous, redundant, and amodal information can direct attention to meaningful events, which may be beneficial for learning. In addition, naming an object and touching or moving or rather showing it are redundant information across the auditory and visual senses and therefore make the relation between the name and the object salient. Such reasoning has led researchers to no longer distnguish motherese in the vocal mode and that in the manual mode, but to include both in the category of multimodal motherese (Gogate et al. 2000). Coming from the perspective of motionese research, Brand and Tapscott (2007) explored the possibility that utterances enhanced the perception of action in 9- to 11-month-old hearing infants and found that co-occurring infant-directed speech and motionese facilitated the segmentation of the observed action into smaller units.

Reinforcing the notion of multimodal motherese, Nagai and Rohlfing (2009) found that when interacting with infants, as opposed to adults, parents attempt to employ a number of strategies in order to increase the saliency of the objects and their initial and final states. These strategies include suppressing their own body movements before starting to execute the task, and generating additional movements on the object, such as tapping it on the table. The resulting behavior is qualitatively different from adult-adult interaction in observable parameters, such as the pace or the smoothness of the movement. It is shown that infants also benefit from this behavior. Koterba and Iverson (2009) showed that hearing infants exposed to a higher repetition of demonstration exhibit longer bangs and shakes of objects whereas infants exposed to a lower repetition spent more time for turning and rotating objects. Ugur et al. (2015) developed a bottom-up architecture of visual saliency that can be used in an infant like learning robot. In this architecture, similar to an infant, the robot tends to find the salient regions in caregiver's action, when they are highlighted by multimodal motherese. Moreover, the saliency preference of the robot also motivates humans to use such motherese as if they were interacting with a human infant, thereby completing the loop. Due to the limited attention mechanism of the robot, human participants tried to teach the task (cup-stacking) by approaching toward the robot and introducing the object in the proximity of the robot. In general, participants amplified their movements and made pauses as if to give the robot a chance to understand the scene. Such delimiting pauses and exaggerations were those also observed in sign motherese as well as in spoken motherese.

The advantage infants enjoy with exposure to multimodal motherese could not be restricted to language learning, but be extended further to the other aspect of child development, e.g., to learning of social understanding that has apparently been a topic of great interest. While one hallmark of social understanding is the acquisition of a theory of mind between 3 and 5 years of age, which is the ability to predict and explain social behavior on the basis of mental state, children acquire certain social cognitive abilities that are predictive of their later social understanding already in infancy (Rakoczy 2012). Such an early competence is the ability to encode human actions as goal-directed. Woodward (1998) showed 6-month-old infants events in which a person reached for one of two toys. After habituation to this event, infants looked significantly longer on test trials in which a person grasped the other toy (new goal trials) than when the person reached for the old toy, but the path of the grasping was changed (old goal trials). Because infants show this preference only for human goal-directed action, and not for the motion path of a mechanical claw, the looking-time patterns indicate that infants encode actions in terms of agent–goal relations. The ability to explain behavior by ascribing goals to an agent, which infants develop during the second half of their first year of life, has turned out to be an important early precursor of later mental state understanding. In fact, Aschersleben, Hofer, and Jovanovic (2008) found a positive correlation between 6-month-old infants' decrement of attention to goal-directed action and their false belief understanding at 4 years of age.

Licata et al. (2014) investigated the relationship between mother–child interaction quality and infants' ability to interpret actions as goal-directed at 7 months assessed by the testing described above. Interaction quality was assessed in a free play interaction using two distinct methods: one assessed the overall affective quality (emotional availability), and one focused on the mother's proclivity to treat her infant as an intentional agent with autonomous thoughts, intentions, emotions, and desires (mind-mindedness). There are good reasons to expect a positive association between both emotional availability and mind-mindedness, and infants' understanding of intentional action. Maternal emotional availability might be beneficial for infants' goal encoding, as learning especially is promoted in a warm and sensitive environment, in which a child can explore his/her environment while being supported by the mother. Maternal mind-mindedness, on the other hand, could be related to infant's goal encoding, as mind-minded mothers may have infants who are better at interpreting human actions as goal-directed because these mothers allow their infants to experience themselves as self-efficient by verbally commenting on their mental states and appropriately responding to them.

Analyses revealed that only maternal emotional availability assessed by the Emotional Availability Scales (EMS) (Biringen 2008), and not maternal mind-mindedness assessed according to the protocol developed by Meins and Fernyhough (2010), was related to infants' goal-encoding ability. The link remained stable even when controlling for child temperament, working memory, and maternal education.

EAS is a method that does not quantify distinct behaviors but analyzes the interactional style of the dyad. It is an emotion-focused measure that refers to the overall affective quality of the relationship. The construct of emotional availability is multidimensional, as it comprises different dimensions of caregiving. Licata et al. argue that the fostering of maternal emotional availability for infants' goal-encoding skills could be mediated through specific ways of the mother engaging in goal-directed behavior with her infant (such as multimodal motherse, and sign motherese), which in turn could support infants' goal sensitivity.

Infant-directed action represented by sign motherese could serve itself as a basis on which infants are enhanced to get access to the meaning of action by helping them to recognize goals and intentions that motivate action. An emotionally available mother could have several attributes that characterize such motherese: an emotionally available mother shows much positive affect and interacts with her child, showing turn-taking, eye contact, and joint attention. It is highly possible to assume that maternal emotional availability also enhances infants' attention and helps them to learn to recognize goals and intentions that motivate action. One reason for this relation could be that magnified emotions amplify information about intentionality and help infants recognize goal achievement (Brand et al. 2002), which is necessary to encode human actions as goal-directed.

More recently, in fact, the findings confirming such reasoning have been reported by a neuroendocrine study employing a computational method on video vignettes of human parent–infant interaction including 32 fathers that were administered oxytocin or a placebo in a crossover experimental design (Weisman et al. 2018). The role of such nine-amino acid peptide hormone in the initiation and maintenance of affiliative bonds and parental repertoire has been elucidated in humans as well as in nonhuman animals (Feldman 2012). Acute administration of oxytocin to a parent was found to enhance the care's, but at the same time also the infant's physiological and behavioral readiness for dyadic social engagement (Weisman et al. 2012). Yet, so far, the exact cues that are involved in this affiliative transmission process have remained unclear.

Weisman et al. revealed the fact that oxytocin administration markedly increased the maximum distance between the father and the infant but at the same time drove the father to reach their closest proximity to the infant earlier, as compared with the placebo condition. Moreover, a father's head tended to move faster and to reach its highest velocity sooner under oxytocin influence. Fathers showed greater acceleration which also tended to vary more. Finally, infant oxytocin increase following interaction with the father correlated with the father's maximum acceleration parameter.

Given that the behavioral variables assessed in the study obviously fall under the definition of motherese in the manual mode, overall implications of its results for deaf studies of cognition and learning should be profound. It implies that such modality is susceptible to intervention in key neuroendocrine pathway that is central to the initiation and maintenance of parental repertoire, namely, the oxyticin system. Moreover, the modality has manifested itself as a consequence of the evolution of the cross-generational transmission of parenting in humans. Future studies should examine whether oxytocin shapes proximity and motion among deaf mothers interacting with their infants, using a signed language, and whether the communication patterns shaped under the hormone affects the infants' subsequent cognitive development.

## 1.8 Conclusions

The reasoning that I have documented so far indicates that, developmentally, the early manual action performed by deaf infants, like non-cry sounds produced by hearing, infants do not transmit meanings, but rather reveal fitness or express states. We make no assumption that very young infants intend to communicate with such action. Moreover, this is also observed in hearing infants. On the other hand, adults who receive the sounds are provided with broad reception/interpretation capabilities that allow them to provide differentiated feedback to signals if they have acquired sign languages. Such reasoning could lead us to notice the importance of considering how selection pressures might engender increase in the tendency of an infant to produce spontaneous meaning less manual action for arguing language evolution at the manual mode, and that in fact, evolution of increased infant spontaneous sign action began with developmental steps in individual deaf infants under the selective pressure of their own deaf caregivers.

In all, during modern human development, infant linguistic capabilities emerge at least partly in response to social interaction either at the vocal mode or at the manual mode, where caregivers react to linguistic capabilities of infants in accordance with a scaffolding principle requiring parental discernment and intuitive parenting to reinforce linguistic exploration and learning. Both endogenous inclination of infants to explore the linguistic space and interactive feedback from caregivers thus foster growth in linguistic capability either at the vocal mode or at the manual mode.

If this assumption about selection pressures for spontaneous action either at the vocal mode or at the manual mode are valid, it follows that hominin parents would also have been selected to become aware of the fitness reflected in infant action and capable of responding to those indicators with selective care and reinforcement of the action. The author considers this to have been illustrated so far in the phenomenon that is called "motherese." And the phenomenon of motionese implies the underlying evolutionarily antecedents for that, by which infants, whether hearing or deaf, are enabled to learn language independent of the auditory capability. In this sense, the gestural theory of language origins can be said to exert relevancy to exactly the same degree to that the vocal theory of language origins can exert relevancy.

## References

Armstrong DF, Stokoe WC, Wilcox SE (1995) Gesture and the nature of language. Cambridge University Press, Cambridge

Aschersleben G, Hofer T, Jovanovic B (2008) The link between infant attention to goal-directed action and later theory of mind abilities. Dev Sci 11:862–868

Bahrick LE, Lickliter (2000) Intersendory redundancy guides attenional selectivity and perceptual learning in infancy. Dev Psychol 35:190–201

Biringen Z (2008) The emotional availability (EA) scales. Unpublished Coding Manual

Brand RJ, Shallcross W (2008) Infants prefer motionese to adult-directed action. Dev Sci 11:853–861
Brand RJ, Tapscott S (2007) Acoustic packaging of action sequences by infants. Infancy 11:321–332
Brand RJ, Baldwin DA, Ashburn LA (2002) Evidence for "motionese": modifications in mothers' infant-directed action. Dev Sci 5:71–87
Chong S, Werker J, Russell J, Caroll J (2003) Three facial expressions mothers direct to their infants. Infant Child Dev 12:211–232
Cooper RP, Aslin RN (1990) Prefernce for infant-directed speech in the first month after birth. Child Dev 61:1584–1595
Csibra G, Gergely G (2006) Social learning and social cognition: the case of pedagogy. In: Processes of change in brain and cognitive development. Attention and performance, vol 21. Oxford University Press, New York, pp 249–274
DeCasper AJ, Lecanuet JP, Busnel MC, Garnier-Deferre C, Maugeais R (1994) Fetal reactions to recurrent maternal speech. Infant Behav Dev 17:159–164
Erting CJ, Prezioso C, O'Grandy Hynes M (1990) The interactional context of deaf mother-infant communication. In: Voltera V, Erting CJ (eds) From gesture to language in hearing and deaf children. Springer, Berlin, pp 97–106
Feldman R (2012) Oxytocin and social affiliation in humans. Horm Behav 61:380–391
Ferguson CA (1964) Baby talk in sic languages. Am Anthropol 66:103–114
Fernald A (1985) Four-month-old infants prefer to listen to motherese. Infant Behav Dev 8:181–195
Fernald A, Simon T (1984) Expanded intonation contours in mothers' speech to newborns. Dev Psychol 20:104–113
Gogate LJ, Bahrick LE, Watson JD (2000) A study of multimodal motherese; The role of temporal synchrony between verbal lebels and gestures. Child Dev 71:878–894
Goodwyn SW, Acredolo LP (1993) Symbolic gesture versus word: is there a modality advantage for onset of symbol use? Child Dev 64:688–701
Griffith PL (1985) Mode-sqitching and mode-finding in a hearing child of deaf parents. Sign Lang Stud 48:195–222
Gusella J, Roman M, Muir D (1986) Experimental manipulation of mother-infant actions. Paper presented at the international conference of infant studies, Los Angels, September
Iverson P, Kuhl PK (1995) Mapping the perceptual magnet effect for speech using signal detection theory and multidimensional scaling. J Acoust Soc Am 97:553–562
Kawai M (1963) On the newly-acquired behaviors of the natural troop of Japanese monkeys on Koshima Island. Primates 4:113–115
Klima ES, Bellugi U (1979) The signs of language. Harvard University Press, Cambridge, MA
Koterba EA, Iverson JM (2009) Investigating motionese: the effect of infant-directed action on infants' attention and object exploration. Infant Behav Dev 32:437–444
Kuhl PK, Andruski JE, Chistovich IA, Chistovich LA, Kozhevnikova EV, Ryskina VL, Stolyarova EI, Sundberg U, Lacerda F (1997) Cross-language analysis of phonetic units in language addressed to infants. Science 277:684–686
Licata M, Paulus M, Thoeremer C, Kristen S, Woodward AL, Socian B (2014) Mother-infant interaction quality and infants' ability to encode actions as goal-directed. Soc Dev 23:340–356
Masataka N (1992) Motherese in a signed language. Infant Behav Dev 15:453–460
Masataka N (1996) Perception of motherese in a signed language by 6-month-old deaf infants. Dev Psychol 32:874–879
Masataka N (1998) Perception of motherese in Japanese Sign Language by 6-month-old hearing infants. Dev Psychol 34:241–246
Masataka N, Koda H, Urasopon N, Watanabe K (2009) Free-ranging macaque mothers exaggerate tool-using bahevior when observed by offspring. PLoS One 4:e4768
Masataka N (2003) The onset of language. Cambridge University Press, Cambridge
Meins E, Fernyhough C (2010) Mind-mindedness coding manual. Unpublished Coding Manual

Nagai Y, Rohlfing KJ (2009) Computational analysis of motionese toward scaffolding robot action learning. IEEE Trans Auton Ment Disord 1:44–54

Newport E (1981) Constrains on structure: evidence from ASL and language learning. In: Collins W (ed) Minnesota symposia on child psychology. Erlbaum, Hillsdale, pp 65–128

Oller DK, Griebel U, Warlaumont AS (2016) Vocal development as a guide to modeling the evolution of language. Top Cogn Sci 8:382–392

Panneton RK (1985) Prenatal experience with melodies: effect on postnatal auditory preference in human newborns. Ph.D. dissertation, University of North Calolina, Greeensboro

Pettito LA, Marentette PF (1991) Babbling in the manual mode: evidence for the ontogeny of language. Science 251:1493–1496

Prinz PM, Prinz EA (1979) Simultaneous acquisition of ASL and spoken English: Phase I: Early lexical development. Sign Lang Stud 25:283–296

Querleu D, Renard K (1981) Les perceptions auditives du foetus humain [Auditory perception of the human fetus]. Journal de Gynecologie Obsterique et Biologie de la Reproduction 10:307–314

Querleu D, Renard K, Crepin G (1981) Perception auditive et reactive foetale aux stimulations sonores. Journal de Gynecologie Obsterique et Biologie de la Reproduction 10:307–314

Rakoczy H (2012) Do infants have a theory of mind? J Dev Psychol 30:59–74

Rohlfing KJ, Fritsch J, Wrede B, Jungmann T (2006) How can multimodal cues from child-directed interaction reduce learning complexity in robots? Adv Robot 22:1183–1199

Saint-Georges C, Chetouani M, Cassel R, Apicella F, Mahdhaoul A, Muratoni F, Laznik M-C, Cohen D (2013) Motherese in interaction: at the cross-road of emotion and cognition? (A systematic review). PLoS One 8:e78103

Sullivan JW, Horowitz FD (1983) Infant intermodal perception and maternal multimodal stimulation: implication for language development. In: Liositt LP, Rovee-Collier CK (eds) Advances in infancy research. Ablex, Norwood, pp 184–239

Takei W (2001) How do deaf infants attain first signs? Dev Sci 4:71–78

Trainor LJ (1996) Infant preferences for infant-directed versus noninfant-directed play songs and lullabies. Infant Behav Dev 19:83–92

Ugur E, Nagai Y, Celikkanat H, Oztop E (2015) Parental scaffolding as a bootstrapping mechanism for learning grasp affordance and imitation skills. Robotics 33:1163–1180

Weisman O, Zagoory-Sharon O, Feldman R (2012) Oxytocin administration to parent enhances infant physiological and behavioral readiness for social engagement. Biol Psychiatry 72:982–989

Weisman O, Delaherche E, Rondeau M, Chetouani D, Feldman R (2018) Oxytocin shapes parental motion during father-infant interaction. Biol Lett 9:21030828

Werker JF, McLeod PJ (1989) Infant preference for both male and female infant-directed talk: a developmental study of attentional and affective responsiveness. Can J Psychol 43:320–346

Werker JF, Pegg JE, McLeod PJ (1994) A cross-language investigation of infant preference for infant-directed communication. Infant Behav Dev 17:323–333

Williamson RA, Brand RJ (2014) Child-directed action promotes 2-year-olds' imitation. J Exp Child Psychol 118:119–126

Woodward AI (1998) Infants selectively encode the goal object of an actor's reach. Cognition 69:1–34

# Chapter 2
# Primate Vocal Anatomy and Physiology: Similarities and Differences Between Humans and Nonhuman Primates

**Takeshi Nishimura**

**Abstract** Our understanding of the evolution of human speech has been expanded by an increasing knowledge of vocal anatomy and physiology in nonhuman primates. Various vocal repertoires have been described in many species based on acoustic evidence. Comparative approaches in vocal anatomy and physiology provide evidence unveiling the acoustic mechanisms for these different vocalizations. Such empirical studies have shown many similarities serving speech faculties, while they also provide evidence supporting the underlying differences between nonhuman primate vocalization and human speech. The vocal apparatuses act under the constraint of their anatomy, and various associations of anatomy and physiology are found in primates, including humans. Efforts to unveil these variations promise a better understanding of primate origins and of the possible evolutionary history of human speech.

**Keywords** Source-filter theory · Vocal folds · Laryngeal air sac · Tongue · Supralaryngeal vocal tract · Descent of the larynx

## 2.1 Introduction

The origin and evolution of language have attracted many scholars in many fields, e.g., linguistics, anthropology, archeology, biology, cognitive science, genetics. Biology, including biological anthropology, has continued to challenge this issue, while paleoanthropologists have faced a major obstacle in that languages do not fossilize (Nishimura 2008). In fact, various vocal repertoires are found in nonhuman primates and birds (McComb and Semple 2005), and their forms and functions have been part of major scientific efforts for understanding the evolutionary processes leading to human language.

T. Nishimura (✉)
Primate Research Institute, Kyoto University, Inuyama, Aichi, Japan
e-mail: nishimura.takeshi.2r@kyoto-u.ac.jp



N. Masataka (ed.), *The Origins of Language Revisited*,
https://doi.org/10.1007/978-981-15-4250-3_2

Speech is simply a medium for communication by language, and it is not the same as language *per se* (Fitch et al. 2005; Hauser et al. 2002). This means that speech is comparable to other forms of vocalization among nonhuman primates and other animals. Nevertheless, it should be noted that no human population lacks verbal communication, and in fact, speech faculties are among the important components for the current forms of human language (Fitch et al. 2005; Hauser et al. 2002). Speech is one of the most efficient mechanisms for language communication, in which humans produce many distinct sounds quickly and sequentially even in a short single exhalation (Greenberg et al. 2003; Lieberman 1984). Although humans might have adopted alternative media for language communication in their history (McNeill 2012; Corballis 2002), the evolution of speech was one of the major factors in the evolution of the language with which we are now endowed, both before and after its origin. Such prominent acoustic properties of speech are achieved through sophisticated manipulation of the peripheral organs, the so-called vocal apparatuses. In this, the pulmonary apparatus generates the power source; the vocal folds in the larynx generate the sound source; and the tongue, jaw, larynx, soft palate, and lips generate the radiated acoustic phenomena constituting voice (Titze 1994; Fitch 2000a; Lieberman 1984). Thus, one of the ways to explore the evolution of speech has been through comparative studies on the anatomy and physiology of the vocal apparatuses in nonhuman primates as the cradle of evolution of speech. Here, I survey current knowledge on the similarities and differences in vocal anatomy and physiology between nonhuman primates and humans. The aim of this chapter is to outline the challenges and prospects facing current and future studies in helping to unveil the evolution of human speech.

## 2.2 Shared Basic Vocal Physiology

The source–filter theory is one principle that well explains the acoustic and physiological mechanisms of speech production in humans (Fig. 2.1) (Titze 1994; Chiba and Kajiyama 1941; Fant 1960). This principle also explains vocalizations in nonhuman primates (Koda et al. 2012; Koda et al. 2015; Riede et al. 2005), while additional mechanisms can be found to generate distinct types of vocalizations among them, as seen in other animals (Riede et al. 2017; Charlton et al. 2013). This principle is also shared in the vocalizations of rodents, songbirds, and crocodiles (Nowicki 1987; Pasch et al. 2017; Reber et al. 2015), meaning that this model represents an evolutionarily antique and universal feature of vocalizations in primates.

The radiated sound is induced by the airflow exhaled in response to the compression of pulmonary volume (Fig. 2.1). It is rarely induced by inhaled airflow in human speech but often in vocalizations for some species of nonhuman primates, for example, in the "pant–hoots" of chimpanzees (Fedurek et al. 2013; Hewitt et al. 2002). The airflow is regulated by the contraction or relaxation of thoracic muscles including the diaphragm and by passive recoil forces within the pulmonary cavity.

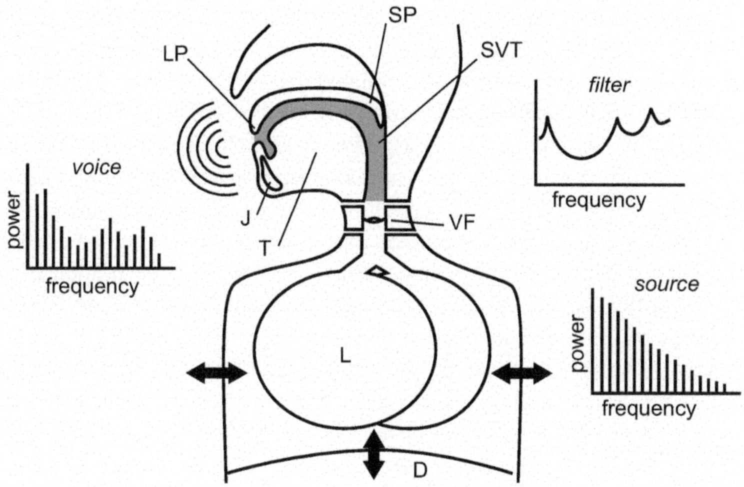


**Fig. 2.1** Diagram for the source–filter theory in speech and vocalizations (Adopted from Nishimura et al. (2008) with some changes). Abbreviations: *D* diaphragm, *J* jaw, *L* lung, *LP* lips, *SP* soft palate, *SVT* supralaryngeal vocal tract, *T* tongue, *VF* vocal folds

The airflow runs cranially through the glottis, to induce cyclical vibrations of the vocal folds and generate the sound source (Fig. 2.1). This physical and acoustic process is termed phonation. The vibration induced passively by airflow results in self-sustained oscillation, described by the myoelastic-aerodynamic (MEAD) model (Van Den Berg 1958; Titze 1980). This mechanism is also found in the avian syrinx (Elemans et al. 2015). The bilateral vocal folds within the larynx make contact and separate from each other periodically, to generate opening and closing phases of the glottis. The acoustic properties of the source are often characterized by acoustic periodicity, sound pressure levels, and the harmonic structure (Herbst 2016; Titze 1994). These mainly determine the musical aspects of radiated sounds such as pitch, volume, and tone. The location and relative power of the fundamental frequency ($f_0$) and its higher harmonics are important properties of the generated voiced sounds. Nevertheless, even if sounds are generated having the same $f_0$ value, other features of the source vary independently and dependently with each other to produce different sounds we perceive (Herbst 2016). The physical process of phonation is complicated, and many aspects of the process are examined in humans by multidisciplinary approaches, including direct measurements, computer simulation, and experimental studies using physical models (Herbst 2016; Titze 1994).

The sound source makes the air filling the supralaryngeal vocal tract (SVT) resonate in accordance with its resonance properties, which are dictated by the volumetric topology of the SVT forming a cavity from the glottis to the mouth (Chiba and Kajiyama 1941; Fant 1960; Titze 1994; Stevens 1972). This means that the SVT serves as a filter to amplify some harmonics of the source near to its inherent resonant frequencies and to suppress others (Fig. 2.1). This process is termed articulation. The volumetric topology of the SVT is modified by the coordinated actions of the tongue and hyoid, jaw, larynx, soft palate, and lips (Lieberman 1984; Crelin 1976; Titze 1994). Voiced sounds along with some frequency bands of the amplified harmonics—formants—are radiated from the mouth. The distribution pattern of the formant positions determines the kind of voiced sounds we perceive, such as vowels in humans (Fant 1960; Titze 1994; Stevens 1972). The physical process of articulation is well known compared with that of phonation (Titze 1994; Stevens 1972).

Source–filter independency is another important feature of human speech. This means that the property of the sound source is only slightly influenced by the resonant properties of the SVT (Chiba and Kajiyama 1941; Fant 1960; Titze 1994). The $f_0$ position is independently changeable from the SVT acoustics in human speech. This is in contrast to rigid source–filter interactions, where vibrations of the vocal folds are inevitably influenced by the acoustic and resonant properties of the SVT, which primarily determine the $f_0$ position (Fletcher and Rossing 1998). Such a strong interaction is seen in some extreme cases of human singing and in some wind instruments (e.g., the trombone). It reduces the flexibility and stability of producing a series of distinct sounds in human speech.

## 2.3 Helium Experiments

The source–filter theory has been tested by examining the acoustics of voices in helium gas, which are modified from natural acoustics, in animal vocalizations (Reber et al. 2015; Madsen et al. 2012; Yamada and Okanoya 2003; Ballintijn and ten Cate 1998; Rand and Dudley 1993; Nowicki 1987; Pasch et al. 2017). The resonance properties of the SVT are influenced by the speed of sound, which is affected by air density, temperature, and pressure. Under helium-enriched conditions, the speed of sound is increased and every resonance frequency is also increased (Nowicki 1987). For vocalizers using SVT filtering, the acoustic features of their voices are inevitably influenced by such increased sound speed, and the positions of all the formants are shifted to higher frequencies (Rand and Dudley 1993; Nowicki 1987). In the case of source–filter independence, such a condition does not shift the $f_0$, whereas only formants are shifted. By contrast, in the case of a strong source–filter interaction, the speed of sound should shift $f_0$ up to a similar degree as among the formants. Experiments using inhaled helium gas have been performed to determine the acoustic functions of given vocal apparatuses in several animals. For example, there are few filtering effects in the vocal sac that is inflated under the jaw in frogs (Rand and Dudley 1993); source–filter independence also applies to songbirds (Nowicki 1987) and alligators (Reber et al. 2015); and moderate source–filter interactions apply to some forms of vocalization in grasshopper mice (Pasch et al. 2017).

Studies on nonhuman primates examined so far all support the view that their varied forms of vocalizations are explained by source–filter independence, as seen in human speech. Hylobatids produce distinct calls termed “songs.” These are loud, high-pitched, pure tone-like, and melodious sounds, which can be heard for over 2 km even in the dense tropical forest of Southeast Asia (Marshall Jr. and Marshall 1976; Geissmann 2002). Helium-breathing experiments demonstrated that white-handed gibbons (*Hylobates lar*) produce their voices in accordance with source–filter independence, probably by tuning the first formant to $f_0$, similar to human soprano singing (Koda et al. 2012). Common marmosets (*Callithrix jacchus*) produce high-pitched "phee" calls with an $f_0$ of >7000 Hz in tropical forests of South America (Bezerra and Souto 2008). These whistle-like sounds are also produced in accordance with source–filter independency probably with a subtle degree of interaction as seen in human speech (Koda et al. 2015). Thus, helium experiments have confirmed that the source–filter principle is common to vocalizations in nonhuman primates and to speech in humans. Humans share the source–filter principle of vocal physiology with nonhuman primates, and we manipulate the vocal apparatuses sophisticatedly for quick and sequential productions of distinct voices in speech.

## 2.4 Vocal Folds and Vocal Membranes

The exhalation phase follows the equally long inhalation phase in tidal respiration at rest, where the air pressure peaks at the onset and then decreases gradually in both phases. By contrast, the long exhalation phase is interrupted by an instantaneous inhalation in speech, where the air pressure is maintained as steady during exhalation (Titze 1994). Such a steady pressure is generated by voluntary controls of the pulmonary muscles involved in respiration, while respiration must be controlled involuntarily for tidal respiration in daily life.

The long and steady exhaled airflow runs up through the glottis, to generate a steady sound source. The human vocal folds are medially and vertically thick (Kurita et al. 1983; Ishii et al. 1999). During modal phonation of speech, the closing phase starts by contact between the lower margins of the thick vocal folds (Fig. 2.2a) (Story 2002). The contact area increases upward, and the upper margins—the free edges of the vocal folds—contact each other. The folds start to separate from each other simultaneously at the lower margins, and they then separate fully at the upper margins, to initialize the open phase of the glottis. The glottal area formed between the bilateral vocal folds increases laterally and then decreases, and the open phase ends by the beginning of contact between the lower margins of the vocal folds to initialize the closure phase. Such wave motions of the mucosa of the vocal folds are well demonstrated by the two-mass model, where each of the vocal folds is composed of upper and lower masses coupled by a linear spring (Story and Titze 1995; Ishizaka and Flanagan 1972). The relationships between the sub- and supraglottal pressures, in addition to the physical properties of the vocal folds, determine the mode of action of the mucosal waves (Story 2002; Titze 1994).

Hirano (1974) proposed that vocal folds, acting as oscillators, are composed of five layers that are histologically distinct in adult humans: the surficial epithelium; the superficial, intermediate, and deep layers of the lamina propria; and the

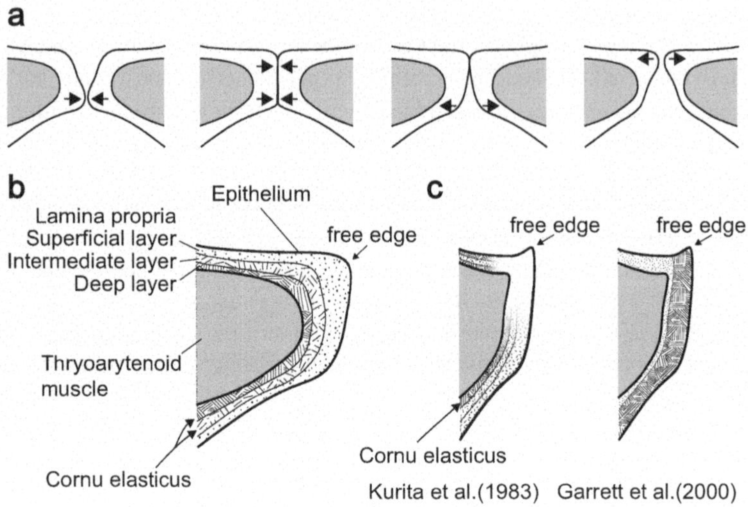


**Fig. 2.2** Vocal fold vibration and anatomy: (**a**) Diagram for the cycle of vocal fold vibration at modal phonation in humans (After Story (2002)), (**b**) vocal fold structure in humans, and (**c**) vocal fold structures observed in macaques

thyroarytenoid muscles (Fig. 2.2b). The superficial layer of the lamina propria coarsely has both collagen and elastin fibers, the intermediate layer is rich in elastin fibers, and the deep layer is rich in collagen fibers (Hirano 1974; Hirano 1975; Kurita et al. 1983; Garrett et al. 2000). The conus elasticus continues to the intermediate and deep layers, and the two layers comprise what is called a vocal ligament along the free edge of the vocal folds (Fig. 2.2b). The layered structure is subdivided into two components—the body and cover—in terms of an oscillator (Titze 1994; Hirano and Kakita 1985; Hirano 1974). The body corresponds to the thyroarytenoid muscle, while the cover corresponds to the epithelium and superficial layer. The intermediate and deep layers are transitional between them and envelop and keep the form of the medially thick thyroarytenoid muscle. The cover slides or glides over the rigid body to generate mucosal waves for producing normal modal phonation in human speech (Story and Titze 1995; Hirano 1974).

The histology of the vocal folds has been examined in detail in few taxa of nonhuman primates. While there are some controversies in terms of the definitions of the layer(s) of the vocal folds, the layered structure in macaques differs from that of humans (Fig. 2.2c). The thyroarytenoid muscle and the lamina propria are thinner in macaques (Ishii et al. 1999; Kurita et al. 1983). The histological compositions are different in the lamina propria of the upper-lateral and medial parts of the vocal folds in macaques (Ishii et al. 1999; Garrett et al. 2000; Kurita et al. 1983; Riede 2010). There seem to be no uniform fibrous layers that fully envelop the thyroarytenoid muscle in macaques. In the upper-lateral part—the floor of the laryngeal ventricle—Kurita et al. (1983) identified three layers in macaques. The superficial and deep layers comprise a network of elastin and collagen fibers, and the intermediate layer includes mostly adipose tissue. By contrast, Garrett et al. (2000) identified no distinct layered structure but instead found dense ground substance there. Two or three layers have been identified in the lower-medial part of the vocal folds (Garrett et al. 2000; Ishii et al. 1999; Kurita et al. 1983; Riede 2010). Kurita et al. (1983) identified three layers: the superficial layer is composed of dense collagen fibers, the intermediate layer is composed of adipose and loose fibrous tissue, and the deep layer is formed by a network of elastin and collagen fibers. Ishii et al. (1999) and Garrett et al. (2000) detected two layers: the superficial layer is rich in both collagen and elastin fibers, and the deep layer is composed of dense ground substance. Riede (2010) also detected adipose tissue in the deep region, while he did not in any layered structure there. Regardless of differences in diagnosis, the superficial region is densely fibrous in macaques, while the deep region is coarsely fibrous. This is quite distinct from humans where the superficial layer is poor in both elastin and collagen fibers. Ishii et al. (1999) found that the elastic fibers are densely distributed around the free edges of the vocal folds in macaques. This was confirmed by the figures published by Riede (2010). In addition, Kurita et al. (1983) indicated that the conus elasticus connects to the deep layer that they detected. These findings suggest that the conus elasticus continues to the deep region, and then such a fibrous layer continues upward to the superficial region, to end around the free edges of the vocal folds. The lack of consensus about a layered structure in this region might reflect that this structure varies along the longitudinal axis of the medial part. Thus, such histological compositions cannot be easily subdivided to delineate two functional components: the body and cover. Such fibrous superficial regions of the free edges and medial parts of vocal folds in macaques probably have physical properties different from those seen in humans where they generate mucosal waves. In fact, the physical properties of the lamina propria of the vocal folds differ between macaques and humans, and the vocal folds can potentially generate many different modes of mucosal waves in macaques (Riede 2010). The properties of vocal fold vibrations in macaques probably reflect such anatomical features. Their vocal fold vibrations are similar to those of human children (Herbst et al. 2018) who have not yet developed the clear layered structure seen in adults (Hirano et al. 1983). The histological compositions and their physical properties need to be examined in other taxa for our better understanding of the evolutionary history of human-like vocal folds.

The vocal membranes or lips have been identified in a number of nonhuman primates, while humans do not have this feature or have its vestige (Mergell et al. 1999; Fitch and Hauser 1995; Starck and Schneider 1960; Harrison and Harrison 1995). The variation and phylogeny of this feature are to be examined in primates. On histological sections, they are often regarded as an upward extension from the vocal folds (Fig. 2.3). A simulation study suggested that they help lower subglottal pressure to give increased efficiency in producing loud and high-pitched calls (Mergell et al. 1999; Fitch 2000a). The morphology of the vocal membranes varies, and this feature can be regarded as the body of vibration and/or the free edges of the vocal folds in some species among nonhuman primates, e.g., strepsirrhines and platyrrhines (Figs. 2.3, 2.4). Although the vocal ligament is not superficial in

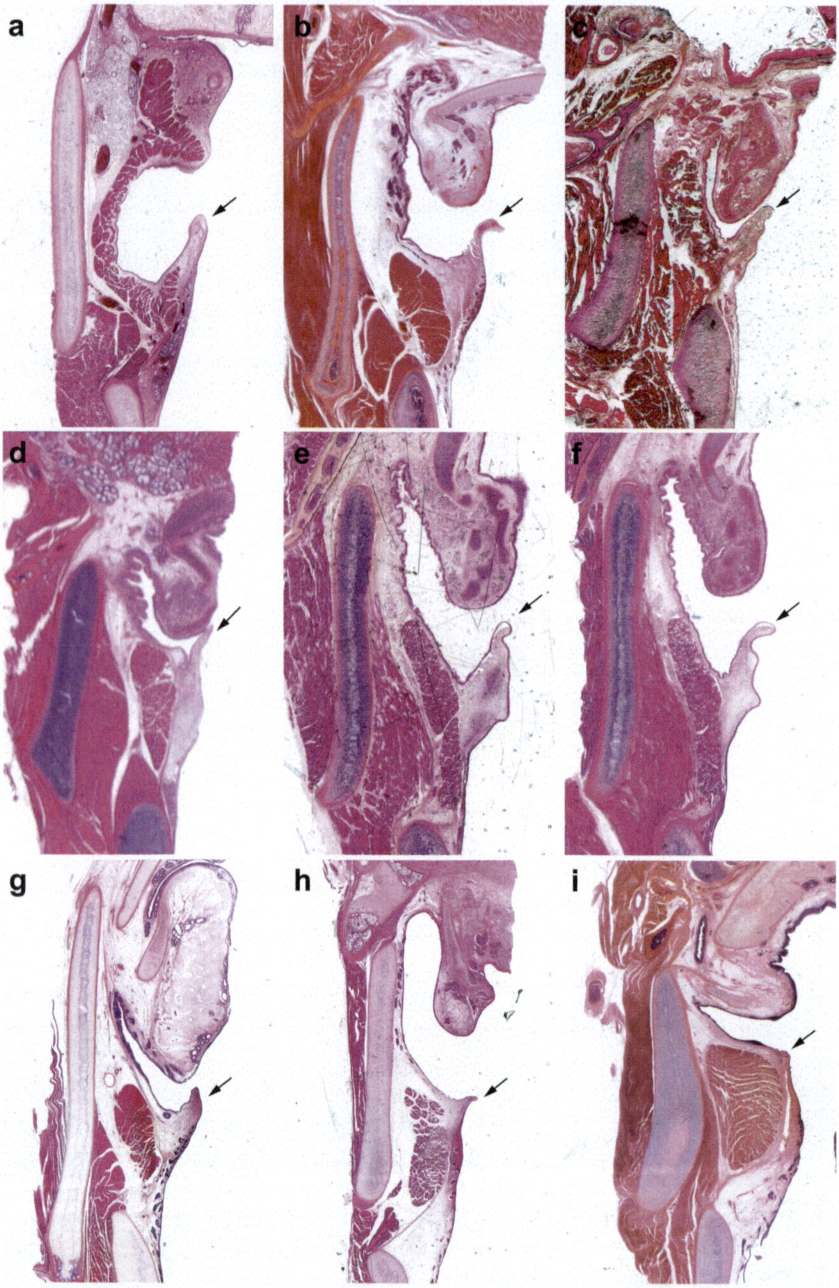


**Fig. 2.3** Histological sections of the vocal fold in nonhuman primates from Hayama's Collection (Hayama 1970). (**a**) *Lemur catta*, female; (**b**) *Cebus apella*, female; (**c**) *Saimiri sciureus*, female; (**d**) *Callithrix jacchus*, male; (**e**) *Saguinus oedipus*, female; (**f**) *Cebuella pygmaea*, male; (**g**) *Ateles paniscus*, female; (**h**) *Macaca sylvanus*, male; (**i**) *Macaca fuscata fuscata*, female; (**j**) *Chlorocebus aethiops*, male; (**k**) *Cercocebus atys*, male; (**l**) *Piliocolobus badius*, male; (**m**) *Semnopithecus entellus*, female; and (**n**) *Trachypithecus obscurus*, male. See also Fig. 2.5. The size and shape of vocal membranes (arrows) are varied among nonhuman primates

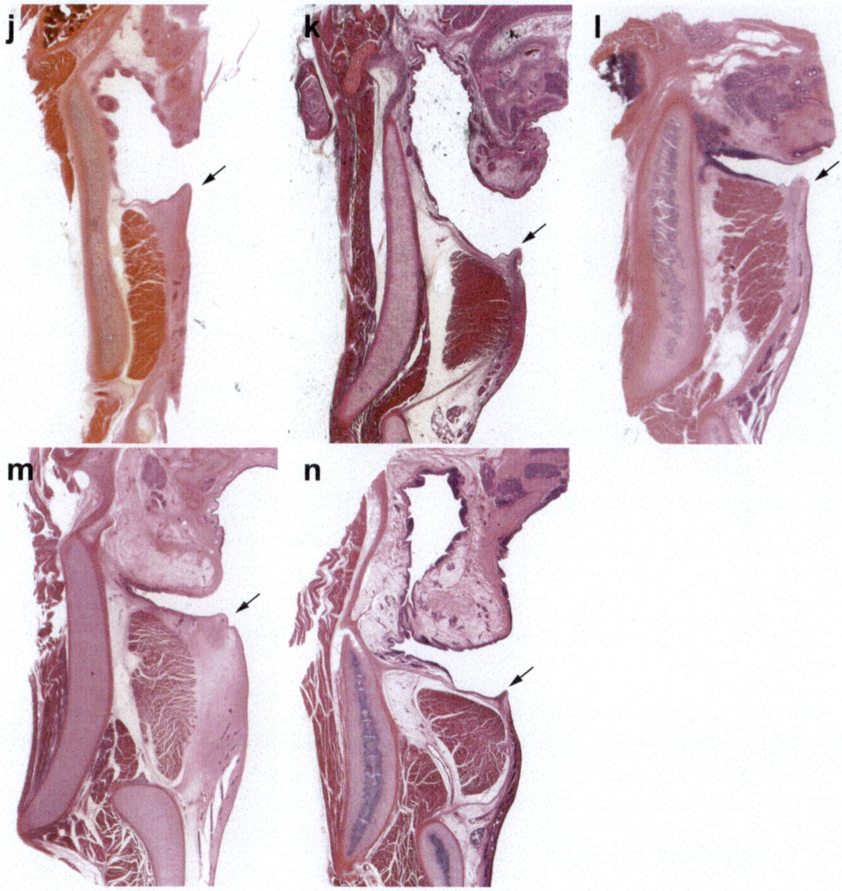


**Fig. 2.3** (continued)

humans, having no, histological studies in macaques suggest that the vocal membranes that extend from the superficial layer are composed of dense fibrous tissue, i.e., vocal ligament. The actions of these membranes have not been examined during vocalizing *in vivo* in nonhuman primates. Further histological and morphological studies need to be performed to increase knowledge on the properties of these structures as oscillators and in their acoustic contributions to calls in nonhuman primates.

This region has not attracted much attention, but the phonatory mechanisms have been examined in some taxa of nonhuman primates (Garcia and Herbst 2018; Herbst et al. 2018; Herbst and Dunn 2018; Herbst 2016). Increasing knowledge on the anatomy and physiology of the vocal folds provides new insights for understanding the evolution of human phonation.

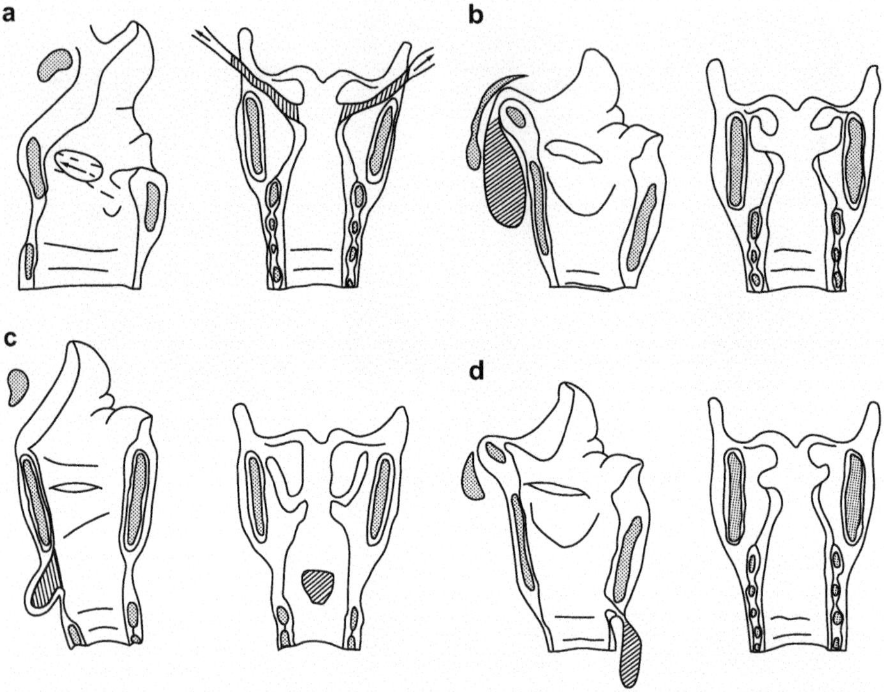


**Fig. 2.4** Four types of the laryngeal air sac in nonhuman primates: (**a**) lateral ventricular sac, (**b**) subhyoidal sac, (**c**) infraglottal sac, and (**d**) dorsal sac. (Adopted from Hayama (1970) after tracing the original figures)

## 2.5 Laryngeal Air Sac

Some primates show an addition and/or modification to existing apparatuses for producing their species-distinct vocalizations. Howler monkeys (genus *Alouatta*) from Central and South America are well known to have an enlarged and modified hyoid and larynx, which are believed to contribute to producing their loud, noisy, and low-frequency roars (Starck and Schneider 1960; Schon 1971; Dunn et al. 2015). The laryngeal air sac is also often cited for its potential contributions to the acoustics of calls made by nonhuman primates (Fitch 2000a; Dunn 2018; Hewitt et al. 2002). Siamangs (*Symphalangus syndactylus*) from Southeast Asia have a large air sac in the ventral region of the neck under the jaw (Mott 1924; Nemai and Keleman 1933; Starck and Schneider 1960; Hayama 1970) that is inflated just before they begin their loud calls (Marshall Jr. and Marshall 1976). Such inflation is also found in a similar region during vocalizing in putty-nosed guenons (*Cercopithecus nictitans*) from Central Africa and Japanese macaques (*Macaca fuscata*) from the Japanese archipelago of East Asia (Itani 1963; Gautier and Gautier 1977; Gautier 1971). Acoustic experiments and simulations using a physical model have supported the view that these sacs are probably used to amplify the voice rather than to modify its formant pattern (Riede et al. 2008; de Boer 2009). This view is supported by experiments involving surgical removal of the sac, which reduces the loudness of vocalization in putty-nosed guenons (Gautier 1971; Gautier and Gautier 1977).

Variations in the anatomy of the laryngeal sac have been reported in several papers (Starck and Schneider 1960; Bartels 1904; Negus 1949) and have been summarized by Hayama (1970) and Hewitt et al. (2002). Four types of laryngeal sacs in nonhuman primates are usually described in terms of the location of their openings to the laryngeal region (Fig. 2.4) (Starck and Schneider 1960; Hayama 1970; Hewitt et al. 2002) as follows:

1. The lateral ventricular sac or the *Saccus ventriculi laryngis lateralis* (VL, Hayama 1970) extends upward from the bilateral laryngeal ventricles to exit the larynx. Bilateral extended tubes fuse with each other in the ventral region of the neck and then expand cranially and/or caudally.
2. The subhyoidal sac or the *Saccus laryngis medianus superior anterior* (MSA, Hayama 1970) opens above the anterior commissure of the bilateral vocal folds at the base of the epiglottis and extends cranially along the middorsal region of the thyroid cartilage. The extended tube usually expands and occupies the dorsal region of the cup-shaped hyoid body and then extends caudally along the ventral region of the neck between the skin and the larynx and trachea.
3. The infraglottal sac or the *Saccus laryngis medianus inferior anterior* (MIA, Hayama 1970) opens at the midventral surface between the thyroid and cricoid cartilages and extends caudally along the ventral region of the neck between the sternohyoid muscle and trachea.
4. The dorsal sac or the *Saccus laryngis medianus inferior posterior* (MIP, Hayama 1970) opens at the middorsal surface between the cricoid cartilage and the first tracheal ring and extends between the trachea and the esophagus.

These differences in opening locations strongly suggest that these four types represent different features in terms of evolutionary homology, even if they share same acoustic effects. A lateral ventricular sac has been identified in *Galago*, *Otolemur*, *Cebus/Sapajus*, and *Callicebus* species, some species of colobines, and siamangs (*Symphalangus syndactylus*) and the great apes (Fig. 2.5). It appears only rarely in *Ateles* and *Aotus* species and the other hylobatids and has not been described in the cercopithecids. Alternatively, a subhyoidal sac is usually found in the cercopithecids (Fig. 2.6). It is also found in *Galago* and *Otolemur* species (Fig. 2.5) but has not been reported in the platyrrhines or hominoids. An infraglottal sac is found among the callitrichids but rarely in *Ateles*, while a dorsal sac is found in the strepsirrhines and atelids (Fig. 2.7). Whereas *Galago*, *Otolemur*, and *Ateles* species probably have multiple types of sacs, the other species of nonhuman primates usually have one type. The presence or absence of sacs varies even in the same taxon among the platyrrhines, strepsirrhines, and hylobatids excluding *Symphalangus*. Such confusion probably reflects wide variations between and within species. In addition, changes during growth might also add confusion. In fact, the sac grows to increase its volume caudally along the neck after birth in male and female chimpanzees (Nishimura et al. 2007).

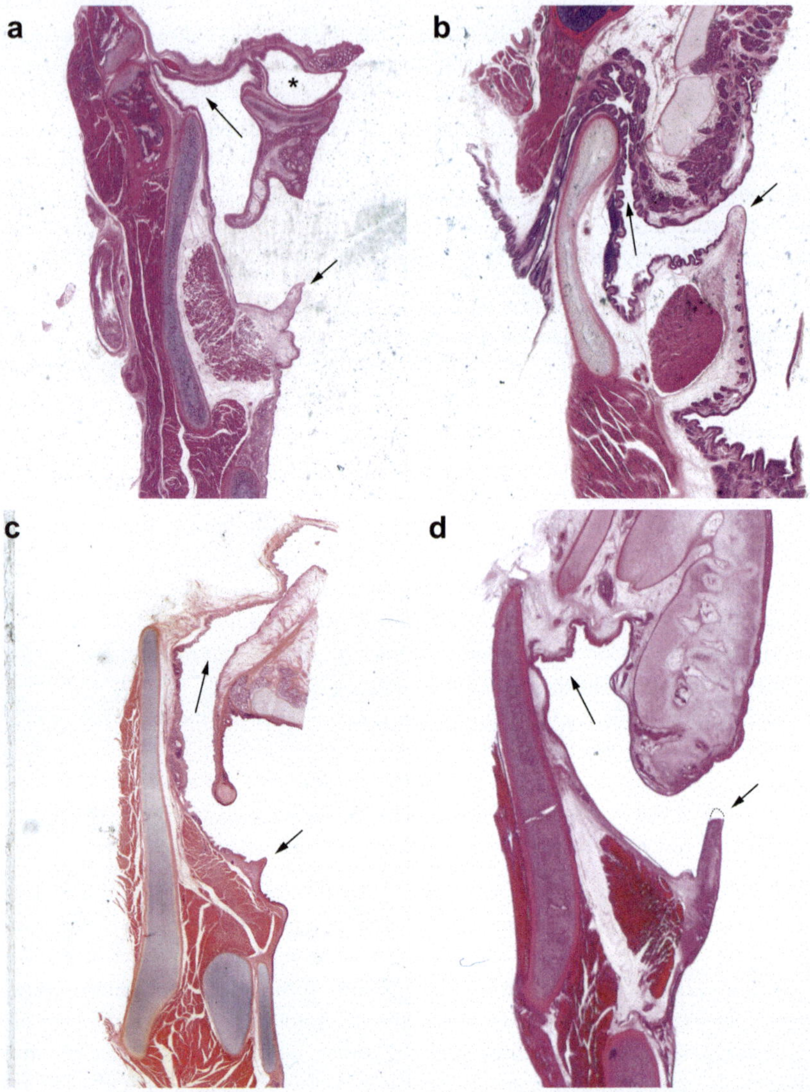


**Fig. 2.5** Histological sections of the lateral ventricular sac in nonhuman primates from Hayama's Collection (Hayama 1970). (**a**) *Galago senegalensis*, male; (**b**) *Aotus trivirgatus*, female; (**c**) *Symphalangus syndactylus*, male; and (**d**) *Pan troglodytes*, male. The sac extends out of the larynx from the laryngeal ventricle (long arrows). * indicates the subhyoidal sac. The vocal membranes (short arrows) are varied

In addition to the differences in opening location, variations in the volume of the sac are expected to vary its acoustic effects. Hayama (1970) expanded our knowledge of this topic by examining specimens from 64 species of nonhuman primates

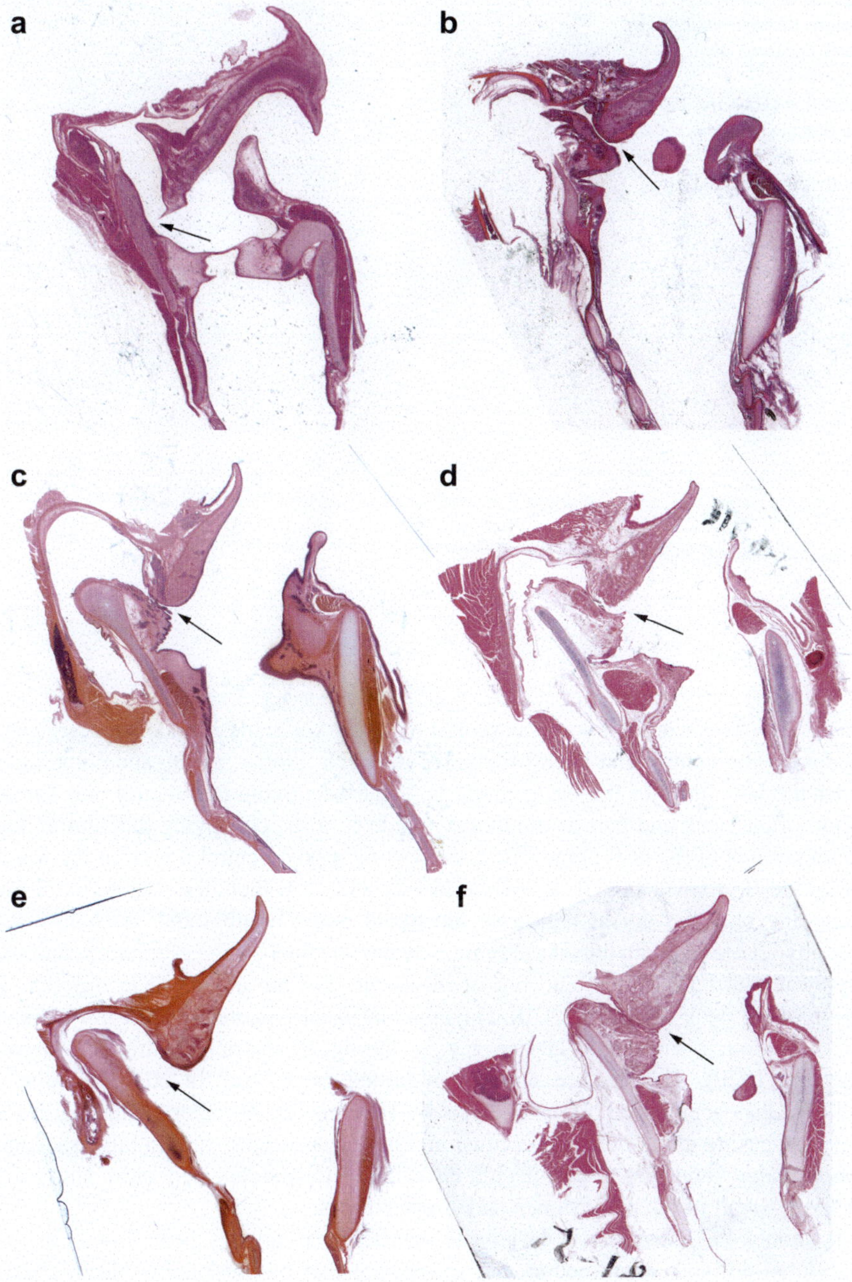


**Fig. 2.6** Histological sections of the subhyoidal sac in nonhuman primates from Hayama's Collection (Hayama 1970). (**a**) *Galago senegalensis*, male; (**b**) *Trachypithecus obscurus*, female; (**c**) *Semnopithecus entellus*, male; (**d**) *Cercocebus atys*, male; (**e**) *Macaca fuscata fuscata*, male; and (**f**) *Papio anubis*, female, juvenile. The subhyoidal sac opens above the anterior commissure of the bilateral vocal folds at the base of the epiglottis (arrow)

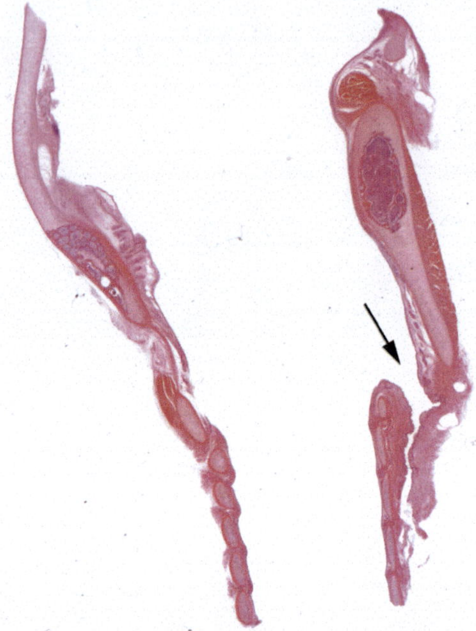

**Fig. 2.7** Histological sections of the dorsal sac in *Ateles paniscus* from Hayama's Collection (Hayama 1970). The dorsal sac opens between the cricoid cartilage and the first tracheal ring (arrow)

(Table 2.1). The size variation was scored by using six grades: grade A, a large sac occupying the ventral and lateral region of the neck, thorax, axilla, and sometimes under the jaw and into the back; grade B, a sac occupying the ventral and lateral region of the neck and part of the thorax; grade C, a sac occupying the ventral and lateral regions of the neck; grade D, a sac occupying the ventral region of the neck; and grade E, a small pouch out of the larynx. Grade E includes a subhyoidal sac occupying only the dorsal region of the hyoid body, a subglottal sac below the sternohyoid muscle, or a dorsal sac lying between the trachea and esophagus, and the laryngeal ventricles that extend out of the larynx but are not fused in the ventral region of the neck (Hayama 1970). Hayama (1970) also scored grade F for a small recess or pouch within the larynx, e.g., a laryngeal ventricle within the larynx (Hayama 1970). This feature was not regarded as a "sac" in other literatures. Table 2.1 has been based on the analysis by Hayama (1970), after excluding grade F of all types, to avoid potential confusion. Great apes usually have a huge sac, and *Symphalangus* and the cercopithecids have large- to middle-sized sacs. These are relatively small in the platyrrhines and strepsirrhines.

It should be noted that a laryngeal ventricle has always been found in the nonhuman primates examined so far, suggesting that its presence is plesimorphic for primates (Hayama 1970). The lateral ventricular sac is open to the laryngeal ventricle, and the former can be regarded as an extension of the latter (Hayama 1970). The lateral ventricular sac probably helps to amplify voiced sounds (de Boer 2009; Riede et al. 2008), and the size variation ranging from a small ventricle to a

**Table 2.1** Size variation of the laryngeal air sac after Hayama (1970)

| | Laryngeal air sac | | | | | | | | Laryngeal ventricle | | |
|---|---|---|---|---|---|---|---|---|---|---|---|
| | Male | | | | Female | | | | | | |
| Species | MSA | MIA | MIP | VL | MSA | MIA | MIP | VL | Male | Female | Remarks |
| *Tupia glis* | E | – | – | – | E | – | – | – | 3 | 3 | |
| *Lemur catta* | – | – | – | – | – | – | – | – | 2 | 2 | |
| *Eulemur mongoz* | – | – | – | – | – | – | – | – | 2 | 2 | |
| *Loris tardigradus* | – | – | – | – | – | – | – | – | 1 | 1 | |
| *Nycticebus caucang* | – | – | – | – | – | – | – | – | 1 | 1 | |
| *Galago senegalensis* | E | – | – | E | E | – | – | E | 4 | 4 | |
| *Galagoides demidoff* | E | – | – | E | E | – | – | E | 4 | 4 | *Galago denidovii* |
| *Callithrix jacchus* | – | – | – | – | – | – | – | – | 3 | 3 | |
| *Callithrix argentata* | – | – | – | – | – | – | – | – | 3 | 3 | |
| *Callithrix penicillata* | – | – | – | – | – | – | – | – | 3 | 3 | |
| *Cebuella pygmaeus* | – | – | – | – | – | – | – | – | 3 | 3 | |
| *Saguinus midas* | – | – | – | – | – | – | – | – | 3 | 3 | *Saguinus tamarin* |
| *Saguinus oedipus* | – | – | – | – | – | – | – | – | 3 | 3 | |
| *Leontopithecus rosalia* | – | – | – | – | – | – | – | – | 3 | 3 | *Leontideus rosalia* |
| *Aotus trivirgatus* | – | – | – | E | – | – | – | E | - | - | |
| *Cebus capucinus* | – | – | – | E | – | – | – | E | 4 | 4 | *Cebus capchinus* |
| *Cebus apella* | – | – | – | E | – | – | – | E | 4 | 4 | |
| *Saimiri sciureus* | – | – | – | – | – | – | – | – | 3 | 3 | |
| *Ateles paniscus* | – | – | E | – | – | – | E | – | 3 | 3 | |
| *Ateles belzebuth* | – | – | – | – | – | E | – | E | 3 | 3 | |
| *Ateles geoffroyi* | – | – | E | – | – | – | E | – | 3 | 3 | |
| *Lagothrix lagothricha* | – | – | – | – | – | – | – | – | 3 | 3 | |
| *Macaca sylvanus* | B | – | – | – | | | | | 3 | | |

(continued)

**Table 2.1** (continued)

| Species | Laryngeal air sac | | | | | | | | Laryngeal ventricle | | Remarks |
|---|---|---|---|---|---|---|---|---|---|---|---|
| | Male | | | | Female | | | | | | |
| | MSA | MIA | MIP | VL | MSA | MIA | MIP | VL | Male | Female | |
| *Macaca sinica* | E | – | – | – | E | – | – | – | 3 | 2 | |
| *Macaca radiata* | E | – | – | – | E | – | – | – | 3 | 2 | |
| *Macaca silenus* | C | – | – | – | D | – | – | – | 3 | 2 | |
| *Macaca nemestrina* | B | – | – | – | C | – | – | – | 3 | 2 | |
| *Macaca fascicularis* | E | – | – | – | E | – | – | – | 3 | 2 | |
| *Macaca mulatta* | D | – | – | – | E | – | – | – | 3 | 2 | |
| *Macaca cyclopis* | D | – | – | – | E | – | – | – | 3 | 2 | |
| *Macaca arctoides* | B | – | – | – | C | – | – | – | 3 | 2 | *Macaca speciosa* |
| *Macaca fuscata fuscata* | B | – | – | – | C | – | – | – | 3 | 2 | |
| *Macaca fuscata yakui* | B | – | – | – | C | – | – | – | 3 | 2 | |
| *Macaca maurus* | B | – | – | – | C | – | – | – | 3 | 2 | |
| *Macaca nigra* | B | – | – | – | | | | | 3 | | *Cercopithecus niger* |
| *Papio anubis* | B | – | – | – | C | – | – | – | 3 | 2 | |
| *Papio cynocephalus* | B | – | – | – | C | – | – | – | 3 | 2 | |
| *Papio hamadryas* | B | – | – | – | C | – | – | – | 3 | 2 | |
| *Mandrillus sphinx* | B | – | – | – | C | – | – | – | 3 | 2 | |
| *Mandrillus leucophaeus* | B | – | – | – | C | – | – | – | 3 | 2 | |
| *Theropitheucs gelada* | B | – | – | – | | | | | 3 | | |
| *Cercocebus torquatus* | B | – | – | – | C | – | – | – | 3 | 2 | |
| *Cercocebus atys* | B | – | – | – | C | – | – | – | 3 | 2 | |
| *Cercocebus galeritus chrysgaster* | B | – | – | – | | | | | 3 | | |
| *Chlorocebus aethiops* | E | – | – | – | E | – | – | – | 3 | 2 | |
| *Cercopithecus cephus* | C | – | – | – | D | – | – | – | 3 | 2 | |

| | | | | | | | | | | | |
|---|---|---|---|---|---|---|---|---|---|---|---|
| *Cercopithecus diana* | C | – | – | – | D | – | – | – | 3 | 2 | *Cercopithecus ciana* |
| *Cercopithecus lhoesti* | C | – | – | – | D | – | – | – | 3 | 2 | |
| *Cercopithecus mitis* | C | – | – | – | D | – | – | – | 3 | 2 | |
| *Cercopithecus mona* | D | – | – | – | E | – | – | – | 3 | 2 | |
| *Cercopithecus neglectus* | C | – | – | – | D | – | – | – | 3 | 2 | |
| *Cercopithecus ascanius* | C | – | – | – | D | – | – | – | 3 | 2 | |
| *Erythrocebus patas* | D | – | – | – | – | – | – | – | 3 | 2 | |
| *Semnopithecus entellus* | B | – | – | – | C | – | – | – | 3 | 2 | *Presbytis entellus* |
| *Trachypithecus obscurus* | D | – | – | – | E | – | – | – | 3 | 2 | *Presbytis obscurus* |
| *Piliocolobus badius* | – | – | – | E | – | – | – | – | 4 | 4 | *Colobus badius* |
| *Hylobates lar* | – | – | – | – | – | – | – | – | 4 | 4 | |
| *Nomascus concolor* | – | – | – | E | – | – | – | E | | | *Hylobates concolar* |
| *Symphalangus syndactylus* | – | – | – | B | – | – | – | B | | | |
| *Pongo pygmaeus* | – | – | – | A | – | – | – | A | | | |
| *Pan troglodytes* | – | – | – | A | – | – | – | A | | | |
| *Gorilla gorilla gorilla* | – | – | – | A | – | – | – | A | | | |
| *Gorila gorilla beringei* | – | – | – | A | – | – | – | A | | | |
| *Homo sapiens* | – | – | – | – | – | – | – | – | 3 | 3 | |

This table is corrected from Hayama (1970): Grade F of the sac in the original article is removed, because it is not regarded as a laryngeal air sac here; the scientific names of some species are corrected and the names cited in the original article are in remarks

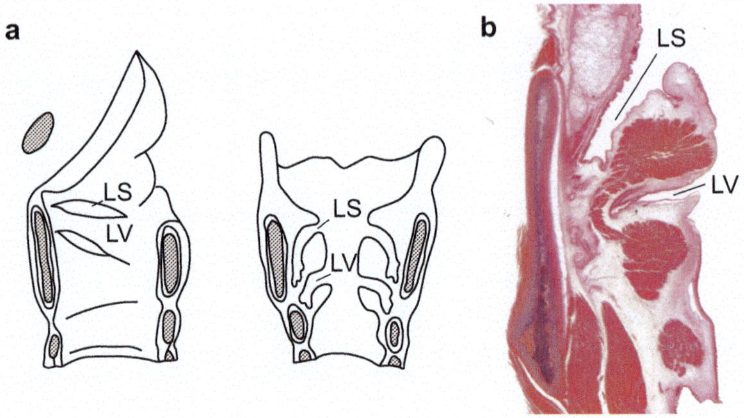


**Fig. 2.8** Another recess in laryngeal region. (**a**) *Saccus laryngis lateralis superior* (LS, Hayama 1970), adopted from Hayama (1970) after tracing the original figure. (**b**) A histological section of LS in *Eulemur mongoz*, female, from Hayama's Collection (Hayama 1970). LS is found superior to the laryngeal ventricle (VL)

huge sac is also likely to contribute to varying acoustic effects. Hayama (1970) scored the size of the laryngeal ventricle as follows: grade 1, small and extending laterally to the medial surface of the thyroid cartilage and a little cranially or caudally from there; grade 2, extending cranially but not reaching the upper rim of the thyroid cartilage; grade 3, extending to the upper rim of the thyroid cartilage; and grade 4, extending well over the upper rim of the thyroid cartilage to end in a tube-like structure at the base of the epiglottis. Grade 4 almost corresponds to grade E mentioned above. The human lateral ventricular sac is classed as grade 3, while nonhuman primates are usually classed higher (Table 2.1, Hayama 1970). Smaller ventricles are often found in strepsirrhines and female individuals among cercopithecids (Hayama 1970). In addition, another small recess, termed the *Saccus laryngis lateralis superior* ("LS" in Hayama 1970), is also found superior to the laryngeal ventricle in some species of strepsirrhines (Fig. 2.8, Hayama 1970). Thus, the larynx has a variety of accessory recesses around the glottis in nonhuman primates, some of which are absent or are rarely formed in humans.

The lateral laryngeal sacs found in great apes interfere with producing the distinct voices required for human speech (de Boer 2009). Instead, the lateral ventricular and subhyoidal sacs probably have the other acoustic effects, including the amplification of calls (Riede et al. 2008; de Boer 2009). The same effect is not always guaranteed for every type of laryngeal sac. In fact, two types of sac open above the glottis, while the other two open below it. This suggests that they have different effects on the air pressure or resonance (if any) of sub-/supraglottal regions. Future examinations are likely to uncover hitherto unknown acoustic contributions of these accessory apparatuses in terms of location, volume, and flexibility (inflatability), in nonhuman primates.

## 2.6 Supralaryngeal Vocal Tract and Tongue

Humans have distinct anatomical features in terms of the SVT and tongue (Lieberman 1984; Fitch 2000a; Takemoto 2001). The SVT is composed of horizontal oral and vertical pharyngeal cavities (Fig. 2.9) (Crelin 1976; Lieberman 1984; Titze 1994). The two cavities are almost equally long and are located almost perpendicular to each other, and a narrow channel termed the oropharyngeal isthmus is formed between the two cavities in humans (Fig. 2.9) (Crelin 1976; Lieberman 1984; Titze 1994). This configuration has been described as a "two-tube" model of the SVT (Titze 1994). It allows the horizontal and vertical cavities to act acoustically as independent resonant tubes, in addition to the whole tube of the SVT (Titze 1994). By contrast, static anatomical studies have shown that the pharyngeal cavity is shorter than the oral cavity and that the two cavities are located almost parallel to each other in nonhuman primates (Fig. 2.9) (Lieberman 1984; Laitman and Reidenberg 1997). A narrow isthmus is not found in nonhuman primates. Their static SVT is shaped like a "single tube" that is slightly curved cranially.

The SVT is contoured ventrally by the tongue surface. The tongue is globular in humans, while it is flat and horizontally cylindrical in nonhuman primates (Fig. 2.9) (Laitman and Reidenberg 1997; Nishimura 2005; Nishimura et al. 2008; Takemoto 2008; Lieberman 1984; Takemoto 2001). The tongue musculature is covered by the epithelium. Movements of such a nonskeletal organ constrained by a flexible outer layer are explained by the muscular hydrostat theory (Kier and Smith 1985; Smith and Kier 1989). The human tongue consists of three directed intrinsic muscle fibers, arranged perpendicular to, in parallel with, and wrapped helically or obliquely around the longitudinal axis of the tongue (Takemoto 2001). The perpendicular muscles are found centrally, whereas the other two components are found peripherally (Takemoto 2001). Contraction of the perpendicular fibers reduces the diameter and increases the length of the tongue to protrude it because the volume of each fiber component is constant (Kier and Smith 1985; Smith and Kier 1989; Takemoto 2001). Contraction of the parallel fibers increases the diameter of the tongue and reduces its length by acting antagonistically against the perpendicular fibers (Kier and Smith 1985; Smith and Kier 1989; Takemoto 2001). Contraction of the helically wrapped fibers twists the tongue around its longitudinal axis (Kier and Smith 1985; Smith and Kier 1989; Takemoto 2001). This model of the actions of the human tongue muscles successfully explains how the three directed intrinsic muscle fibers arranged within the globular tongue allow for the deformations of the tongue surface with a high degree of freedom by subtle actions of intrinsic muscles (Takemoto 2001).

The composition of the tongue musculature is common to chimpanzees and humans (Takemoto 2008). Although there have been few systematic studies of the tongue musculature in the other nonhuman primates, many species, including macaques and baboons, probably share it (Doran 1975; Sokoloff 2000; Sonntag 1925). Nevertheless, their muscle fibers are arranged within their horizontally cylindrical tongue, and their mutual orientations differ from those in the human

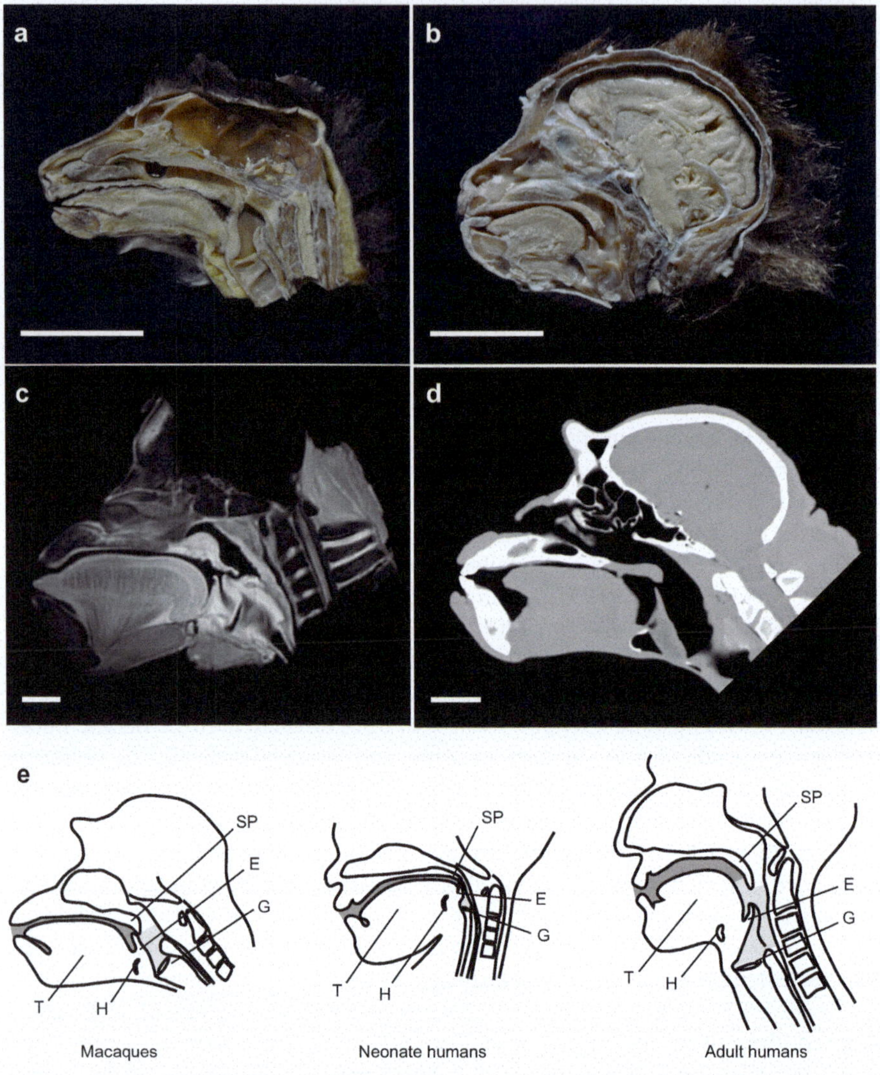


**Fig. 2.9** Midsagittal sections of the supralaryngeal vocal tract in primates. (**a**) *Eulemur fulvus*, section of an embalmed specimen; (**b**) *Macaca fascicularis*, section of an embalmed specimen; (**c**) *Pongo pygmaeus*, scan of magnetic resonance imaging; (**d**) *Pan troglodytes*, scan of computed tomography; and (**e**) diagrams in macaques, neonate humans, and adult humans (Adopted from Nishimura et al. (2008) with some changes). Abbreviations: *E* epiglottis, *G* glottis, *H* hyoid, *SP* soft palate, *T* tongue. Scale is 3 cm

tongue (Takemoto 2008). The primary actions of such tongues are restricted to protrusion and retrusion, and this prevents them from undergoing the flexible deformation seen in the human tongue (Takemoto 2008).

## 2.7 Developmental Descent of the Larynx

The differences in the shape of the adult SVT and tongue between humans and nonhuman primates arose in evolution of developmental changes in the positions of the hyoid and larynx and in the anteroposterior length of the jaw. Human babies show an SVT configuration and tongue posture that resemble those seen in adult nonhuman primates (Fig. 2.9c) (Negus 1949; Sasaki et al. 1977). The positions of the hyoid and larynx at rest are close to the soft palate, and the epiglottis is located in the nasopharynx: described as an "intranarial" position (Fig. 2.9c) (Negus 1949; Sasaki et al. 1977). The hyoid and larynx descend along the cervical curvature in the first 9 years of human life (Fitch and Giedd 1999; Sasaki et al. 1977; Lieberman et al. 2001). The epiglottis attached to the thyroid cartilage is detached from the soft palate, and the tongue base, attached to the cranial surface of the hyoid body, is pulled downward into the pharynx (Fig. 2.9c). This descent of the larynx lengthens the vertical pharyngeal cavity and makes it face the posterior surface of vertically thick tongue. On the other hand, humans do not develop a snout, and our face remains short and flat even in adults. The horizontal oral cavity is lengthened during infancy but is almost complete to reach an adult length in the early juvenile stage (Lieberman et al. 2001; Fitch and Giedd 1999). Such derived growth patterns in laryngeal position and face bring about the two-tube SVT and globular tongue configuration of humans.

Nonhuman primates demonstrate different growth patterns of the pharyngeal and oral cavities, while the pharyngeal cavity per se is also lengthened during growth. Chimpanzees (*Pan troglodytes*) show detachment of the epiglottis from the soft palate during infancy as seen in humans, to make the pharyngeal cavity face the posterior surface of the tongue (Nishimura 2005; Nishimura et al. 2006), while Japanese macaques do not show such detachment (Fig. 2.10) (Nishimura et al. 2008; Flugel and Rohen 1991; Laitman et al. 1977). The thyroid cartilage descends against the hyoid in chimpanzees but not in Japanese macaques, precluding the descent of the epiglottis against the soft palate (Fig. 2.10) (Nishimura et al. 2008). These features come about because the thyroid cartilage is connected to the hyoid by a moderately long ligament and they both certainly act independently from each other in hominoids including humans, while they are almost articulated with each other in other primates (Nishimura 2003). Regardless of the descent of the larynx, the pharyngeal cavity is quite short relative to the oral cavity in adult chimpanzees (Fig. 2.9) (Nishimura 2005; Nishimura et al. 2006). By contrast, nonhuman primates develop a long snout. In fact, the face continues to grow long after the early juvenile stage to lengthen the horizontal cavity after the completion of laryngeal descent in chimpanzees (Nishimura 2005; Nishimura et al. 2006; Nishimura et al. 2008). Thus, regardless of the presence or absence of developmental descent of the larynx, inherited growth patterns of face ensure the development of a single-tube SVT in nonhuman primates.

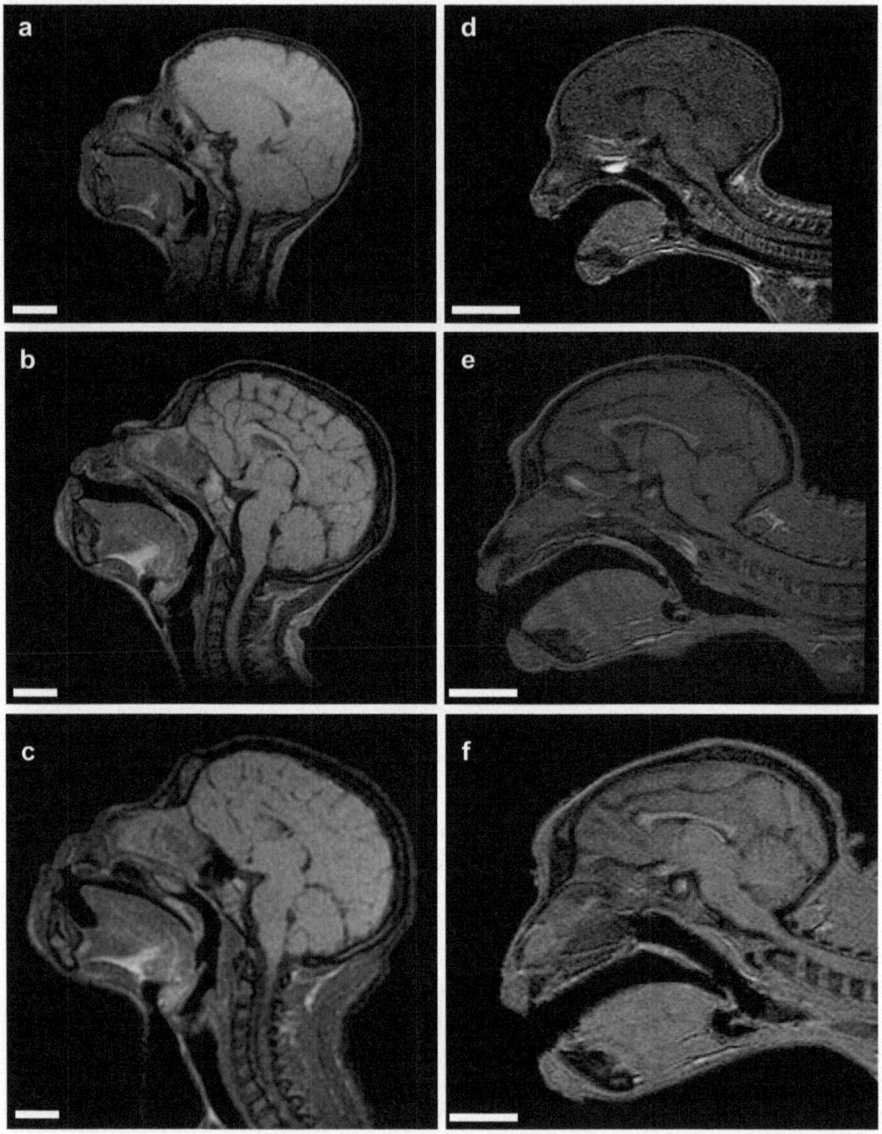


**Fig. 2.10** Growth of the SVT in chimpanzees and Japanese macaques. (Left) Midsagittal magnetic resonance images of a female chimpanzee (*Pan troglodytes*) at (**a**) 4 months of age, (**b**) 2 years of age, and (**c**) 4 years of age. (Adopted from Nishimura et al. (2006) with some changes. Scale is 2.0 cm). (Right) Midsagittal magnetic resonance images of Japanese macaques (*Macaca fuscata fuscata*) at (**a**) 1 month of age in a male subject, (**b**) 13 months of age in a male subject, and (**c**) 25 months of age in a male subject. (Adopted from Nishimura et al. (2008) with some changes. Scale is 2.0 cm)

## 2.8 Mobility of the Tongue and Larynx

Humans use the highly deformable tongue to modify the SVT topology for producing speech sounds. The descended hyoid-larynx secures a long pharyngeal cavity without any particular actions during speech production in adult humans, and they move in a rather restricted range during speech production (Hiiemae and Palmer 2003; Hiiemae et al. 2002). Such small actions are not attributed to any anatomical restrictions, and in fact, they move in a larger range for mastication and are pulled cranially quickly for swallowing (Hiiemae et al. 2002; Hiiemae and Palmer 2003; Ekberg and Sigurjonsson 1982). During speech production, the human hyoid moves slightly anteriorly. It remains there as an anchor for the tongue, and the varied combinations of contraction and relaxation in each of the tongue's intrinsic muscles deform the surface topology of the globular tongue. This means that the sophisticated deformation of the tongue *per se* shifts the position of the narrow channel between the two cavities anteroposteriorly and changes the length and topology of the two cavities (Nishimura 2018). Thus, such a highly deformative and globular tongue is a primal component to facilitate the quick and sequential shifts of varied SVT topology during human speech.

Nonhuman primates probably realize sequential and varied deformations of the SVT topology during their vocalizations. The static anatomical distinction between humans and nonhuman primates means that investigators have underestimated the plasticity of SVT modifications in nonhuman primates (e.g., Lieberman et al. 1969). Helium experiments have now shown that gibbons change the sound source property and the filter effects of the SVT independently from each other in acoustic terms (Koda et al. 2012). This finding indicates that their melodious songs are generated by shifting the $f_0$ to a higher frequency and by synchronizing this with similar changes in the first resonance frequency of the SVT, to maintain a formant tuning with pitch (Koda et al. 2012). They are required to perform highly coordinated actions by increasing the vibration rate of the vocal folds and deforming the SVT topology by enlarging the mouth and/or shortening its length. In addition, highly coordinated respiratory actions are also required. Diana monkeys (*Cercopithecus diana*) from West Africa demonstrate a formant transition in that the formant pattern changes in a single alarm call (Riede and Zuberbuhler 2003; Riede et al. 2005). Putty-nosed guenons (*Cercopithecus nictitans*) from West Africa combine distinct calls into a third structure having another meaningful message, which is different from each of the original calls (Arnold and Zuberbuhler 2006; Zuberbuhler 2006). Campbell's monkeys (*Cercopithecus campbelli*) from West Africa combine an alarm call followed by a suffix call to broaden the meaning of the original call (Ouattara et al. 2009). These formant transitions and combinations require a sequential deformation of the SVT (Riede et al. 2005). Further, macaque species (*Macaca*) from Asia and restricted parts of North Africa probably skillfully assume varied SVT topologies and produce a range of resonance patterns that are larger than thought previously (Fitch et al. 2016). Thus, these findings strongly suggest that a wide range of SVT topology *per se* is not unique to humans and rather the variation of such a faculty probably underlies each distinct vocal repertoire in any given taxon of nonhuman primates.

The SVT topology is probably modified by coordinated actions in vocal apparatuses in nonhuman primates, while these actions are not always the same as those performed during speech production in humans (Nishimura 2018). In fact, the horizontally cylindrical tongue physically prevents nonhuman primates from coordinated actions of the tongue muscles for deformation of the tongue surface as seen in humans (Takemoto 2008; Lieberman 1968). Alternatively, nonhuman primates and other mammals are often found to move the hyoid and larynx downward just before beginning vocalizing (Fitch 2000b; Fitch and Reby 2001; Fitch et al. 2016). This action is one of the options for modifying the topology of the SVT, by pulling the tongue base downward into the pharynx to modify the tongue's posture and extend and modify the vertical cavity including pharyngeal and laryngeal cavities. In fact, it reduces the resonance frequencies by extending the SVT length in red deer (*Cervus elaphus*) (Fitch and Reby 2001). Such an active lowering of the hyoid and larynx also occurs in human singing to produce "singing formants" that emphasize resonance in the posterior pharyngeal cavity (Sundberg 1974). Thus, the dynamic actions of associated apparatus, as seen in hyoidal and laryngeal lowering, enable variations in the SVT topology in nonhuman primates having a rather static tongue.

## 2.9 Conclusions

Recent empirical studies have challenged the traditional view that nonhuman primates have anatomical and physiological constraints in retaining stereotyped postures of the vocal apparatuses. This misconception stemmed in part from a lack of knowledge of comparative vocal anatomy and physiology in nonhuman primates. Primates principally share the same underlying physiology and anatomy in the vocal apparatus involved in each process of the source–filter theory. Nevertheless, nonhuman primates also show wide variations and many differences from humans in terms of respiration control, histology of the vocal folds, accessory sacs and recesses in the laryngeal region, and morphology of the tongue and SVT. These features probably indicate that acoustic mechanisms are widely diversified for vocalization among nonhuman primates. Each of such diversified faculties can perform a part of acoustic aspects of human speech in various ways with effort. In fact, sequential transitions of calls and the production of a wide range of call sounds are both found among nonhuman primates. It should be noted that humans are able to produce multiple distinct sounds quickly even in a single exhalation. This is indispensable for fluent language communication without much effort. This faculty probably remains as the "Rubicon" (the point of no return) for the evolution of speech in humans as derived from vocalizations in nonhuman primates. Crossing it could be achieved in part by the evolutionary shifts in vocal anatomy and physiology surveyed here. Alternative associations of vocal anatomy and physiology need to be examined in nonhuman primates to explore the evolutionary processes leading to the quick and sequential productions of distinct voices through the long evolutionary history of primates.

**Acknowledgments** I am grateful to the late Dr. Sugio Hayama for his systematic work cited here and to my colleagues for their collaborations in earlier studies. I thank the Japan Monkey Centre and the Primate Research Institute of Kyoto University for providing specimens and histological sections and Mses Sumiko Tsubouchi and Atsuko Kataoka for scanning the histological sections. I also thank Prof. Nobuo Masataka, the editor, for inviting me to contribute to this volume. This work was supported by JSPS KAKENHI Grant Numbers 19H01002 and 18H03503.

## References

Arnold K, Zuberbuhler K (2006) Language evolution: semantic combinations in primate calls. Nature 441(7091):303. https://doi.org/10.1038/441303a

Ballintijn MR, ten Cate C (1998) Sound production in the collared dove: a test of the 'whistle' hypothesis. J Exp Biol 201(10):1637–1649

Bartels P (1904) Über die Nebenräume der Kehlkopfhöhle. Beiträge zur vergleichenden und zur Rassen-Anatomie. Z Morphol Anthropol 8(1):11–61

Bezerra BM, Souto A (2008) Structure and usage of the vocal repertoire of Callithrix jacchus. Int J Primatol 29(3):671–701. https://doi.org/10.1007/s10764-008-9250-0

Charlton BD, Frey R, McKinnon AJ, Fritsch G, Fitch WT, Reby D (2013) Koalas use a novel vocal organ to produce unusually low-pitched mating calls. Curr Biol 23(23):R1035–R1036. https://doi.org/10.1016/j.cub.2013.10.069

Chiba T, Kajiyama M (1941) The vowel: its nature and structure. Tokyo-Kaiseikan, Tokyo

Corballis MC (2002) From hand to mouth: the origins of language. Princeton University Press, Princeton

Crelin ES (1976) The human vocal tract: anatomy, function, development, and evolution. Vantage Press, New York

de Boer B (2009) Acoustic analysis of primate air sacs and their effect on vocalization. J Acoust Soc Am 126(6):3329–3343. https://doi.org/10.1121/1.3257544

Doran GA (1975) Review of the evolution and phylogeny of the mammalian tongue. Acta Anat (Basel) 91(1):118–129

Dunn JC (2018) Sexual selection and the loss of laryngeal air sacs during the evolution of speech. Anthropol Sci 126(1):29–34. https://doi.org/10.1537/ase.180309

Dunn JC, Halenar LB, Davies TG, Cristobal-Azkarate J, Reby D, Sykes D, Dengg S, Fitch WT, Knapp LA (2015) Evolutionary trade-off between vocal tract and testes dimensions in howler monkeys. Curr Biol 25(21):2839–2844. https://doi.org/10.1016/j.cub.2015.09.029

Ekberg O, Sigurjonsson SV (1982) Movement of the epiglottis during deglutition. A cineradiographic study. Gastrointest Radiol 7(2):101–107

Elemans CP, Rasmussen JH, Herbst CT, During DN, Zollinger SA, Brumm H, Srivastava K, Svane N, Ding M, Larsen ON, Sober SJ, Svec JG (2015) Universal mechanisms of sound production and control in birds and mammals. Nat Commun 6:8978. https://doi.org/10.1038/ncomms9978

Fant G (1960) Acoustic theory of speech production. Mouton, The Hague

Fedurek P, Schel AM, Slocombe KE (2013) The acoustic structure of chimpanzee pant-hooting facilitates chorusing. Behav Ecol Sociobiol 67(11):1781–1789. https://doi.org/10.1007/s00265-013-1585-7

Fitch WT (2000a) The evolution of speech: a comparative review. Trends Cogn Sci 4(7):258–267

Fitch WT (2000b) The phonetic potential of nonhuman vocal tracts: comparative cineradiographic observations of vocalizing animals. Phonetica 57(2–4):205–218. doi:28474

Fitch WT, Giedd J (1999) Morphology and development of the human vocal tract: a study using magnetic resonance imaging. J Acoust Soc Am 106(3):1511–1522. https://doi.org/10.1121/1.427148

Fitch WT, Hauser MD (1995) Vocal production in nonhuman primates: acoustics, phylogeny, and functional constraints on "Honest" advertisement. Am J Primatol 37:191–219

Fitch WT, Reby D (2001) The descended larynx is not uniquely human. Proc R Soc Lond B Biol Sci 268(1477):1669–1675

Fitch WT, Hauser MD, Chomsky N (2005) The evolution of the language faculty: clarifications and implications. Cognition 97(2):179–210;. discussion 211-125. https://doi.org/10.1016/j.cognition.2005.02.005

Fitch WT, de Boer B, Mathur N, Ghazanfar AA (2016) Monkey vocal tracts are speech-ready. Sci Adv 2(12):e1600723. https://doi.org/10.1126/sciadv.1600723

Fletcher NH, Rossing TD (1998) The physics of musical instruments, 2nd edn. Springer, New York

Flugel C, Rohen JW (1991) The craniofacial proportions and laryngeal position in monkeys and man of different ages (a morphometric study based on CT-scans and radiographs). Mech Ageing Dev 61(1):65–83

Garcia M, Herbst CT (2018) Excised larynx experimentation: history, current developments, and prospects for bioacoustic research. Anthropol Sci 126(1):9–17. https://doi.org/10.1537/ase.171216

Garrett CG, Coleman JR, Reinisch L (2000) Comparative histology and vibration of the vocal folds: implications for experimental studies in microlaryngeal surgery. Laryngoscope 110 (5):814–824. https://doi.org/10.1097/00005537-200005000-00011

Gautier JP (1971) Etude morphologique et Fonctionnelle des annexes extra-laryngees des cercopithecinae; liaison avec les cris d'espacement. Biol Gabonica 7(2):229–267

Gautier JP, Gautier A (1977) Communication in old world monkeys. In: Sebeok TA (ed) How animals communicate. Indiana University Press, Bloomington, pp 890–964

Geissmann T (2002) Duet-splitting and the evolution of gibbon songs. Biol Rev Camb Philos Soc 77(1):57–76

Greenberg S, Carvey H, Hitchcock L, Chang SY (2003) Temporal properties of spontaneous speech - a syllable-centric perspective. J Phon 31(3–4):465–485. https://doi.org/10.1016/j.wocn.2003.09.005

Harrison DFN, Harrison DFN (1995) The anatomy and physiology of the mammalian larynx. Cambridge University Press, Cambridge

Hauser MD, Chomsky N, Fitch WT (2002) The faculty of language: what is it, who has it, and how did it evolve? Science 298(5598):1569–1579

Hayama S (1970) The *Saccus laryngis* in primates. J Anthropol Soc Nippon 78:274–298 (in Japanese with English abstract)

Herbst CT (2016) Biophysics of vocal production in mammals. In: Suthers RA, Fitch WT, Fay RR, Popper AN (eds) Vertebrate sound production and acoustic communication. Springer, Cham, pp 159–189. https://doi.org/10.1007/978-3-319-27721-9_6

Herbst CT, Dunn JC (2018) Non-invasive documentation of primate voice production using electroglottography. Anthropol Sci 126(1):19–27. https://doi.org/10.1537/ase.180201

Herbst CT, Koda H, Kunieda T, Suzuki J, Garcia M, Fitch WT, Nishimura T (2018) Japanese macaque phonatory physiology. J Exp Biol 221(Pt 12):jeb171801. https://doi.org/10.1242/jeb.171801

Hewitt G, MacLarnon A, Jones KE (2002) The functions of laryngeal air sacs in primates: a new hypothesis. Folia Primatol (Basel) 73(2–3):70–94

Hiiemae KM, Palmer JB (2003) Tongue movements in feeding and speech. Crit Rev Oral Biol Med 14(6):413–429

Hiiemae KM, Palmer JB, Medicis SW, Hegener J, Jackson BS, Lieberman DE (2002) Hyoid and tongue surface movements in speaking and eating. Arch Oral Biol 47(1):11–27

Hirano M (1974) Morphological structure of vocal cord as a vibrator and its variations. Folia Phoniatr (Basel) 26(2):89–94. https://doi.org/10.1159/000263771

Hirano M (1975) Phonosurgery: basic and clinical investigations. Otol (Fukuoka) 21:239–242
Hirano M, Kakita Y (1985) Cover-body theory of vocal fold vibration. In: Daniloff RG (ed) Speech science. College-Hill Press, San Diego, pp 1–46
Hirano M, Kurita S, Nakashima T (1983) Growth, development, and aging of humna vocal folds. In: Bless DM, Abbs JH (eds) Vocal fold physiology: contemporary research and clinical issues. College-Hill Press, San Diego, pp 22–43
Ishii K, Yamashita K, Akita M (1999) Fibrous structure of connective tissue in the vocal fold of the Japanese monkey (Macaca fuscata). Okajimas Folia Anat Jpn 76(2–3):107–115
Ishizaka K, Flanagan JL (1972) Synthesis of voiced sounds from a 2-mass model of the vocal cords. Bell Syst Tech J 51(6):1233–1268
Itani J (1963) Vocal communication of the wild Japanese monkey. Primates 4(2):11–66
Kier WM, Smith KK (1985) Tongues, tentacles and trunks - the biomechanics of movement in muscular-hydrostats. Zool J Linnean Soc 83(4):307–324. https://doi.org/10.1111/j.1096-3642.1985.tb01178.x
Koda H, Nishimura T, Tokuda IT, Oyakawa C, Nihonmatsu T, Masataka N (2012) Soprano singing in gibbons. Am J Phys Anthropol 149(3):347–355. https://doi.org/10.1002/ajpa.22124
Koda H, Tokuda IT, Wakita M, Ito T, Nishimura T (2015) The source-filter theory of whistle-like calls in marmosets: acoustic analysis and simulation of helium-modulated voices. J Acoust Soc Am 137(6):3068–3076. https://doi.org/10.1121/1.4921607
Kurita S, Nagata K, Hirano M (1983) A comparative study of the layer structure of the vocal fold. In: Bless DM, Abbs JH (eds) Vocal fold physiology: contemporary research and clinical issues. College-Hill Press, San Diego, pp 3–21
Laitman JT, Reidenberg JS (1997) The human aerodigestive tract and gastroesophageal reflux: an evolutionary perspective. Am J Med 103(5A):2S–8S
Laitman JT, Crelin ES, Conlogue GJ (1977) The function of the epiglottis in monkey and man. Yale J Biol Med 50(1):43–48
Lieberman P (1968) Primate vocalizations and human linguistic ability. J Acoust Soc Am 44 (6):1574–1584
Lieberman P (1984) The biology and evolution of language. Harvard University Press, Cambrige, MA
Lieberman PH, Klatt DH, Wilson WH (1969) Vocal tract limitations on the vowel repertoires of rhesus monkey and other nonhuman primates. Science 164(3884):1185–1187
Lieberman DE, McCarthy RC, Hiiemae KM, Palmer JB (2001) Ontogeny of postnatal hyoid and larynx descent in humans. Arch Oral Biol 46(2):117–128
Madsen PT, Jensen FH, Carder D, Ridgway S (2012) Dolphin whistles: a functional misnomer revealed by heliox breathing. Biol Lett 8(2):211–213. https://doi.org/10.1098/rsbl.2011.0701
Marshall JT Jr, Marshall ER (1976) Gibbons and their territorial songs. Science 193 (4249):235–237. https://doi.org/10.1126/science.193.4249.235
McComb K, Semple S (2005) Coevolution of vocal communication and sociality in primates. Biol Lett 1(4):381–385. https://doi.org/10.1098/rsbl.2005.0366
McNeill D (2012) How language began: gesture and speech in human evolution. Cambridge University Press, Cambridge
Mergell P, Fitch WT, Herzel H (1999) Modeling the role of nonhuman vocal membranes in phonation. J Acoust Soc Am 105(3):2020–2028
Mott F (1924) A study by serial sections of the structure of the larynx of *Hylobates syndactylus* (Siamang gibbon). Proc Zool Soc London 94(4):1161–1170
Negus VE (1949) The comparative anatomy and physiology of the larynx. William Heinemann Medical Books, London
Nemai J, Keleman G (1933) Beitrage zur Kenntnis des Gibbonkehlkopfes. Z Anat Entwicklungsgesch 100:512–520
Nishimura T (2003) Comparative morphology of the hyo-laryngeal complex in anthropoids: two steps in the evolution of the descent of the larynx. Primates 44(1):41–49. https://doi.org/10.1007/s10329-002-0005-9

Nishimura T (2005) Developmental changes in the shape of the supralaryngeal vocal tract in chimpanzees. Am J Phys Anthropol 126(2):193–204. https://doi.org/10.1002/ajpa.20112
Nishimura T (2008) Understanding the dynamics of primate vocalization and its implications for the evolution of human speech. In: Masataka N (ed) The origins of language. Springer, Tokyo, pp 111–131
Nishimura T (2018) The descended larynx and the descending larynx. Anthropol Sci 126(1):3–8. https://doi.org/10.1537/ase.180301
Nishimura T, Mikami A, Suzuki J, Matsuzawa T (2006) Descent of the hyoid in chimpanzees: evolution of face flattening and speech. J Hum Evol 51(3):244–254. https://doi.org/10.1016/j.jhevol.2006.03.005
Nishimura T, Mikami A, Suzuki J, Matsuzawa T (2007) Development of the laryngeal air sac in chimpanzees. Int J Primatol 28(2):483–492. https://doi.org/10.1007/s10764-007-9127-7
Nishimura T, Oishi T, Suzuki J, Matsuda K, Takahashi T (2008) Development of the supralaryngeal vocal tract in Japanese macaques: implications for the evolution of the descent of the larynx. Am J Phys Anthropol 135(2):182–194. https://doi.org/10.1002/ajpa.20719
Nowicki S (1987) Vocal tract resonances in oscine bird sound production: evidence from birdsongs in a helium atmosphere. Nature 325(6099):53–55. https://doi.org/10.1038/325053a0
Ouattara K, Lemasson A, Zuberbuhler K (2009) Campbell's monkeys use affixation to alter call meaning. PLoS One 4(11):e7808. https://doi.org/10.1371/journal.pone.0007808. doi:ARTN e7808
Pasch B, Tokuda IT, Riede T (2017) Grasshopper mice employ distinct vocal production mechanisms in different social contexts. Proc R Soc London Ser B Biol Sci 284(1859):20171158. https://doi.org/10.1098/rspb.2017.1158
Rand AS, Dudley R (1993) Frogs in helium - the anuran vocal sac is not a cavity resonator. Physiol Zool 66(5):793–806
Reber SA, Nishimura T, Janisch J, Robertson M, Fitch WT (2015) A Chinese alligator in heliox: formant frequencies in a crocodilian. J Exp Biol 218(Pt 15):2442–2447. https://doi.org/10.1242/jeb.119552
Riede T (2010) Elasticity and stress relaxation of rhesus monkey (Macaca mulatta) vocal folds. J Exp Biol 213(Pt 17):2924–2932. https://doi.org/10.1242/jeb.044404
Riede T, Zuberbuhler K (2003) The relationship between acoustic structure and semantic information in Diana monkey alarm vocalization. J Acoust Soc Am 114(2):1132–1142
Riede T, Bronson E, Hatzikirou H, Zuberbuhler K (2005) Vocal production mechanisms in a non-human primate: morphological data and a model. J Hum Evol 48(1):85–96. https://doi.org/10.1016/j.jhevol.2004.10.002
Riede T, Tokuda IT, Munger JB, Thomson SL (2008) Mammalian laryngseal air sacs add variability to the vocal tract impedance: physical and computational modeling. J Acoust Soc Am 124(1):634–647. https://doi.org/10.1121/1.2924125
Riede T, Borgard HL, Pasch B (2017) Laryngeal airway reconstruction indicates that rodent ultrasonic vocalizations are produced by an edge-tone mechanism. R Soc Open Sci 4 (11):170976. https://doi.org/10.1098/rsos.170976
Sasaki CT, Levine PA, Laitman JT, Crelin ES Jr (1977) Postnatal descent of the epiglottis in man. A preliminary report. Arch Otolaryngol 103(3):169–171
Schon MA (1971) The anatomy of the resonating mechanism in howling monkeys. Folia Primatol 15(1):117–132
Smith KK, Kier WM (1989) Trunks, tongues, and tentacles – moving with skeletons of muscle. Am Sci 77(1):29–35
Sokoloff AJ (2000) Localization and contractile properties of intrinsic longitudinal motor units of the rat tongue. J Neurophysiol 84(2):827–835
Sonntag CF (1925) The comparative anatomy of the tongues of the Mammalia. XII. Summary, classification and phylogeny. Proc Zool Soc London 1925:701–U781
Starck D, Schneider R (1960) Respirationsorgane. A. Larynx. In: Hofer HO, Schultz AH, Starck D (eds) Primatologia, vol 3–2. Karger, Basel, pp 423–587

Stevens KN (1972) The quantal nature of speech: evidence from articulatory-acoustic data. In: David EE, Denes PB (eds) Human communication: a united view. McGrew Hill, New York, pp 51–66

Story BH (2002) An overview of the physiology, physics and modeling of the sound source for vowels. Acoust Sci Technol 23(4):195–206. https://doi.org/10.1250/ast.23.195

Story BH, Titze IR (1995) Voice simulation with a body-cover model of the vocal folds. J Acoust Soc Am 97(2):1249–1260. https://doi.org/10.1121/1.412234

Sundberg J (1974) Articulatory interpretation of the "singing formant". J Acoust Soc Am 55 (4):838–844

Takemoto H (2001) Morphological analyses of the human tongue musculature for three-dimensional modeling. J Speech Lang Hear Res 44(1):95–107

Takemoto H (2008) Morphological analyses and 3D modeling of the tongue musculature of the chimpanzee (Pan troglodytes). Am J Primatol 70(10):966–975. https://doi.org/10.1002/ajp.20589

Titze IR (1980) Comments on the myoelastic-aerodynamic theory of phonation. J Speech Hear Res 23(3):495–510. https://doi.org/10.1044/jshr.2303.495

Titze IR (1994) Principles of voice production. Prentice Hall, Englewood Cliffs

Van Den Berg J (1958) Myoelastic-aerodynamic theory of voice production. J Speech Hear Res 1 (3):227–244

Yamada H, Okanoya K (2003) Song syntax changes in Bengalese finches singing in a helium atmosphere. Neuroreport 14(13):1725–1729. https://doi.org/10.1097/01.wnr.0000087731.58565.29

Zuberbuhler K (2006) Language evolution: the origin of meaning in primates. Curr Biol 16(4): R123–R125. https://doi.org/10.1016/j.cub.2006.02.003

# Chapter 3
# Integrations of Multiple Abilities Underlying the Vocal Evolutions in Primates

**Hiroki Koda**

**Abstract** Language is cognitive systems unique to humans, and other animals never showed such equivalent one. Recent studies have strongly suggested that the language would not be originated solely from one unique ability, but rather it would emerge as a consequence of the integrations of multiple abilities which would be commonly shared with nonhuman animals. These ideas suggest the importance of the comparative approaches for the cognitions and communications between humans and nonhuman animals. Particularly, speech or vocal communications are typical examples of the basic biological components which are all necessary for language emergences. The evolutionary pathway from primate vocal systems to human communication via language systems has been always paid a special attention by many evolutionary biologists. Here I review to focus on the recent progress of the studies in the vocal evolution in the primate lineages, mainly for the two dimensions in the primate vocalizations, i.e., vocal controllability and speech homologous facial expressions in monkeys. The limited ability of vocal production control distinguishes us from other primates. Recent experiments, however, have begun to reveal voluntary vocal control ability in nonhuman primates, showing successful attempts of the vocal operant conditionings. These accumulating findings have concluded the functional expansion of cognitive motor control from hand to mouth, suggesting the possible evolutionary history where the motor cortex expansions from forelimb to larynx would occur in the human evolution. Besides, orofacial action is a crucial component for the speech movements, which are characterized by facial actions of ~5 Hz oscillations of lip, mouth, or jaw movements. A recent promising candidate homologue for these facial actions is a lip-smacking, a facial display of primates, which seems to be independent of speech. Interestingly, such facial actions are also characterized by stable 5 Hz oscillation patterns of jaw or mandible actions, matching that of speech. Recent studies have confirmed the parallel development and kinematics between speech and lip-smacking actions, suggesting a common neural mechanism for the central pattern generator underlying orofacial movements,

H. Koda (✉)
Primate Research Institute, Kyoto University, Kyoto, Japan
e-mail: koda.hiroki.7a@kyoto-u.ac.jp



N. Masataka (ed.), *The Origins of Language Revisited*,
https://doi.org/10.1007/978-981-15-4250-3_3

which would evolve to speech with a sensory foundation for perceptual saliency particular to 5 Hz rhythms widely observed in primate species. Thus, these stepwise acquisitions of multiple independent components such as vocal controllability or facial actions would be all necessary evolutionary events before speech emergence, and the integrations of the multiple components would be a key to finally establish human speech.

**Keywords** Vocal controllability · Facial expression · Motor cortex · Mandible action

Language is a cognitive system unique to humans; other animals have never shown an equivalent one. Recent studies have strongly suggested that language does not originate solely from one unique ability but rather as a consequence of the integration of multiple abilities, which are commonly shared with nonhuman animals. This suggests the importance of comparative approaches for cognition and communication between humans and nonhuman animals. In particular, speech or vocal communications are typical examples of basic biological components, which are all necessary for language formation. The evolutionary pathway, from primate vocal systems to human communication via language systems, has always been paid a special attention by many evolutionary biologists. Here, I review the focus on the recent progress of studies in the vocal evolution in primate lineages, mainly for two dimensions in primate vocalizations: vocal controllability and speech homologous facial expressions in monkeys. The limited ability of vocal production control distinguishes us from other primates. Recent experiments, however, have begun to reveal voluntary vocal control ability in nonhuman primates, showing successful attempts of vocal operant conditionings. Findings such as these, which are constantly being revealed, have concluded that the functional expansion of cognitive motor control from hand to mouth suggests the possible evolutionary history where motor cortex expansions, from forelimb to larynx, would occur in human evolution. Besides, orofacial action is a crucial component in speech movements, which involve facial actions of ~5 Hz oscillations of the lip, mouth, or jaw. A recent promising candidate for the homologue of such facial actions is lip-smacking—a facial display of primates, which seems to be independent of speech. Interestingly, such facial actions are also characterized by stable 5 Hz oscillation patterns of jaw or mandible action, matching that of speech. Recent studies have confirmed that the parallel development and kinematics between speech and lip-smacking actions suggest a common neural mechanism for the central pattern generator underlying such orofacial movements. Thus, stepwise acquisitions of multiple independent components, such as vocal controllability or facial actions, would all be necessary evolutionary steps before the emergence of speech, and the integration of multiple components would be a key to finally establishing human-like speech.

## 3.1 Human Speech and Monkey Vocalizations

We believe that language is a unique cognitive ability, which biologically divides us from any other animal species. Humans, as is commonly held, are the only animals which have language. However, consensus among scholars as to the definition of language ability is poor. Some scholars theorized that mental systems may combine two single target objects to operate as a combined unit or in a recursive way: the "merge" operation (linguistics of generative grammar) (e.g., Chomsky 1998) and cognitive linguistics of the "theory of mind" (where mental systems of one agent may understand/predict the mental states of others) (e.g., Tomasello 2009). Recent discussions regarding the evolution of language argue that language is an evolutionary result of the integration of multiple biological components observed in modern humans (Hauser et al. 2002; Okanoya 2007; Fitch 2010, 2017). Hence, some of these components may be observed in nonhuman animals which could be tested in attempts to discover the biological mechanisms underlying such convergent or homologous traits of language components and hypothesize as to the evolutionary causations for these.

Per these views of the integration of language components, speech is a (sensory-) motor component which is commonly used in human communication via language. Apparently, speech is not a prerequisite for the emergence of language; gestural communications with rules of syntax spontaneously emerge in cases of auditory deficit. Thus, some researchers do not emphasize the importance of speech ability in the evolutionary pathway of the emergence of language (Corballis 2002), but rather prefer to focus only on the core components of recursive operations (e.g., Bolhuis et al. 2014). However, we know that communication via language of the auditory-vocal domain is most common. Accordingly, speech should be considered an essential generating aspect regarding the origins of language. Additionally, speech is clearly a difference between human and nonhuman primates, including apes. Traditional attempts to search for language abilities among nonhuman primates (i.e., "great ape language," since such as Kellogg 1968; Hayes and Hayes 1951), attempts to teach the great ape language by using speech, had failed. This failure is due to several biological reasons: nonhuman primates do not have the "sufficient" vocal apparatus for speech sound articulation (e.g., Nishimura et al. 2003), nor "sufficient" vocal controllability (e.g., Jürgens 2002), nor "sufficient" sound perceptions (e.g., Kojima 2003), and so forth. This means that many kinds of sub-components are phylogenetically insufficient for speech.

Supposition is made that the ability of speech is a sub-component of language faculty (particularly of sensory-motor aspect). I propose that speech is a (sensory-) motor ability, unique to humans, forming an integrated complex of sub-sub-abilities, which are required for evolution of this motor system in primate lineages. In this chapter, I consistently argue that the concepts of *components* and *integration* are keys to understand the evolutionary pathway from monkey vocalizations to human speech, which is shown to be analogous with recent theoretical views of human language evolution.

## 3.2 Volitional Controllability Embedded in Speech Ability

Speech is partially characterized by our motor ability to modify the acoustic properties in the intentional way. For example, we can produce the sound of "speech," which is acoustically encoded as /spí:tʃ/, when we intend to produce such. By contrast, apes and monkeys never vocally produce /spí:tʃ/, even if trained intensively from birth, as I described above. What mechanisms are involved in speech? A component underlying these motor abilities is the volitional control of vocal production. It has long been held that this constitutes a main gap in the ability between humans and nonhuman primates. By establishing the biological mechanisms and origins underlying the volitional control of vocal production in the primate lineage, we may reconstruct the evolutionary pathway from the "primitive" form of monkey vocalization to the "modern" form of human speech. When and why did the ancestral species of *Homo sapiens* acquire these abilities? By deeply reviewing recent evidence of the vocal training of monkeys, I will attempt to establish the origins of volitional control of vocal production.

What is volitional control? It is the ability to explicitly control motor action in a goal-directed way over the actions mainly governed by the autonomic, reflexive, or emotional systems (e.g., Bari and Robbins 2013). Accordingly, it could be involved in any kind of motor action, including vocalizations and forelimb or hindlimb actions. Primates—mammals well-adapted and evolved to arboreal life—especially develop forelimb control and higher levels of volitional control of manual actions, such as hand grasping and finger pinching (e.g., lemurs climbing trees, social grooming in macaques, or apes using tools). To greater or lesser extents, all volitional motor controls are neuroanatomically involved in the cortical network of the motor cortex, being the main network of the brain (Jürgens 2002; Hammerschmidt and Fischer 2008). For example, hand dexterity in primates is assumed to be the result of the evolution of the motor system's cortical networks (Lemon 2008). In contrast, the voluntary control of vocalization in nonhuman primates is extremely limited, as described above. Actually, this is well supported by neuroanatomical evidence for nonhuman primates' motor circuits for vocal production; that is, the direct cortico-motoneuronal connection from the primary motor cortex to motoneurons responsible for each motor action component (Jürgens 2002; Hammerschmidt and Fischer 2008). Direct cortico-motoneuronal connections from the forelimb to motoneurons exist, but those of the larynx lack in nonhuman primates. In humans, both direct pathways responsible for the vocal and manual action units exist. This is one of the primary and direct reasons why our speech system is behaviorally and phylogenetically isolated from those of nonhuman primates. Instead, nonhuman primates' dominant neural circuits responsible for vocal production is believed to be the subcortical networks consisting of the anterior cingulate cortex (ACC), periaqueductal gray (PAG), or brainstem (Jürgens 2002; Hammerschmidt and Fischer 2008). Thus, the present general consensus is the system dichotomy for the vocal action control between humans and nonhuman primates (e.g., Hage and Nieder 2016; Hage 2018a; Simonyan and Horwitz 2011; Simonyan 2014).

Importantly, the system dichotomy is consistent with the findings of human vocal ontogeny, paralleled with the vocal system phylogeny. Undoubtedly, humans begin their vocalizations from crying, not from "speech." The basic framework for vocal development in humans, from prelinguistic infancy to the post-linguistic child, has supposedly followed two main courses: the developmental systems of crying and then coo-babbling (Locke 1995; Oller and Griebel 2008; Oller 2000; Barr et al. 2000). Just after birth, infant vocalization begins with crying, closely related to their emotional states, such as discomfort, distress, or pain. Then, babies start to produce an acoustically resonated vocalization called "cooing" or "babbling," both regarded as onsets to the development of language (speech sounds), where voluntary vocal control matures through infant-mother interactions (Masataka 2003). Importantly, the development of cooing or babbling to speech is fundamentally independent of how crying develops. Cries develop in terms of their acoustic properties (resonance, rhythms, or variations) but have been believed to occur dominantly under emotional states (Barr et al. 2000). By contrast, the cooing-babbling system develops with more directed ways of emotionally independent control in vocal production, i.e., volitional control. Furthermore, crying in human babies shares the neural motor system with vocalization in nonhuman primates (Newman 2007). Thus, the system dichotomy has been observed at ontogenetic levels (Hage and Nieder 2016). These dualisms of vocal developments, crying and cooing, have been believed to be uniquely human. Consequently, nonhuman primate vocalizations are homologous of infant cries and completely independent of speech.

## 3.3 Dissecting the Ability of Voluntary Motor Control

Primate vocalizations are basically divided into two systems: (1) emotion-dependent cries and (2) intention-driven calls, both of which seem to rely on the presence of cortico-motoneuronal connections in the vocal domain. However, this simple system dichotomy is challenged by the evidence of vocal training successes in certain nonhuman primates, which have been investigated to date. Traditionally, operant conditioning has been the gold standard to test the ability of voluntary vocal control in nonhuman primates (Sutton et al. 1973, 1974, 1981a, b; Larson and Kistler 1984; Aitken and Wilson 1979; Aitken 1981; Pierce 1985; Coudé et al. 2011; Hage et al. 2013, 2016; Hage and Nieder 2013a; Hage 2018a; Koda et al. 2007, 2018; Hihara et al. 2003). The evidence suggests that, despite difficulties, it may still be possible that nonhuman primates have the ability to be vocally conditioned. In fact, most findings finally showed success of voluntary vocal control. What was occurring when emotional-dependent vocalizations of ancestral primates gradually or suddenly transitioned to controllable cognitive ones, characteristic of human speech? The simple system dichotomy is insufficient to answer those complex questions.

The timing control of vocalizations, at least, would be conditional for nonhuman primates, but the differential conditioning of vocal characters is very difficult, suggesting that a timing control of vocalizations could be within a range of goal-directed actions. The most recent experiments have begun to report successes of the conditioning of different types of vocalization (for the most relevant paper, see Hage et al. 2013; Hage 2018a), but other dimensions of vocal control—such as pitch—are still thought to be difficult to condition. Thus, such goal-directed vocal control is not on an achievement of single mechanisms (i.e., voluntary or involuntary), but rather due to an integration of several motor abilities. This should be separated into multiple components, as with similar views of language faculty. As a general principle of motor learnings, the goal-directed and voluntary motor controls are achieved with the integration of multiple aspects of motor controls (e.g., Paus 2001; Bari and Robbins 2013; Jahanshahi et al. 2015). This includes various kinds of motor actions. Particularly, in previous attempts to shape voluntary motor controls, nonhuman primates were consistently required to discern preceding stimuli (cues) and to explicitly control their actions at different levels—to prepare, inhibit, disinhibit, and execute required motor actions. There are likely some process-specific limitations for motor planning in nonhuman primate vocalizations. After our recent study of action learning comparisons between vocal and manual actions of Japanese macaques (Koda et al. 2018), I aim to discuss what components play an important role in switching from emotional-dependent to cognitive-driven vocal production in nonhuman primates.

Recently, we trained Japanese macaques to execute motor actions of vocalization or touching, via a visual stimulus in order to systematically compare learning patterns (Koda et al. 2018). Generally, the results suggested that macaques have an ability to control both these motor actions in a goal-directed way. While both vocalization and touching are seemingly "homologue" goal-directed behaviors, they could be achieved by the completely different learning process of motor control. Firstly, the training durations were substantially different: touching actions were rapid, but vocalizations were slow. More relevantly, motor inhibitions (which here mean the ability to inhibit motor actions when required to stop or not execute the action) were almost perfectly learned in the touching actions, but not in the vocalizations. Additionally, the generalization patterns were different between touching and vocalizations: The former was flexibly executed in response to sudden changes of cues, but the latter were not, and indeed the macaques failed to execute vocalizations due to violations of expectations for cue timing. Thus, touching action (manual action) is voluntarily controlled in most levels of the motor control, to inhibit, disinhibit, and execute. In contrast, all the different levels of motor components were not completed or not well coordinated in the conditioned vocalizations, compared with manual actions, such as touching. Here I supposed that motor preparation and inhibitory control were thus not well matured in nonhuman primate vocalizations.

## 3.4 Steps from Involuntary to Voluntary Actions

My idea might be inspired by experimental evidence with several accumulated observations to be overlooked. For example, evidence that vocal conditioning is not always achieved for all monkeys is relevant for this discussion. Actually, the researchers who tried vocal conditioning always selected the subject before starting the training, by well observing whether the monkeys spontaneously vocalized much frequently (Pierce 1985). This suggests several basic requirements for the achievement of voluntary vocal control. First, spontaneous vocalizations should be provoked and be triggered easily by change of emotional states or external stimulations much before learning vocal initiation control. Otherwise, the animal has no opportunities to learn when they should execute or inhibit a vocal action. Recent evidence on motor development in humans has emphasized the importance of such spontaneous motor actions for the acquisition of goal-directed movements (e.g., Watanabe et al. 2011; Robertson et al. 2001, 2007). In the general principals of motor development, such as limb control, goal-directed behavior is not observed just after birth, but undoubtedly progressively develops later through experiences such as trial and error in manipulating body parts. Infants never start to perform adult-like goal-directed actions, such as reaching or grasping; they simply show immature spontaneous movements characterized as excessive and multiple actions, without skilled coordination and sometimes termed "general movements" (GMs) (Hadders-Algra 2002, 2004, 2007; Prechtl and Hopkins 1986). Through brain maturation, particularly in the prefrontal area, and with sensory-motor feedback interactions to external environments, spontaneous movements will be regulated and switch to goal-directed actions, mainly governed by the cognitive voluntary control (Kanemaru et al. 2012). Ontogenetically, these GMs would be considered as precursors for goal-directed behaviors, giving sufficient opportunities for learning cognitive motor control, such as inhibition, disinhibition, or execution (Watanabe et al. 2011). Likewise, spontaneous occurrences of vocalizations would be prerequisites before learning when they should inhibit or execute the vocalizations. Nonhuman primates could be trained to meet with the "criterions," showing higher rates of spontaneous vocalizations observed as "vocal GMs."

Nonhuman primates who are motivated enough to vocalize would, arguably, learn to prepare vocalizations until presentations of visual stimuli, at different times, may simultaneously enhance the differentiation of contexts where the animal may vocalize or not. Consequently, they would decouple vocalizations from emotion via such processes of inhibitory control, leading to the achievement of voluntary vocal control. This would be a key process switching from emotional-coupled movements represented as GMs to cognitive-driven, goal-directed behaviors. When looking at vocal development in primates, classical studies in alarm call development of vervet monkeys support this decoupling process (Seyfarth et al. 1980). When immature vervet monkeys first emit an alarm call to other animals, including predators and non-threatening animals, they gradually learned to tune the specific alarm call at the predators and to suppress it toward non-threatening animals

(Seyfarth et al. 1980). Recent models of vocal development in marmosets also demonstrated that immature marmosets emitted babbling-like variable vocalizations and progressively strengthened the link of the specific call with the relevant specific context while weakening those calls with other inappropriate contexts, leading to the development of appropriate vocal usage in their vocal communication (Takahashi et al. 2015). In our training on Japanese macaques, we observed extremely high rates of vocalizations in early phases of training (Koda et al. 2018). These findings suggest that increases of call rates, an essential component for the achievement of vocal control in nonhuman primates, preceded voluntary vocal control.

These possible processes for decoupling vocalizations from emotion is observed in human infant cries (Barr et al. 2000), characterizing unique developments in human primates. Such cries have been proposed to also be shared with nonhuman primate vocalizations in the neural mechanisms underlying their production, which is believed to be independent of subsequent vocal development for spoken language and homologue with nonhuman primate vocalizations (Newman 2007). However, human crying is more unique than nonhuman primate vocalizations, in terms of the excessiveness of crying, sometimes observed as colic—a common clinical syndrome, where infant cries for more than 3 h a day, for 3 days a week, and for 3 or more weeks (Wessel et al. 1954). These excessive vocal productions are never found in other primates and mammals, making them a unique dimension of human vocal ontogeny (Soltis 2004). To date, the reasons why human infants cry so easily have remained unclear, especially regarding both its mechanisms and evolutionary causation, but excessive crying would provide an advantage for infant humans to learn when they should maintain or stop crying. This may lead to emotional communication via vocalization, which is an evolutionary basis of vocal communications before the emergence of language communication. Finally, vocalizations would be cognitively controlled in a top-down fashion and in a goal-directed way to develop the coo-babbling and speech sounds, which would indeed be the system duality observed only in humans in the primate lineage.

## 3.5 From Hand to Mouth: Expansions of Motor Abilities

Phylogenetically, when does the system duality for vocal actions emerge in primate evolution? The "from hand to mouth" hypothesis by psychologist Michael Corballis may hold the answer (Corballis 2002). Corballis supposed that the gestural origins of communication signals in language lie with an extinct ancestral species in the human lineage. This supposition concurs with a recent hypothesis that functional brain expansions of the motor cortex provide voluntary motor control abilities at different levels of motor patterns in primates and humans, further emphasizing system duality of cognitive and emotional control in vocal productions (Simonyan and Horwitz 2011; Simonyan 2014; Hage and Nieder 2016). Analogously, when looking at the dexterity of the forelimb, hands, or fingers and voluntary actions in mammals, the extent of dexterity is well correlated with expansions of direct cortico-motoneuronal connections, an essential system for the dexterity of manual actions, in the motor control areas of the forelimb, hands, and fingers (Lemon 2008; Kuypers 1981; Lemon and Griffiths 2005; Bortoff and Strick 1993; Nakajima et al. 2000; Porter and Lemon 1993; McKiernan et al. 1998). Likewise, humans acquire voluntary vocal control due to an expansion of the direct cortico-motoneuronal connections in human laryngeal motor cortex (LMC), which is lacking in nonhuman primate LMC (Rathelot and Strick 2006, 2009). This suggests an evolutionary shift occurred only after extinct hominid ancestors. The parallel phenomena of motor cortex expansions innovated the novel functions of top-down cognitive control for motor execution in primate evolution: leaving one to hypothesize that motor innovations might have first occurred in forelimb control between nonprimates and primates and next in vocal control between nonhuman primates and humans.

## 3.6 Lip-Smacking Is a Facial Root

Speech does not emerge solely by the acquisition of voluntary vocal control. Independent of the ability of voluntary vocal control, facial action patterns would be prepared, enabling humans to generate unique motor actions, consisting of consonants and vowels, and rapid changes in acoustic characters even in single utterances were exhibited. Discussions of voluntary controllability of vocalizations mainly involved in the laryngeal controls including the vocal folds, while those of facial actions seen in speech would be related through articulations (Fitch 2010). I reviewed the idea of speech action evolutions: the evolution of speech actions partially originates from facial expressions, not from vocal actions.

Without doubt, the rapid change of consonance and vowel production in a speech stream requires the ability of rapid oscillated action control of the lip, mouth, tongue, and jaw movements to modify the resonance properties, together with voicing determined by the vocal fold (Lindblom and Sundberg 1971). In contrast, such fluent, rapid, and flexible actions in nonhuman primates have rarely occurred in a comparable manner with their vocalizations. Here we again find fundamental gaps in rhythmic actions between human speech and nonhuman primate vocalizations, independent of the ability of voluntary motor control. Origins of this speech rhythmic action can be traced back to the early innovative hypothesis of "frame/content theory," proposed by Peter F. MacNeilage (MacNeilage 1998). This theory focused on speech sound features consisting of two rhythmic parts: (1) the "syllabic frame," where the cycle constitutes the syllable, and (2) the "segmental content," where open and closed phases are segments (vowels and consonants). MacNeilage hypothesized that these communication-related frames perhaps first evolved when the ingestion-related cycles of mandibular oscillation took on communicative significance as lip-smacks, tongue-smacks, and teeth chatters—displays which are prominent in many nonhuman primates (from the abstract in (MacNeilage 1998)). When this idea was first proposed, overlapping facial actions of monkeys and the "frame" of speech were regarded as merely a phenomenon of correlation and not confirmed by empirical evidence. Recently, however, some typical facial actions in several primates, such as lip-smacking or teeth chattering, have been widely accepted as plausible candidates for historical precursors of speech rhythmic actions (e.g., Ghazanfar and Takahashi 2014).

The first recent empirical candidates of lip-smacking actions in macaques have mainly been led by Asif Ghazanfar and his colleagues (Ghazanfar and Takahashi 2014). Lip-smacking is generally characterized by vertical movements of the jaw and lips, with lip protrusion, which functions as a face-to-face communication signal directed at other monkeys (Preuschoft and van Hooff 1995). Despite great divergence of speech sounds among multiple languages, oscillatory rhythms of orofacial movements exhibit constant and convergent rhythms of 5 Hz, independent of multiple cultures and languages (Chandrasekaran et al. 2009). This suggests that constant 5 Hz orofacial movements are generated by a common neural mechanism behind the divergence in language. Importantly, stable 5 Hz orofacial movements are also found in the lip-smacking macaques. They are now hypothesizing that this orofacial movement in macaque facial actions is homologous of those of speech, based on their recent findings (Ghazanfar and Takahashi 2014). First, when observing inner actions during lip-smacking by real-time X-ray cineradiography, they found similar oscillatory dynamics in coordinated motions of the hyoid bones, tongue, jaw, and lips between human speech and macaque lip-smacking (Ghazanfar et al. 2012). More interestingly, these action dynamics were distinguished from those of other related orofacial actions such as chewing (Ghazanfar et al. 2012). This parallel between speech and lip-smacking is also supported by developmental evidence (Morrill et al. 2012). Lip-smacking was observed in the early stage development of immature macaques, who expressed "immature" lip-smacking, and its oscillatory rhythms were slower than those of the matured lip-smacking of adult macaques. This developmental "elevation" in the action rhythms from slow to rapid rates, interestingly, overlaps with the development of speech orofacial actions, which are found in the development from cooing to babbling in humans. These developmental changes are not observed, however, with the action of chewing (Morrill et al. 2012). The physiological evidence does support the specializations of lip-smacking actions isolated from other orofacial actions such as chewing or sucking. The electromyography recording of the dynamic coordination of monkeys' mimetic musculature revealed apparent gaps between lip-smacking and ingestive movements (chewing or sucking), suggesting unique foundations underlying lip-smacking (Shepherd et al. 2012).

## 3.7 Speech Action Embedded in Primate Facial Expressions

Overall, the key feature is lip-smacking oscillation rhythms, and its overlaps with speech facial actions, consistently suggesting homology. Importantly, this homology between speech and lip-smacking does not indicate that speech facial actions would have directly evolved from facial action of vocalizations. Rather, it suggests a central pattern generator underlying both speech and lip-smacking orofacial movements, which existed in a common ancestor between humans and macaques. Consequently, human speech evolved directly from lip-smacking-like movements that were used for communication in an extinct common primate ancestor. Considering the hypothesis that 5 Hz oscillations of a central pattern generator were embedded in the brain systems of extinct primate ancestors, orofacial movements exhibiting the same rate of oscillation may not only be restricted to speech or typical lip-smacking but may also be found in extensive forms of orofacial displays in primates in various social contexts, exhibiting variable features among species.

In fact, these predictions were supported by variant forms of facial actions in other primates. Vocalizations in geladas (*Theropithecus gelada*) are an example of primates beyond genus *Macaca* (Bergman 2013). Their vocalizations exhibit similar 5 Hz pulsed patterns, and the pulses were generated by orofacial actions. Arguably, lip-smacking-like facial actions actively contributed to rhythm generations in primate vocalization and in terms of the similar manner of the sound generation in speech. Another example is our recent finding of teeth chattering, a unique facial expression of stump-tailed macaques (*Macaca arctoides*) (Toyoda et al. 2017). Such teeth chattering was referred to as a subtype of lip-smacking (i.e., grin-lip-smack) in the genus *Macaca* (Blurton Jones and Trollope 1968). However, this teeth chattering lacks lip protrusion, which is observed in common lip-smacking (Blurton Jones and Trollope 1968). Instead, teeth chattering shows the bared teeth display. The contexts of these facial expressions are different: lip-smacking is used in face-to-face interaction, but teeth chattering was observed during courtship displays of the male mounting the female (Blurton Jones and Trollope 1968). Despite different actions and contexts, teeth chattering exhibited a 5 Hz rhythm in facial actions, paralleling the lip-smacking action in other macaques, indicating oscillatory rhythm convergence beyond the differences in species, action forms, and social contexts (Toyoda et al. 2017).

What mechanisms should be assumed to explain these convergent phenomena? Currently, a plausible and reasonable idea is that central pattern generators (CPGs) are specialized for these orofacial movements in primate lineages (Toyoda et al. 2017; Ghazanfar and Takahashi 2014). The CPGs were originally proposed as motor neural mechanisms to explain the temporal regulation of stereotypic rhythmic actions. Given that stereotypic rhythms execute without any voluntary control, as seen in walking or swimming, GPCs might be located in the brainstem. These autonomic and rhythmic actions are also found in orofacial domains, such as chewing, drinking, and respiration (Moore et al. 2014). Most of these actions are necessary for homeostasis and are fundamentally achieved by the coordination of multiple action units (including the lips, tongue, jaw, mandible, or larynx), which are regulated by motor neurons projected from the cortical (motor cortex) or subcortical (cerebellum or brainstem) areas. Regardless of functional differences, the motor regulation systems should partly overlap for chewing, drinking, respiration, vocalizing, and lip-smacking. In primate evolution, an extinct primate ancestor might have modified or newly installed CPGs specialized for the orofacial actions, independent of ingestive and respiratory movements, to further use as communication signals.

What is the evolutionary trajectory of the emergence of specialized CPGs? At the very least, similar (homologous) facial actions have not been reported in nonprimate mammals. How far back can we go to a common ancestor of nonhuman primates or of nonprimate mammals for such facial expressions? What evolutionary event (s) triggered these stable orofacial actions? When looking for similar facial actions widely in primate lineages, lip-smacking equivalent facial expressions have been reported in some New World monkeys, but not in prosimians (Ghazanfar and Takahashi 2014). This may suggest that the evolutionary shift from prosimians to simians might be a crucial event for the acquisition of novel motor systems regulating orofacial rhythms. However, I acknowledge that it may be misleading to overextend the "5 Hz rhythm theory." For example, in our observations in stump-tailed macaques, we reported similar 5 Hz rhythms in "panting" vocalizations, but we simultaneously performed careful discussion for its homologous origins (Toyoda et al. 2017). Even if vocal or orofacial actions showed similar 5 Hz rhythms in some species, it may be a mere coincidence of the particular rhythms, not originating from homologue mechanisms. In this view, vocal rhythms, mainly determined by the larynx, would be treated as having a different origin to the orofacial movements discussed here. Likewise, lip-smacking would not be a perfect explanation for speech action evolution. Indeed, lip-smacking is reported in chimpanzees, but the rhythms of their lip-smacking are likely different from those of macaques or indeed those of human speech (personal communication, Michio Nakamura). Thus, we need more careful comparisons and discussion to reconstruct the evolutionary history of speech actions.

## 3.8 Tentative Conclusion: The Integration of Components for the Emergence of Speech

Speech is a component of the faculty of language and is itself about as complex as language. As I reviewed above, several possible candidate key mechanisms would improve theorization of emotional-dependent and nonoperational animal vocalizations up to cognitively driven and acoustically rich human speech, i.e., a direct cortico-motoneuronal connection or CPG specialization for rapid and communication-oriented orofacial actions. Of course, these two mechanisms should not be considered as the only possible candidates. Another theorization may be respiration rhythms for all vocal motor actions. Importantly, however, integrated processes of multiple components for the evolution of our speech ability, where multiple components emerge originally independent of the adaptive functions for modern speech, are integrated to evolve into the present speech ability. One example of a step in this evolution might be how vocal organs evolved, because vocal anatomy primarily determines the possible variations of acoustic features (Fitch et al. 2016).

In another level, the phonation-respiration systems should have a key role here. The rhythmic facial actions found in lip-smacking may have been integrated with phonation-respiration systems, to segment a single phonated utterance to more temporally short units in human lineage but rarely in nonhuman primates. The brain systems for voluntary control of vocal actions covering all components should have evolved only in humans, which ultimately separate us from nonhuman primates. It remains unclear as to when, how, and why each of the multiple components emerged and were integrated. These perspectives in evolutionary modifications triggering novel biological functions will contribute to the elucidation of the ultimate questions regarding not only the evolution of speech but also that of language.

**Acknowledgments** This review is based mostly on my recent collaborative studies and partially financially supported by MEXT/JSPS KAKENHIs (17H06380, 18H03503, 19H01002). I would personally like to thank my recent collaborators: Takeshi Nishimura, Takumi Kunieda, Aru Toyoda, Takashi Morita, Tamaki Maruhashi, and Suchinda Malaivijitnond for helping me organize the projects. Special thanks to Kazuo Okanoya for his guidance of the most parts of my current research ideas.

## References

Aitken PG (1981) Cortical control of conditioned and spontaneous vocal behavior in rhesus monkeys. Brain Lang 13(1):171–184

Aitken PG, Wilson WA (1979) Discriminative vocal conditioning in rhesus monkeys: evidence for volitional control? Brain Lang 8(2):227–240

Bari A, Robbins TW (2013) Inhibition and impulsivity: behavioral and neural basis of response control. Prog Neurobiol 108(September):44–79. https://doi.org/10.1016/j.pneurobio.2013.06.005. Elsevier Ltd

Barr RG, Hopkins B, James A. (James Andrew) Green (2000) Crying as a sign, a symptom and a signal: clinical, emotional and developmental aspects of infant and toddler crying. Cambridge University Press, Cambridge

Bergman TJ (2013) Speech-like vocalized lip-smacking in geladas. https://doi.org/10.1016/j.cub.2013.02.038

Blurton Jones NG, Trollope J (1968) Social behaviour of stump-tailed macaques in captivity. Primates 9(4):365–393. Springer

Bolhuis JJ, Tattersall I, Chomsky N, Berwick RC (2014) How could language have evolved? PLoS Biol 12(8):e1001934. https://doi.org/10.1371/journal.pbio.1001934

Bortoff GA, Strick PL (1993) Corticospinal terminations in two new-world primates: further evidence that corticomotoneuronal connections provide part of the neural substrate for manual dexterity. J Neurosci 13(12):5105–5118. https://doi.org/10.1523/JNEUROSCI.13-12-05105.1993

Chandrasekaran C, Trubanova A, Stillittano S, Caplier A, Ghazanfar AA (2009) The natural statistics of audiovisual speech. PLoS Comput Biol 5(7):e1000436. https://doi.org/10.1371/journal.pcbi.1000436

Chomsky N (1998) Minimalist inquiries: the framework, MIT occasional papers in linguistics 15. Distributed by MIT Working Papers in Lingustics, Cambridge, MA, pp 1–56

Corballis MC (2002) From hand to mouth: the origins of language. Princeton University Press, Princeton

Coudé G, Ferrari PF, Rodà F, Maranesi M, Borelli E, Veroni V, Monti F, Rozzi S, Fogassi L (2011) Neurons controlling voluntary vocalization in the macaque ventral premotor cortex. PLoS One 6 (11):e26822. https://doi.org/10.1371/journal.pone.0026822
Fitch WT (2010) The evolution of language. Cambridge University Press, Cambridge. https://doi.org/10.1017/CBO9780511817779
Fitch WT (2017) Empirical approaches to the study of language evolution. Psychon Bull Rev 24:3–33. https://doi.org/10.3758/s13423-017-1236-5
Fitch WT, de Boer B, Mathur N, Ghazanfar AA (2016) Monkey vocal tracts are speech-ready. Sci Adv 2(12):e1600723
Ghazanfar AA, Takahashi DY (2014) The evolution of speech: vision, rhythm, cooperation. Trends Cogn Sci 18(10):543–553. Elsevier Ltd. https://doi.org/10.1016/j.tics.2014.06.004
Ghazanfar AA, Takahashi DY, Mathur N, Tecumseh Fitch W (2012) Cineradiography of monkey lip-smacking reveals putative precursors of speech dynamics. Curr Biol 22(13):1176–1182
Hadders-Algra M (2002) Variability in infant motor behavior: a hallmark of the healthy nervous system. Infant Behav Dev 25(4):433–451. https://doi.org/10.1016/S0163-6383(02)00144-3
Hadders-Algra M (2004) General movements: a window for early identification of children at high risk for developmental disorders. J Pediatr 145(2):S12–S18. https://doi.org/10.1016/j.jpeds.2004.05.017
Hadders-Algra M (2007) Putative neural substrate of normal and abnormal general movements. Neurosci Biobehav Rev 31:1181–1190. https://doi.org/10.1016/j.neubiorev.2007.04.009
Hage SR (2018a) Dual neural network model of speech and language evolution: new insights on flexibility of vocal production systems and involvement of frontal cortex. Curr Opin Behav Sci 21(June):80–87. https://doi.org/10.1016/j.cobeha.2018.02.010
Hage SR (2018b) Dual neural network model of speech and language evolution: new insights on flexibility of vocal production systems and involvement of frontal cortex. Curr Opin Behav Sci 21(June):80–87. https://doi.org/10.1016/j.cobeha.2018.02.010
Hage SR, Nieder A (2013a) Single neurons in monkey prefrontal cortex encode volitional initiation of vocalizations. Nat Commun 4(September):2409. https://doi.org/10.1038/ncomms3409. Nature Publishing Group
Hage SR, Nieder A (2013b) Single neurons in monkey prefrontal cortex encode volitional initiation of vocalizations. Nat Commun 4(September):2409. https://doi.org/10.1038/ncomms3409. Nature Publishing Group
Hage SR, Nieder A (2016a) Dual neural network model for the evolution of speech and language. Trends Neurosci 39(12):813–829. https://doi.org/10.1016/j.tins.2016.10.006
Hage SR, Gavrilov N, Nieder A (2013) Cognitive control of distinct vocalizations in rhesus monkeys. J Cogn Neurosci 25(10):1692–1701. http://www.ncbi.nlm.nih.gov/pubmed/23691983
Hage SR, Gavrilov N, Nieder A (2016) Developmental changes of cognitive vocal control in monkeys. J Exp Biol 219:1744–1749. https://doi.org/10.1242/jeb.137653
Hage SR, Nieder A (2016b) Dual neural network model for the evolution of speech and language. Trends Neurosci 39(12):813–829. https://doi.org/10.1016/j.tins.2016.10.006
Hammerschmidt K, Fischer J (2008) Constraints in primate vocal production. In: The Evolution of Communicative Creativity: From Fixed Signals to Contextual Flexibility, pp 93–119
Hauser MD, Chomsky N, Tecumseh Fitch W (2002) The faculty of language: what is it, who has it, and how did it evolve? Science 298(5598):1569–1579. https://doi.org/10.1126/science.298.5598.1569
Hayes KJ, Hayes C (1951) The intellectual development of a home-raised chimpanzee. Proc Am Philos Soc 95(2):105–109. JSTOR
Hihara S, Yamada H, Iriki A, Okanoya K (2003) Spontaneous vocal differentiation of coo-calls for tools and food in Japanese monkeys. Neurosci Res 45(4):383–389. https://doi.org/10.1016/S0168-0102(03)00011-7

Jahanshahi M, Obeso I, Rothwell JC, Obeso JA (2015) A fronto-striato-subthalamic-pallidal network for goal-directed and habitual inhibition. Nat Rev Neurosci 16(12):719–732. http://www.ncbi.nlm.nih.gov/pubmed/26530468

Jürgens U (2002) Neural pathways underlying vocal control. Neurosci Biobehav Rev 26 (2):235–258. http://www.ncbi.nlm.nih.gov/pubmed/11856561

Kanemaru N, Watanabe H, Taga G (2012) Increasing selectivity of interlimb coordination during spontaneous movements in 2- to 4-month-old infants. Exp Brain Res 218(1):49–61. https://doi.org/10.1007/s00221-012-3001-3

Kellogg WN (1968) Communication and language in the home-raised chimpanzee. Science 162 (3852):423–427. https://doi.org/10.1126/science.162.3852.423

Koda H, Kunieda T, Nishimura T (2018) From hand to mouth: monkeys require greater effort in motor preparation for voluntary control of vocalization than for manual actions. R Soc Open Sci 5(11):180879. https://doi.org/10.1098/rsos.180879

Koda H, Oyakawa C, Kato A, Masataka N (2007) Experimental evidence for the volitional control of vocal production in an immature gibbon. Behaviour 144:681–692

Kojima S (2003) Search for the origins of human speech, a: auditory and vocal functions of the chimpanzee. Kyoto University Press, Kyoto

Kuypers, HGJM (1981) Anatomy of the descending pathways. Comprehensive Physiology. Wiley Online Library

Larson CR, Kistler MK (1984) Periaqueductal gray neuronal activity associated with laryngeal EMG and vocalization in the awake monkey. Neurosci Lett 46(3):261–266

Lemon RN (2008) Descending pathways in motor control. Annu Rev Neurosci 31(1):195–218. https://doi.org/10.1146/annurev.neuro.31.060407.125547

Lemon RN, Griffiths J (2005) Comparing the function of the corticospinal system in different species: organizational differences for motor specialization? Muscle Nerve 32(3):261–279. https://doi.org/10.1002/mus.20333

Lindblom BEF, Sundberg JEF (1971) Acoustical consequences of lip, tongue, jaw, and larynx movement. J Acoust Soc Am 50:1166

Locke J (1995) The child's path to spoken language. Harvard University Press, Cambridge, MA

MacNeilage PF (1998) The frame/content theory of evolution of speech production. Behav Brain Sci 21(4):499–546. https://doi.org/10.1017/S0140525X98001265. Cambridge University Press

Masataka, Nobuo. 2003. The onset of language. Cambridge University Press Cambridge doi:https://doi.org/10.1017/CBO9780511489754

McKiernan BJ, Marcario JK, Karrer JH, Cheney PD (1998) Corticomotoneuronal postspike effects in shoulder, elbow, wrist, digit, and intrinsic hand muscles during a reach and prehension task. J Neurophysiol 80(4):1961–1980. https://doi.org/10.1152/jn.1998.80.4.1961

Moore JD, Kleinfeld D, Wang F (2014) How the brainstem controls orofacial behaviors comprised of rhythmic actions. Trends Neurosci 37(7):370–380. Elsevier Ltd. https://doi.org/10.1016/j.tins.2014.05.001

Morrill RJ, Paukner A, Ferrari PF, Ghazanfar AA (2012) Monkey lipsmacking develops like the human speech rhythm. Dev Sci 15(4):557–568. https://doi.org/10.1111/j.1467-7687.2012.01149.x

Nakajima K, Maier MA, Kirkwood PA, Lemon RN (2000) Striking differences in transmission of corticospinal excitation to upper limb motoneurons in two primate species. J Neurophysiol 84 (2):698–709. https://doi.org/10.1152/jn.2000.84.2.698

Newman JD (2007) Neural circuits underlying crying and cry responding in mammals. Behav Brain Res 182(2):155–165. https://doi.org/10.1016/j.bbr.2007.02.011

Nishimura T, Mikami A, Suzuki J, Matsuzawa T (2003) Descent of the larynx in chimpanzee infants. Proc Natl Acad Sci 100(12):6930–6933. National Acad Sciences

Okanoya K (2007) Language evolution and an emergent property. Curr Opin Neurobiol 17:271–276. https://doi.org/10.1016/j.conb.2007.03.011

Oller DK (2000) The emergence of the speech capacity. Lawrence Erlbaum Associates Publishers, Mahwah

Oller DK, Griebel U (2008) Evolution of communicative flexibility: complexity, creativity, and adaptability in human and animal communication. MIT Press, Cambridge

Paus T (2001) Primate anterior cingulate cortex: where motor control, drive and cognition interface. Nat Rev Neurosci 2(6):417–424. https://doi.org/10.1038/35077500

Pierce JD (1985) A review of attempts to condition operantly alloprimate vocalizations. Primates 26 (2):202–213. https://doi.org/10.1007/BF02382019. Springer

Porter R, Lemon R (1993) Corticospinal function and voluntary movement. https://doi.org/10.1093/acprof:oso/9780198523758.001.0001

Prechtl HFR, Hopkins B (1986) Developmental transformations of spontaneous movements in early infancy. Early Hum Dev 14(3–4):233–238. https://doi.org/10.1016/0378-3782(86)90184-2

Preuschoft S, van Hooff JARAM (1995) Homologizing primate facial displays: a critical review of methods. Folia Primatol 65(3):121–137. http://www.karger.com/DOI/10.1159/000156878

Rathelot J-A, Strick PL (2006) Muscle representation in the macaque motor cortex: an anatomical perspective. Proc Natl Acad Sci U S A 103(21):8257–8262. https://doi.org/10.1073/pnas.0602933103. National Academy of Sciences

Rathelot J-A, Strick PL (2009) Subdivisions of primary motor cortex based on cortico-motoneuronal cells. Proc Natl Acad Sci U S A 106:918–923. https://doi.org/10.1073/pnas.0808362106

Robertson SS, Bacher LF, Huntington NL (2001) The integration of body movement and attention in young infants. Psychol Sci 12(6):523–526. https://doi.org/10.1111/1467-9280.00396

Robertson SS, Johnson SL, Masnick AM, Weiss SL (2007) Robust coupling of body movement and gaze in young infants. Dev Psychobiol 49(2):208–215. https://doi.org/10.1002/dev.20201. Wiley Subscription Services, Inc., A Wiley Company

Seyfarth RM, Cheney DL, Marler P (1980) Monkey responses to three different alarm calls: evidence of predator classification and semantic communication. Science 210(4471):801–803. http://www.sciencemag.org/content/210/4471/801.short

Shepherd SV, Lanzilotto M, Ghazanfar AA (2012) Facial muscle coordination in monkeys during rhythmic facial expressions and ingestive movements. J Neurosci 32(18):6105–6116. https://doi.org/10.1523/JNEUROSCI.6136-11.2012

Simonyan K (2014) The laryngeal motor cortex: its organization and connectivity. Curr Opin Neurobiol 28C(October):15–21. https://doi.org/10.1016/j.conb.2014.05.006. Elsevier Ltd

Simonyan K, Horwitz B (2011) Laryngeal motor cortex and control of speech in humans. Neuroscientist 17(2):197–208. https://doi.org/10.1177/1073858410386727

Soltis J (2004) The developmental mechanisms and the signal functions of early infant crying. Behav Brain Sci 27(04):477–484. https://doi.org/10.1017/S0140525X0442010X. Cambridge University Press

Sutton D, Larson C, Lindeman RCC (1974) Neocortical and limbic lesion effects on primate phonation. Brain Res 71(1):61–75. https://doi.org/10.1016/0006-8993(74)90191-7

Sutton D, Larson C, Taylor EM, Lindeman RC (1973) Vocalization in rhesus monkeys: conditionability. Brain Res 52(C):225–231

Sutton D, Trachy RE, Lindeman RC (1981a) Primate phonation: unilateral and bilateral cingulate lesion effects. Behav Brain Res 3(1):99–114

Sutton D, Trachy RE, Lindeman RC (1981b) Vocal and nonvocal discriminative performance in monkeys. Brain Lang 14(1):93–105. https://doi.org/10.1016/0093-934X(81)90067-5

Takahashi DY, Fenley AR, Teramoto Y, Narayanan DZ, Borjon JI, Holmes P, Ghazanfar AA (2015) The developmental dynamics of marmoset monkey vocal production. Science 349 (6249):734–738. https://doi.org/10.1126/science.aab1058

Tomasello M (2009) Constructing a language. Harvard university Press, Cambridge

Toyoda A, Maruhashi T, Malaivijitnond S, Koda H (2017) Speech-like orofacial oscillations in stump-tailed macaque (Macaca arctoides) facial and vocal signals. Am J Phys Anthropol 164 (2):435–439. https://doi.org/10.1002/ajpa.23276

Watanabe H, Homae F, Taga G (2011) Developmental emergence of self-referential and inhibition mechanisms of body movements underling felicitous behaviors. J Exp Psychol Hum Percept Perform 37(4):1157–1173. https://doi.org/10.1037/a0021936. American Psychological Association

Wessel MA, Cobb JC, Jackson EB, Harris GS, Detwiler AC (1954) Paroxysmal fussing in infancy, sometimes called colic. Pediatrics 14(5):421–435. http://www.ncbi.nlm.nih.gov/pubmed/13214956

# Chapter 4
# Conversation Among Primate Species

**Loïc Pougnault, Florence Levréro, and Alban Lemasson**

**Abstract** The literature in psychology and sociolinguistic suggests that human interlocutors, when conversing, virtually sign a sort of contract that defines the exchange rules in both structural and social domains. These rules make the messages more understandable and the interaction more predictable, but they may also act as a social bond regulator. These rules can be very basic such as speech overlap avoidance, respect of response delays, turn-taking and vocal accommodation to the context and interlocutor's social status. Interestingly, these rules are universally spread among human cultures questioning their biological basis and motivating the search for possible parallels with our primate cousins. Here, we will review the available literature on monkeys and apes. We will describe the different forms of vocal interactions, the temporal rules underlying these coordinated interactions, the non-random social selection of interlocutors and the context-dependent acoustic plasticity associated to these exchanges. The fact that primate species are socially varied, in terms of both social structure and social organisation, is another interesting aspect, since different social needs may predict different vocal interaction patterns and conversational rules. For example, duets, choruses and dyadic exchanges are not randomly distributed in the primate phylogeny and may even show different

---

Florence Levréro and Alban Lemasson contributed equally with all other contributors.

---

L. Pougnault
Univ Rennes, Normandie Univ, CNRS, EthoS (Éthologie animale et humaine) – UMR 6552, F-35000, Rennes, France

Université de Lyon/Saint-Etienne, CNRS, Equipe Neuro-Ethologie Sensorielle, ENES/CRNL, UMR5292, INSERM UMR_S 1028, Saint-Etienne, France

ZooParc de Beauval & Beauval Nature, Saint Aignan, France

F. Levréro
Université de Lyon/Saint-Etienne, CNRS, Equipe Neuro-Ethologie Sensorielle, ENES/CRNL, UMR5292, INSERM UMR_S 1028, Saint-Etienne, France

A. Lemasson (✉)
Univ Rennes, Normandie Univ, CNRS, EthoS (Éthologie animale et humaine) – UMR 6552, F-35000, Rennes, France
e-mail: alban.lemasson@univ-rennes1.fr



N. Masataka (ed.), *The Origins of Language Revisited*,
https://doi.org/10.1007/978-981-15-4250-3_4

functions. Also, age proximity, kin membership, social affinity and hierarchy seem to play species-specific roles. Regarding plasticity, cases of vocal sharing and acoustic matching have been described in some species, notably in contact calls which are the calls the most frequently involved in dyadic exchanges. At last, a few studies also show that these 'primitive' conversational rules are often broken by juveniles and that the appropriate way to vocally interact with others may be socially learned, thus another aspect that do not seem strictly human.

**Keywords** Vocal accommodation · Conversation · Turn-taking · Vocal interaction

## 4.1 Introduction

The psychology and sociolinguistic literature suggests that human interlocutors, when conversing, virtually sign a sort of contract that defines exchange rules concerning both the structural and the social domains. These rules make messages easier to understand and interactions more predictable, and in addition they can act as a social bond regulator. These rules can be basic such as speech overlap avoidance, respect of response delays, turn-taking and vocal accommodation in relation to context and interlocutors' social status. The fact that these rules are universal among human cultures questions their biological basis and motivates the search for parallels with our primate cousins. Here, we review the relative literature concerning monkeys and apes. We present different forms of vocal interactions, the temporal rules underlying coordinated vocal interactions, the non-random social selection of interlocutors and the context-dependent acoustic plasticity associated with these exchanges. The fact that primates live in various social systems in terms of both social structure and social organisation is another interesting aspect. Different social needs may predict different vocal interaction patterns and conversational rules. For example, duets, choruses and dyadic exchanges are not randomly distributed over the primate phylogeny and may even possess different functions. Also, age proximity, kin membership, social affinity and hierarchy seem to play species-specific roles. Plasticity in cases of vocal sharing and acoustic matching has been described for some species, notably for contact calls that are most frequently involved in dyadic exchanges. A few reports show that these 'primitive' conversational rules are often broken by juveniles and that the appropriate way to interact vocally with others can be learned socially, thus revealing an aspect that does not seem to be strictly human.

## 4.2 *Homo sapiens*' Conversational Rules

### *4.2.1 A Universal 'Contract of Communication'*

Informal verbal interactions are the core matrix for human social life (Stivers et al. 2009). Anthropologists, ethnographers, psychologists, ethologists and sociolinguists agree that, even though notable cultural variations exist, comparable conversational basic phenomena appear in unrelated languages (Beach 1990). The universality of conversations has been well illustrated by the French social psychologist Rodolphe Ghiglione in his so-called 'contract of communication' (Ghiglione 1986). When interlocutors are engaged in a conversation, they virtually sign a sort of contract based on shared principles concerning both the temporal and social domains. This contract facilitates vocal coordination between interlocutors and promotes inter-comprehension. Ghiglione (1986) listed several principles, among which 'context relevance' (i.e. evaluation of context as pertinent or not to initiate a conversation), 'reciprocity' (i.e. evaluation of the partner as a valid interlocutor) and 'contract-based temporal rules' (i.e. respect of turn-taking between speakers and speech overlap avoidance).

#### 4.2.1.1 Temporally Ruled Interactions

Turn-taking represents the social core of human interactions (Levinson 2016). It is characterised by a reciprocal exchange of alternating, short and flexible turns between two or more interlocutors (Pika et al. 2018). Conversing is a well-known example of behaviour based on turn-taking (Sacks et al. 1974), and this key temporal coordination can be found in all languages and human societies around the world (Stivers et al. 2009). To engage in turn-taking, interlocutors A and B have to respect three basic rules: (*i*) A and B coordinate their speech temporally and avoid talking simultaneously; (*ii*) at any given moment during a conversation, interlocutors have coordinated roles – one of them sends a signal and the other responds (i.e. following the pattern AB and not AA or BB); (*iii*) during a longer conversation, the two interlocutors alternate (i.e. going from AB to ABA, ABAB, etc.) (Ghiglione 1986; Kerbat-Orecchioni 1990; Levinson 1983).

The optimal timing to take turns depends on the ability to coordinate language comprehension (prediction) and language production (formulation) (Sacks et al. 1974). Because, response delays are short, a speaker must begin formulating his/her response while his/her partner is still talking, and this act of formulation must be based on a prediction about what the partner is going to say (Sacks et al. 1974). The silent gaps between turns must last sufficiently to avoid overlap but be short enough to allow a certain fluidness during the verbal exchange. The brevity of silent gaps, often less than 200 ms, is particular striking because linguistic formulation takes time: preparing to name a picture, for instance, takes at least 600 ms (Indefrey and

Levelt 2004) and can rise to about 1500 ms on average for a sentence (Griffin and Bock 2000).

Moreover, the timing of turn-taking is less precise and less homogeneously distributed among societies than often claimed (Heldner and Edlund 2010). Humans' common response delay varies with their culture (Stivers et al. 2009), and it can be influenced selectively and locally by social aspects of context (Sacks et al. 1974). For instance, the common response delays for subjectively on-time responses of Nordic cultures interlocutors (e.g. Danish 203 ms) are longer than those of Japanese (36 ms) (Stivers et al. 2009). These different response delays could be due to a specific cultural interactional pace or stem from more general differences of the overall tempo of social life (Hall 1959). Despite these differences, the available data are consistent with the presence of a universal and stable system of turn-taking avoiding overlap and minimising silent gaps (Stivers et al. 2009).

#### 4.2.1.2 Socially Guided Interaction Patterns

Human oral communication is typically intentional and directional, targeting a dedicated audience or a particular individual interlocutor. To engage in a conversation, interlocutors are thus not chosen randomly. Affinity and age proximity with a chosen interlocutor are often considered determining factors in both traditional and modern human societies (Bascom 1942). However, humans do not converse exclusively with friends. Humans are hence able to adjust their social distance to the interlocutor during a given conversation due to an acoustic convergence/divergence phenomenon named 'vocal accommodation' (Babel 2010; Chartrand and van Baaren 2009; Gallois et al. 2005). Short-term accommodation (i.e. minutes or hours) can be distinguished from long-term accommodation (i.e. month or years), and long-term accommodation is commonly assumed to be based on repeated short-term accommodation (Nguyen and Delvaux 2015; Trudgill 2008). The communication accommodation theory claims that linguistic short-term convergence and its opposite, divergence, are strategically used by speakers engaged in spoken interactions in order to respectively minimise or maximise their social distance with their interlocutors, so as to reinforce their own social identity (Gallois et al. 2005). In this way, speakers are able to match various kinds of acoustic features, such as fundamental frequencies or vowel spectra, in response to social factors such as attraction (Babel 2012) and strength of the relationship (Babel 2010; Pardo et al. 2012). During long-term accommodation, the change observed in speech-involved attributes is measured over long stretches of dialogue such as accent, speaking rate, intensity, low-frequency band variation, pause frequency and utterance length (Gregory 1990). Convergent characteristics in vocal behaviour reflect a speaker's motivation to increase social integration or identification (McGarva and Warner 2003).

### 4.2.2 *Functions of Conversations*

Humans may converse to transmit a message to a specific person. They can develop different strategies in order to increase the chance of success to continue a given conversation, such as persistence (i.e. signal repetition) and elaboration (i.e. signal variation) (Golinkoff 1986, 1993). However, observational studies of human conversations in relaxed social settings suggest that these consist predominantly of exchanges of social information (mostly concerning personal relationships and experience) (Dunbar et al. 1997). Conversations are thus used to facilitate social integration and social bonding.

Respect of turn-taking during conversation, 'as an orderly distribution of opportunities to participate in social interaction', is considered one of the 'most fundamental preconditions' for a viable social organisation (Schegloff 2002). Apart from just 'politeness', maintaining mutual comprehensibility when participants talk at the same time is obviously difficult (Duncan 1972). When this occurs, overlap is mostly associated with negative and strong personality attributes (Ter Maat et al. 2010) and considered as serious impoliteness by most human societies (Calame-Griaule 1965). There is a fundamental distinction between accidental overlap and deliberate overlap (Kurtić et al. 2013). Conversely, pauses between turns are considered as more agreeable (Ter Maat et al. 2010). Nevertheless, long silence gaps are considered less affiliative, or more distancing, than shorter ones (Roberts et al. 2011). Temporal rules make messages more understandable and interactions more predictable, but they can also act as a social bond regulator.

### 4.2.3 *Ontogeny of Conversational Behaviours*

Maturation and social experience clearly shape efficiency in turn-taking, but turn-taking is definitely found at very early ages. Reissland and Stephenson (1999) compared vocal interactions between mothers and their babies who were either premature (range 26–32 weeks postmenstrual, mean = 30.2 weeks) or term infants (from 38 to 42 weeks postmenstrual, mean = 40.1 weeks). Their conversational turns were investigated at the hospital (1 week old for term infants and 39 weeks postmenstrual for preterm infants) and then at home 2 months later. No differences in vocalisation rates or conversational turns could be evidenced at the hospital. However, when they were 2 months old, mothers of preterm infants responded more to their infants' non-cry vocalisations, and they took every opportunity to do so. Term infants were more likely to respond to their mother's vocalisations. Another report focusing on mother-preterm conversational turns found that both adult word count per hour and infant vocalisation count per hour increased from 32 to 36 weeks postmenstrual age (Caskey et al. 2014). When a parent was present, infants produced significantly more conversational turns per hour at both ages.

Around 3 or 4 months old, infants are more capable to control their vocal production voluntarily. Two structurally different patterns of dyadic vocal interactions between mother and infant emerge: the 'coaction mode' when both members perform the same or similar behaviour at the same time and the 'alternating mode' when mothers often create or shape an antiphonal or alternating pattern of vocalisations between themselves and their infants (Stern et al. 1975). While more overlaps than alternations are found during the second semester of life, the contrary is observed during the third semester (Ginsburg and Kilbourne 1988).

Caregivers play a major role in the development of conversational abilities. They use cues such as stable timing and exaggerated pauses to help infants learn to take turns (Stern et al. 1977; Stern and Gibbon 1979). Longitudinal approaches reveal that the structure of vocal interactions elicited by the mother's behaviour is temporally stable during the first half year of life in face-to-face interactions (Cossette et al. 1986; Stern et al. 1977). At very early stages –before 2 months – infants' vocal production occurs in the form of periodically occurring bursts (Schaffer 1977). Between these bursts, the mother skilfully inserts her own actions, and by doing so, she is attempting to provide some temporal organisation for her interaction with the infant. Mothers' speech to 2-month-old infants is timed in such a way as to leave time for the infant's response. Maternal utterances are brief (between 0.5 s and 1.5 s) followed by pauses lasting around 1 s and usually followed by another utterance. Pauses between utterances that are connected through repetition of form, content or topic rarely exceed 3 s and pauses longer than 3 s generally demarcate the end of mutual engagement episodes (Stern et al. 1977; Stern and Gibbon 1979). The majority of maternal responses to infants aged between 2 and 4 months occur within 1 s after the signal (Keller et al. 1999) and pauses between turns of young infants and mothers range from 500 ms to 1 s (Jaffe et al. 2001).

Of course both caregivers and infants are actors in the acquisition process. The fact that mothers and infants engaged in a social interaction match each other's duration of vocal utterances and pauses suggests mutual regulation of turn-taking (Beebe et al. 1985; Jaffe et al. 2001). This is in line with the theory of social learning that presupposes that infant vocalisations are reinforced by continuous social stimulation. Masataka's (2003) experimental results support the anecdotal evidence that infants between 12 and 18 weeks old are less likely to begin vocalising while their caregivers are speaking than when they are silent (Ginsburg and Kilbourne 1988). Even when mothers stimulate their infants randomly, they maintained their infants' attention, and infants focused their attention on their mothers to respond contingently. Neglected children fail to develop this ability to converse, showing irrelevant turns, interruptions, simultaneous talking and non-contingent responding. Moreover, vocal matching ability appears early during ontogeny. It occurs in 41–57% of infants' non-crying vocalisations, during spontaneous vocal interactions between mothers and their 2-, 3- and 5-month-old infants. With age, matches become more and more complex in number and types of included features such as pitch, pitch modulation, duration and rhythm (Papoušek and Papoušek 1989).

These studies show that prelinguistic infants engage in proto-conversational interactions with their caregivers and that these interactions are well coordinated and well timed (Hilbrink et al. 2015). However, when children begin to interact using language, their fluency at turn-taking declines; gaps in linguistic turn-taking are then longer than previously in nonlinguistic turn-taking (Hilbrink et al. 2015). This development trajectory, from well-timed nonlinguistic interactions to less fluent linguistic interactions, raises the possibility that preschool children (around 9 months old) may in fact lack competence in flexibly coordinating linguistic prediction and formulation (Lindsay et al. 2019). This slowing down occurs when the infants begin to grasp the significance of intentional communication. Levinson (2016) argued that the poor timing of children's conversational turn-taking is not indicative of an underlying lack of competence, but he suggests that children can flexibly coordinate prediction about what a conversational partner will say by formulating early a response, but compared to adults, it is hard for them to encode linguistically the ideas that they want to express, and thus they respond after a longer delay.

### *4.2.4 Underlying Neurobiological Processes*

The cognitive processes governing human behaviour during a verbal interaction are complex and vary with context. We chose to focus on one well-studied conversation context, the face-to-face interaction. Unlike other types of communication, face-to-face interactions involve more continuous turn-taking (Wilson and Wilson 2005). Recently Stephens et al. (2010) found that, when the speaker's voice is aurally presented to the listener, the brain activities of both listener and speaker synchronise. A successful communication notably results in a temporally coupled neural response pattern, a pattern not systematically found if speakers talk a language unknown to listeners. Using the near-infrared spectroscopy developed by Cui et al. (2012) that measures simultaneously the brain activity of two conversing persons, Jiang et al. (2012) examined neural mechanistic features of face-to-face communication compared with back-to-back dialogue, face-to-face monologue and back-to-back monologue. These authors highlighted a simultaneous neural synchronisation in the left inferior frontal cortex – left IFC – (usually important for language production) increased in face-to-face communication. This facilitation could be explained by an action-perception system (Fogassi and Ferrari 2007; Nishitani et al. 2005). The left IFC, in addition to other brain regions, is indeed a site where mirror neurons are located (Rizzolatti and Arbib 1998). These neurons are well known to respond to the observations of an action, to a sound associated with that action or to observations of mouth-communication gestures (Ferrari et al. 2003; Kohler et al. 2002).

## 4.3 Nonhuman Primates' Patterns of Vocal Interactions and Conversational-Like Features

The first part of this chapter shows that, despite obvious cultural variations, language seems 'by nature' interactive. This questions the existence of possible underlying biological bases and encourages scientists to tackle the evolutionary origin of 'conversational' behaviour. For some authors, the ability to converse in a rule-governed way with a particular interlocutor probably appeared somewhere before the common ancestor of modern humans and Neanderthals (600,000 years ago), since all the genetic and physiological prerequisites for speech seems in place (Dediu and Levinson 2013), while they seem lacking in *Homo erectus* (ca 1.6 My) (Levinson and Holler 2014; MacLarnon and Hewitt 2004). During the intervening million years, simple alternation of vocal utterances during vocal exchanges may have provided the framework for linguistic complexity to evolve (Levinson 2016). The temporal properties of turn-taking may have remained fixed, creating the foundations on which more complex linguistic material was progressively added. For instance, more complex linguistic material could be free flowing, intricately varied, rapid sound sequences suitable for the fast transfer of complex, highly flexible communication allowed by fine neurobiological control of the respiratory muscles and complex cognitive processes. Now, language diversity is considered as driven by cultural evolution (Levinson 2016). However, other authors suggest that the rules governing humans' vocal interactions are much more ancestral as reports stress parallels with birds and mammals (Henry et al. 2015) and particularly with nonhuman primates (Lemasson et al. 2011a, b; Snowdon and Cleveland 1984; Sugiura and Masataka 1995).

### 4.3.1 *Definition of Conservation-Like Vocal Exchanges*

The vast majority of nonhuman primates are highly social species that live in groups governed by complex relationships (Dunbar 2009; Grubb et al. 2003; Kappeler and Van Schaik 2002; McComb and Semple 2005). Their social life requires the existence of communication rules respected by all individuals, allowing each individual to share information and to develop and to maintain social bonds with particular group members. Although the level of ability to communicate visually (mimics, gestures, postures) varies greatly among species, all nonhuman primates are vocal communicants (Lemasson 2011). In this context, vocal interactions play a key role in their social communication network (Pika et al. 2018).

Nonhuman primates produce a variety of call types, some dedicated to specific contexts, such as alarm calls, distress calls, copulation calls and food calls. These calls increase the chances of individuals to survive and to reproduce. However, several species also produce so-called 'social/contact' calls, not specifically linked to a given context but typically associated with peaceful and relaxed behaviours. They are produced with high call rates (most often higher than the other call types) and usually involved in socially and temporally coordinated vocal exchanges. We believe that these contact call exchanges are good candidates for 'conversation-like vocal interactions' in primates. They involve a diversity of recurrent vocal partners. Interlocutors are typically familiar individuals belonging to a given social group and can be of any age and sex. Production of conversation-like vocal exchanges is not related to 'primary need' contexts, time of day or year, and they relate more to daily-basis short-distance communication. Thus, these types of interactions differ from vocal interactions related to reproduction in insects (Bailey 2003), amphibians (Forester and Harrison 2008) and even mammals like rodents (Okobi et al. 2019) or deer (Clutton-Brock and Albon 1979; Reby and McComb 2003).

To what degree vocal nonhuman primates' communication is intentional is a matter of current debate. Nevertheless, the debate concerning vocal production (control of acoustic structure of the sound emitted) is more important than that concerning vocal usage (control of utterance of a vocalisation). Several reports show that vocal usage is socially aware. Indeed, the identity of potential receivers (i.e. active interlocutors or passive listeners) influences the vocal behaviour of emitters who can choose to vocalise only when particular conspecifics are present. Audience effect on call rates (but even on call structure to a lesser extent) has been found in a range of primate species. Good evidence shows that the composition of the nearby audience can affect the likelihood of calling. For example, some contact calls are preferentially emitted when approaching specific partners: baboons' grunts and macaques' girneys addressed to affiliated individuals (Silk 2002) and female/low-ranking male chimpanzees' pant-grunts addressed to the higher-ranking male (Laporte and Zuberbühler 2010). However, the audience effect has been observed in a broad range of contexts and for several types of calls (e.g. chimpanzees: Leavens et al. 2004, 2009; Slocombe et al. 2010; macaques: Hauser and Marler 1993; brown capuchins: Pollick et al. 2005; red-capped mangabeys: Bouchet et al. 2010). The social and temporal rules underlying contact call exchanges are good examples to illustrate the level of control of vocal usage by nonhuman primates.

Conversation-like call exchanges are only one type of nonhuman primates' vocal interactions. Authors typically distinguish this behaviour from other vocal patterns like isolated calling (one call is emitted and no other calls can be heard around), repeated calling (the same callers call several times in a row), disorganised phono-responses (one individual produces a call, typically an alarm call, and all the group members respond in an apparent chaotic way), chorusing (two or more individuals overlap their emissions of a given call type) and duetting (two individuals synchronise long series of calls or songs with a stereotyped temporal association). Yoshida and Okanoya (2005) hypothesised that during evolution choruses came first, then duets and vocal exchanges finally emerged. These different calling patterns are however found at different levels of the primate phylogeny, and all the great apes present all these patterns. The two 'non-conversational' interaction patterns (i.e. chorus, duet) suggest abilities of animals to control the timing of their vocal utterances with predictable temporal coordination. For example, male chimpanzees produce a species-typical long-distance call known as pant-hoot. This vocalisation is often emitted during choruses in a food arrival context (Clark and Wrangham 1993) or to maintain contact over long distances (Mitani and Nishida 1993). Pant-hoots start with a build-up introduction that facilitates synchronisation of the climax part of the pant-hoots of the different overlapping callers (Fedurek et al. 2013). Duetting generally concerns vocal interactions between two partners of the opposite sex (Bailey 2003) and are common among nonhuman primates with stable monogamy and territoriality (e.g. tarsiers, indris, Mentawai langurs and gibbons: Fuentes 1998; Geissmann and Orgeldinger 2000; Haimoff 1986; MacKinnon and MacKinnon 1980). Each species of gibbon has its species-specific temporal duet organisation with predictable male/female/couple parts (Geissmann 2002).

### *4.3.2 Temporal Organisation of Conversation-Like Vocal Exchanges*

Campbell's monkeys' conversation-like vocal exchanges have been well studied. The adult males and females of this species form harem groups and live in West African forests. They present different vocal repertoires and patterns of vocal usage. Males typically produce long loud calls to warn about danger that trigger disorganised phono-responses from all females using their own versions of alarm calls (Ouattara et al. 2009a, b). However the social core of intra-harem communication relies on female contact calls. Their vocal interactions are organised in temporally structured ways generally involving only two or three interlocutors responding to each other (Lemasson et al. 2010) and forming the so-called contact call exchanges respecting a turn-taking pattern (Lemasson et al. 2011a). Exchanging partners can be physically close (e.g. grooming) or distant (without any visual contact). The respect of the turn-taking during vocal exchanges has been observed in a broad range of nonhuman primate species (e.g. pygmy marmosets: Snowdon and Cleveland 1984; common marmosets: Takahashi et al. 2013; Japanese macaques: Sugiura 1998; gorillas: Lemasson et al. 2018; bonobos: Levréro et al. 2019). A given caller typically produces a single call and waits for the vocal response of another group member before possibly calling again.

The first important rule observed is that the respondent adjusts the timing of its response by leaving a short silent gap after the initiator's call. This gap respects a maximum duration (about 1–3 s according to species) to ensure call coordination and a minimum silent delay to prevent call overlap (Henry et al. 2015). A study focusing on long periods of common marmosets' dyadic vocal interactions showed vocal cooperation and temporal coordination (Takahashi et al. 2013). Marmosets adjusted their vocal output reciprocally and continuously with each other. Dyads engage in turn-taking by showing that they prevent overlapping call (waiting ~5 s before responding); the timing of their calls is periodically coupled, and they mime each other so that when one speeds up or slows down its call timing, the other will follow.

Individual Campbell's monkeys respond to each other with a short intercall interval of up to 1 s but rarely less than 260 ms (Lemasson et al. 2010). The common response delay is rarely shorter than an average call duration (e.g. Diana's monkeys: Candiotti et al. 2012; squirrel monkeys: Masataka and Biben 1987; Japanese macaques: Sugiura 1993; spider monkeys: Briseno-Jaramillo et al. 2018; gorillas: Lemasson et al. 2018; bonobos: Levréro et al. 2019).

Response delay is however flexible as it varies with context, such as the physical distance between callers. The interval between Japanese macaque interlocutors is longer when group members are scattered, and more importantly, an initiator persists and repeats its call when it fails to get a response, and the farther away the targeted respondent, the longer it waits (Sugiura 2007). Sugiura (1993) also suggested a possible cultural effect on the common response delay as it varies among Japanese macaque' population. Thus, delays are longer in the Yakushima population than in the Ohirayama population. As the original members of the Ohirayama group were translocated from Yakushima island 33 years before this study, this difference suggests possible trans-generational social transmission of the conversational rule in addition to a dialect-like structural modification of the contact (coo) calls recorded in these two populations (Tanaka et al. 2006). Social determinants are also key factors that influence duration of the silence gap between squirrel monkeys' exchanged vocalisations. This gap is significantly shorter when the respondent has a strong affiliative bond with the initiator (Biben et al. 1986; Masataka and Biben 1987). Some species, such as common marmosets (Yamaguchi et al. 2009) and spider monkeys (Briseño-Jaramillo et al. 2018) can exchange two types of contact calls, i.e. marmosets' phees and trills or spider monkeys' whinnies and high whinnies. Adult interlocutors typically match the call type used to respond to one another. Nevertheless, response delay can be call type – dependent. Common marmosets respond faster to a trill vocalisation than to a phee vocalisation, and the delay between repeated trills, a way to elicit a response from conspecifics, is shorter (3 s) than the delay between repeated phees (6 s). Authors suggested that phees are usually longer and louder than trills and would thus be used preferentially by distant interlocutors (Yamaguchi et al. 2009)

Several studies furthered their investigations and showed that the respect of the expected temporal pattern matters for animals, at least adults. Playback experiments showed that adults discriminate spontaneously between artificial vocal exchanges respecting the turn-taking principle (Campbell's monkeys: Lemasson et al. 2011a) and the non-overlap principle (gorillas: Pougnault et al. 2020) and those breaking these rules. In both cases, subjects lost interest in the broadcast when the interaction did not respect the temporal rule. Also, in Japanese macaque, adults that produce repeated calls in a row, leaving no chance for turn-taking responses, are low-ranking males (Lemasson et al. 2013a, b).

These temporal features (i.e. turn-taking and overlap avoidance during vocal interactions) by themselves are not sufficient to distinguish primates' vocal interactions from those of other animal species as similar temporal features have been observed for a broad range of nonprimate species (insects: Hartbauer et al. 2005, Hedwig 2006; amphibians: Schwartz 1987, Zelick and Narins 1985; birds: Hyman

2003, Mennill et al. 2003, Vehrencamp et al. 2014; mammals: Behr et al. 2009, Carter et al. 2008, Ghazanfar et al. 2002, Goll et al. 2017, Miller and Wang 2006). What makes the vocal exchanges of some primate species particular is that they present the necessary combination of temporal and social dimensions that define 'conversation-like' vocal exchanges.

### *4.3.3 Social Organisation of Conversation-Like Vocal Exchanges*

The 'social bonding hypothesis' proposed by Dunbar (1996) highlighted the key role played by oral interactions during the evolution of primates. This theory suggested that, in large social groups and in groups with solid bonds, the function of gestural grooming was partially transferred to vocal interactions. Gestural grooming plays an important role in forming new bonds and strengthening existing ones in all primate species (Dunbar 1991). As group size and bonding network increased, conversation-like oral exchanges may have developed since it then became impossible to allocate sufficient time to groom all partners physically, so vocal exchanges could play a 'grooming-at-distance' function (Dunbar 1996).

In line with Dunbar's predictions, two recent studies confirmed that preferred contact call exchange partners are the same as physical grooming partners and that conversation-like exchanges can be considered grooming-at-distance (Japanese macaques: Arlet et al. 2015; spider monkeys: Briseño-Jaramillo et al. 2018). Moreover, female Japanese macaques that groomed each other more also responded to each other' calls more. Thus, the exchange partner, as for humans, is not randomly nor opportunistically chosen. These findings are supported by several other studies reporting that individuals involved in a positive relationship are more likely to exchange calls (e.g. cotton-top tamarins: Jordan, et al. 2004, pygmy marmosets: Snowdon and Cleveland 1984, bonobos: Levréro et al. 2019, Campbell's monkeys: Lemasson et al. 2006, squirrel monkeys: Soltis et al. 2002).

Interlocutors can be selected according to other factors, such as social status. For Campbell's monkeys, age of individuals is an important regulating factor; elders, regardless of dominance status, elicit more responses from their younger conspecifics despite a lower call production (Lemasson et al. 2010). The hierarchy of dominance in this species is very subtle, and agonism is rare (Lemasson et al. 2006). The attention paid to the age of the interlocutors has also been found for marmosets (Chen et al. 2009). Interestingly an effect of age has been found concerning western lowland gorillas' conversation-like interactions. The closer two gorillas are in age, the more likely they are to exchanges soft calls (Lemasson et al. 2018). Age proximity could play a key role in great apes that are characterised by a long development and longevity (Levréro et al. 2019). Dominance status can matter more for other species. For example, dominant male and female white-faced capuchins receive more vocal responses than subordinates of either sexes (Digweed et al. 2007).

### 4.3.4 Neurobiological Processes Associated with Vocal Exchanges

Although much is known about the acoustic structure of primate calls and the social context in which their vocalisations are usually uttered, our knowledge concerning the neocortical control of vocal interactions is still limited. Information concerning this domain comes mainly from studies of lesions of squirrel monkeys, marmosets or macaques (Aitken 1981; Jürgens et al. 1967; Jürgens and Ploog 1970; Sutton et al. 1974). Nonhuman primates have been thought, for several decades, to have no voluntarily control over their vocal production, a fundamental difference from human language (Jürgens 1995). The first neurological studies on this topic showed that the production of some of squirrel monkeys' vocalisations was related to the limbic system and processed in subcortical areas associated with emotions. Nobody seriously disputes the fact that humans have far greater control over their vocal production apparatus and the evidence that some Old and New World monkeys and apes have limited vocal control (Lemasson et al. 2013a, b).

One line of evidence comes from studies based on conversation-like vocal exchanges. For example, when a marmoset hears a conspecific vocalisation and decides not to answer, the primary auditory cortex is activated (Simões et al. 2010). In contrast, when the individual answers, the ACC (i.e. anterior cingulate cortex), the VLPFC (left ventrolateral prefrontal cortex; note that the VLPFC is a key region for speech control in humans) and the DMPFC (dorsomedial prefrontal cortex) are all activated, suggesting the existence of context-dependent vocal control in nonhuman primates (Jürgens et al. 1996; Simões et al. 2010). Other studies reported activation of the ventral premotor cortex when a subject produces conditioned vocalisations, but not when a subject responds spontaneously to the presence of food (pigtailed macaques: Coudé et al. 2011). Single-cell recordings in the ventrolateral prefrontal cortex of monkeys trained to vocalise (contact calls) in response to visual cues evidenced call-related neurons (rhesus macaques: Hage and Nieder 2013).

A recent study showed a robust correlation between marmosets' cortical activity and conversation-like contact call exchanges and in addition that the magnitude of neuronal responses increased in relation to conversation length. This supports the notion that the neural process is strongly related to the social context of the vocal exchange (Nummela et al. 2017). Authors hypothesised that the change in the frontal motor cortex may reflect a change in social arousal and attention (Miller et al. 2010). This could serve as a sensory gating function to facilitate rapid processing of conspecific vocalisations throughout the auditory system (Miller et al. 2010) and precipitate a cascade of subsequent social decision-making processes (Toarmino et al. 2017).

Additionally, recently Okobi et al. (2019) identified a region of singing mice's motor cortex (the orofacial motor cortex) that influences vocalisation and mediates rapid vocal interaction. Authors showed that the motor cortex is required for adaptive counter-calling but not for the production of the call itself. This is the first demonstration of cortical dependence of a precise vocal interaction in mammals. Also, they demonstrated that the motor cortex is able to adjust dynamically the pacing and duration of call sequences, consistent with changes of these parameters during social interactions.

All these results support the notion that diverse cortical structures are involved in the control of nonhuman primates' vocal communication. Neuronal processes, such as the social context change in the frontal cortex state, may occur only when primates are actively interacting with one another.

### 4.3.5 Acoustic Plasticity Associated with Vocal Exchanges

As stated above, most human conversations deal with exchanges about personal relationships and experience (Dunbar et al. 1997). Interestingly, contact calls are often the call types that present the highest level of acoustic identity coding (Lemasson and Hausberger 2011 but see Keenan et al., sub). Beyond identity, these call types can also code for different sorts of social information through a process of vocal convergence. Over more or less long periods of time, individuals can modify the frequency modulation pattern of their contact call to copy their preferred partners (e.g. pygmy marmosets: Snowdon and Elowson 1999). Female Campbell's monkeys that are more closely affiliated are more likely to share particular vocal variants than conspecifics that are not, regardless of genetic relatedness (Lemasson et al. 2011b). However, an individual's variants are not stable over its entire adult life. Reports show that these changes are triggered by alterations of females' social relationships (Lemasson et al. 2003; Lemasson and Hausberger 2004). Authors showed, using playback experiments, that only current variants (conversely to older ones) elicited immediate call responses (Lemasson et al. 2005). Bonobos, another so-called tolerant species, similarly share variants (Levréro et al. 2019). The greater the age difference between two bonobo callers, the less likely they will share acoustic variants. These authors hypothesised that juvenile bonobos display a 'youngster vocal badge' as they often play with peers and this could favour the development of social bonds during their long immaturity period. However, vocal copying by Japanese macaques, who also often exchange contact (coo) calls but form despotic societies, does not explain social affinity. Instead, subordinates copy dominants' calls (Lemasson et al. 2016). Interestingly, macaque species vary socially, some are tolerant, while others are despotic (Thierry et al. 2000; Thierry 2007). Recently a study confirmed that hierarchy explains acoustic similarity in another despotic species, rhesus macaques, but not in their tolerant cousins, Tonkean macaques (De Marco et al. 2019).

At the time scale of a vocal exchange, acoustic matching can be observed in nonhuman primates. The respondent produces a call that matches the acoustic characteristics of the call used by the initiator, a phenomenon comparable to the vocal accommodation mentioned previously for humans. This has been clearly demonstrated using observations and playback experiments with Japanese macaques (Sugiura 1993, 1998) and bonobos (Levrero, pers. comm.). Female Diana monkeys converge vocally in social cohesion contexts and diverge in spatial cohesion contexts (Candiotti et al. 2012).

These copied acoustic variants can be seen as social badges that individuals use during vocal exchanges to advertise their affinity and to facilitate social integration (Lemasson and Hausberger 2004; Sewall et al. 2016).

Apart from convergence/divergence processes, acoustic plasticity can be found in conversation-like contact call exchanges. During vocal interactions, individuals can use several kinds of signals to elicit vocalisation by their interlocutors or to indicate their desire to interact vocally. Biben et al. (1986) demonstrated that during squirrel monkeys' vocal interactions, low-frequency chucks are more likely to initiate vocal exchanges, whereas high-frequency chucks terminate them. Call combination can be used for similar purposes. Red-capped mangabeys typically add an optional 'Uh' unit to their contact call when they engage in vocal exchanges (Bouchet et al. 2010). Bonobos also use call combination of long-distance vocalisations during interparty movements (Schamberg et al. 2016). When wild bonobos are motivated to move from one foraging subgroup to another, they produce a combination of 'whistle + high-hoot'. They are likely to travel to the other subgroup if their initial vocalisation elicited an answer; otherwise they demonstrate an apparent persistence in the production of 'whistle + high-hoot'. Moreover, individuals can modify the acoustic structure of their call combinations to allow the listener to distinguish between those given spontaneously and those given in a response to another call.

In addition, when a Japanese macaque fails to elicit a vocal response from its conspecifics, he/she can elaborate by emitting a second (acoustically modified) call to emphasise his/her motivation to receive a response (Koda 2004). The second call will be longer with frequency modulations of larger amplitude. Playback experiments confirmed that such modified calls trigger systematically more responses. Squirrel monkeys' (Biben et al. 1986) and common marmosets' (Choi et al. 2015) calls with 'exaggerated acoustic patterns' are used preferentially when the initiator is far from the targeted respondents.

### 4.3.6 Breaking the Conversational Rules by Nonhuman Primates

As in humans, the appropriate way to behave vocally when conversing results from social development in several nonhuman primate species. Immature individuals break social and temporal rules more often than adults, supporting the idea of a possible learning process. For example, immature Japanese macaques and Campbell's monkeys are less likely to adhere to the turn-taking rules than adults (Lemasson et al. 2011b, 2013a, b). Analyses of spontaneous vocal exchanges of both species show that juveniles break the turn-taking rule more often (up to 12 times more frequently) by overlapping conspecifics' calls or by producing several repeated calls. Using playback experiments, these authors showed that, conversely to adults, juvenile Campbell monkeys do not detect or do not care about respecting (or not) turn-taking in these artificial call exchanges (Lemasson et al. 2011a).

Chow et al. (2015) suggested that turn-taking during common marmosets' vocal interactions is learned during ontogeny and that its development is guided by parents' behaviours. These authors found that parents were significantly less likely to produce a response following an interruption of their own call by their offspring than during vocal interactions when no interruption occurred. Interestingly, overlaps during vocal interactions are more frequent between offspring and their parents than between siblings. Additionally, despite only small differences in the duration of phee calls produced by mothers and fathers, offspring are significantly more likely to interrupt their father than their mother. Indeed, whereas juvenile marmosets adopt an adult-like temporal pattern of vocal exchanges with their mothers around 10–-12 months, this is not the case during vocal interactions with their fathers. However, while vocal exchanges with mothers and fathers gradually slowed over the first year of life, the temporal pattern stayed relatively constant during sibling vocal interactions. Chow et al. (2015) suggested that these ontogenic changes are not simply due to a general maturational process but that marmosets learn to adapt the relative timing of their vocal responses based on the specific social context and the identity of their interlocutors.

The hypothesis presented by Chow et al. (2015) could be valid if the quantity of contingent parental responses was correlated with the rate of turn-taking development so that infants with more responsive parents learned to adjust the temporal pattern of their turn-taking faster. Takahashi et al. (2016) investigating parental influence on juvenile marmosets' turn-taking development could not evidence any relationships between maturation rate of vocal turn-taking and overall frequency of contingent parental responses. However, the capacity to grasp the vocal turn-taking temporal pattern increases with age of individuals. During the first post-natal week, marmosets interact only weakly with their parents, but these interactions increase with experience. However, the vocal pattern used by parents is not modified during this period. Takahashi et al. (2016) hypothesised that this was due to self-monitoring by young marmosets rather than parental influence.

A similar developmental pattern is observed for at respect of social and acoustic rules. For example, a recent study showed that free-ranging spider monkey adults respond preferentially to friends and perform systematically call-type matching immature did not and young individuals even less than older immatures (Briseno-Jaramillo et al. 2018). Adults may have either the higher cognitive abilities required for processing such complex information or the social experience necessary for understanding these conversational rules. Subadult Japanese macaques (aged 3–5 years old) were less efficient than adults at respecting the call-matching rule during interindividual exchanges of coo calls (Masataka 2003, Sugiura 1998, Sugiura and Masataka 1995). A playback study ran by Bouchet et al. (2017) confirmed that adults of this species care whether vocal exchanges respect the call-matching, while juveniles do not. A report suggested that a social rule guides western lowland gorillas' contact (grunt) call exchanges, i.e. interlocutors are preferentially selected among group members close in age (Lemasson et al. 2018). A playback experiment broadcasting artificial vocal exchanges respecting or not this social rule showed that older subjects, more experienced, paid more attention to exchanges respecting the age proximity expectation than younger ones (Pougnault et al. 2020).

## 4.4 Conclusion

During recent decades, research focusing on the existence and the developmental acquisition of rules governing nonhuman primates' conversation-like vocal interactions has enlarged perspectives. Current findings bring to light nonhuman primates' ability to take turns during alternated vocal interactions. They select their vocal partner and respect a common response delay to avoid overlap between turns, a delay which seems culturally and socially influenced as in humans. Moreover, nonhuman primates demonstrate vocal elaboration and persistence capacities so as to elicit or to perpetuate a vocal interaction. None of these abilities seem to appear spontaneously but are acquired through social experience during all their ontogeny. Additionally, recent neurobiological arguments support the idea that vocal production during vocal interactions is mediated by neuro-cortical control. Authors suggest volitional control of vocal utterances during social interactions. Human conversation, and more precisely the rules governing interactions, could be homologies (traits shared with other primates) rather than analogies (parallel evolution) (Henry et al. 2015). In any event, these abilities could have been an important step in the evolution of primates' vocal communication.

## References

Aitken PG (1981) Cortical control of conditioned and spontaneous vocal behavior in rhesus monkeys. Brain Lang 13(1):171–184. https://doi.org/10.1016/0093-934X(81)90137-1

Arlet M, Jubin R, Masataka N, Lemasson A (2015) Grooming-at-a-distance by exchanging calls in non-human primates. Biol Lett 11(10):20150711–20150714. https://doi.org/10.1098/rsbl.2015.0711

Babel M (2010) Dialect divergence and convergence in New Zealand English. Lang Soc 39(4):437–456. https://doi.org/10.1017/S0047404510000400

Babel M (2012) Evidence for phonetic and social selectivity in spontaneous phonetic imitation. J Phon 40:177–189. https://doi.org/10.1016/j.wocn.2011.09.001

Bailey WJ (2003) Insect duets: underlying mechanisms and their evolution. Physiol Entomol 28:157–174. https://doi.org/10.1046/j.1365-3032.2003.00337.x

Bascom WR (1942) The principle of seniority in the social structure of the Yoruba. Am Anthropol 44:37–46. https://doi.org/10.1525/aa.1942.44.1.02a00050
Beach WA (1990) Searching for universal features of conversation. Res Lang Soc Interact 24 (1–4):351–368. https://doi.org/10.1080/08351819009389347
Beebe B, Jaffe J, Feldstein S, Mays K, Alson D (1985) Interpersonal timing: the application of an adult dialogue model to mother-infant vocal and kinesic interactions. In: Field T, Fox N (eds) Social perception in infants. Ablex Publishing, Norwood, pp 217–247
Behr O, Knörnschild M, Von Helversen O (2009) Territorial counter-singing in male sac-winged bats saccopteryx bilineata: low-frequency songs trigger a stronger response. Behav Ecol Sociobiol 63(3):433–442. https://doi.org/10.1007/s00265-008-0677-2
Biben M, Symmes D, Masataka N (1986) Temporal and structural analysis of affiliative vocal exchanges in squirrel monkeys (*Saimiri sciureus*). Behaviour 98(1):259–273
Bouchet H, Pellier AS, Blois-Heulin C, Lemasson A (2010) Sex differences in the vocal repertoire of adult red-capped mangabeys (*Cercocebus torquatus*): a multi-level acoustic analysis. Am J Primatol 72(4):360–375. https://doi.org/10.1002/ajp.20791
Bouchet H, Koda H, Lemasson A (2017) Age-dependent change in attention paid to vocal exchange rules in Japanese macaques. Anim Behav 129:81–92. https://doi.org/10.1016/j.anbehav.2017.05.012
Briseno-Jaramillo M, Ramos-Fernández G, Palacios-Romo TM, Sosa-López JR, Lemasson A, Sosa-López JR, Lemasson A (2018) Age and social affinity effects on contact call interactions in free-ranging spider monkeys. Behav Ecol Sociobiol 72:192. https://doi.org/10.1007/s00265-018-2615-2
Calame-Griaule G (1965) Ethnologie et langage. la parole chez les dogon. Gallimard, Paris
Candiotti A, Zuberbuhler K, Lemasson A (2012) Convergence and divergence in Diana monkey vocalizations. Biol Lett 8(3):382–385. https://doi.org/10.1098/rsbl.2011.1182
Carter GG, Skowronski MD, Faure PA, Fenton B (2008) Antiphonal calling allows individual discrimination in white-winged vampire bats. Anim Behav 76:1343–1355. https://doi.org/10.1016/j.anbehav.2008.04.023
Caskey M, Stephens B, Tucker R, Vohr B (2014) Adult talk in the NICU with preterm infants and developmental outcomes. Pediatrics 133(3):e578–e584. https://doi.org/10.1542/peds.2013-0104
Chartrand TL, van Baaren R (2009) Human mimicry. Adv Exp Soc Psychol 41:219–274. https://doi.org/10.1016/S0065-2601(08)00405-X
Chen HC, Kaplan G, Rogers LJ (2009) Contact calls of common marmosets (*Callithrix jacchus*): influence of age of caller on antiphonal calling and other vocal responses. Am J Primatol 71:165–170. https://doi.org/10.1002/ajp.20636
Choi JY, Takahashi DY, Ghazanfar AA (2015) Cooperative vocal control in marmoset monkeys via vocal feedback. J Neurophysiol 114:274–283. https://doi.org/10.1152/jn.00228.2015
Chow CP, Mitchell JF, Miller CT (2015) Vocal turn-taking in a non-human primate is learned during ontogeny. Proc R Soc B Biol Sci 282:20150069. https://doi.org/10.1098/rspb.2015.0069
Clark AP, Wrangham RW (1993) Acoustic analysis of wild chimpanzee pant hoots: do Kibale Forest chimpanzees have an acoustically distinct food arrival pant hoot? Am J Primatol. https://doi.org/10.1002/ajp.1350310203
Clutton-Brock TH, Albon SD (1979) The roaring of Red Deer and the evolution of honest advertisement. Behaviour 69(4):145–170
Cossette L, Malcuit G, Pomerleau A, Julien D (1986) Temporal structure of maternal language directed at infants 3 months old. Can J Psychol 40(4):414–422
Coudé G, Ferrari PF, Rodà F, Maranesi M, Borelli E, Veroni V, Monti F, Fogassi L (2011) Neurons controlling voluntary vocalization in the macaque ventral premotor cortex. PLoS One 6(11): e2682. https://doi.org/10.1371/journal.pone.0026822
Cui X, Bryant DM, Reiss AL (2012) NIRS-based hyperscanning reveals increased interpersonal coherence in superior frontal cortex during cooperation. NeuroImage 59(3):2430–2437. https://doi.org/10.1016/j.neuroimage.2011.09.003

De Marco A, Rebout N, Massiot E, Sanna A, Sterck EHM, Langermans JAM, Cozzolino R, Thierry B, Lemasson A (2019) Differential patterns of vocal similarity in tolerant and intolerant macaques. Behaviour 1(aop):1–25. https://doi.org/10.1163/1568539X-00003562

Dediu D, Levinson SC (2013) On the antiquity of language: the reinterpretation of Neandertal linguistic capacities and its consequences. Front Psychol 4(397):1–17. https://doi.org/10.3389/fpsyg.2013.00397

Digweed SM, Fedigan LM, Rendall D (2007) Who cares who calls? Selective responses to the lost calls of socially dominant group members in the white-faced capuchin (*Cebus capucinus*). Am J Primatol 69:829–835. https://doi.org/10.1002/ajp.20398

Dunbar RIM (1991) Functional significance of social grooming in primates. Folia Primatol 57:121–131

Dunbar RIM (1996) Grooming, gossip and the evolution of language. Faber & Faber, London

Dunbar RIM (2009) The social brain hypothesis and its implications for social evolution. Ann Hum Biol 36(5):562–572. https://doi.org/10.1080/03014460902960289

Dunbar RIM, Marriott A, Duncan NDC (1997) Human conversational behavior. Hum Nat 8 (3):231–246. https://doi.org/10.1007/BF02912493

Duncan S (1972) Some signals and rules for taking speaking turns in conversations. Soc Psychol 23 (2):283–292

Fedurek P, Schel AM, Slocombe KE (2013) The acoustic structure of chimpanzee pant-hooting facilitates chorusing. Behav Ecol Sociobiol 67(11):1781–1789. https://doi.org/10.1007/s00265-013-1585-7

Ferrari PF, Gallese V, Rizzolatti G, Fogassi L (2003) Mirror neurons responding to the observation of ingestive and communicative mouth actions in the monkey ventral premotor cortex. Eur J Neurosci 17(8):1703–1714. https://doi.org/10.1046/j.1460-9568.2003.02601.x

Fogassi L, Ferrari PF (2007) Mirror neurons and the evolution of embodied language. Curr Dir Psychol Sci 16(136). https://doi.org/10.1353/lan.2005.0202

Forester DC, Harrison WK (2008) The significance of persistent vocalisation by the spring peeper, *Pseudacris Crucifer* (Anura: *Hylidae*). Behaviour 103(1/3):1–15. https://doi.org/10.1163/156853989x00303

Fuentes A (1998) Re-evaluating primate monogamy. Am Anthropol 100(4):890–907

Gallois C, Ogay T, Giles H (2005) Communication accommodation theory a look back and a look ahead. In: Gudykunst B (ed) Theorizing about intercultural communication. Sage, Thousand Oaks, pp 121–148

Geissmann T (2002) Duet-splitting and the evolution of gibbon songs. Biol Rev Camb Philos Soc 77:57–76. https://doi.org/10.1017/S1464793101005826

Geissmann T, Orgeldinger M (2000) The relationship between duet songs and pair bonds in siamangs, *Hylobates syndactylus*. Anim Behav 60:805–809. https://doi.org/10.1006/anbe.2000.1540

Ghazanfar AA, Smith-Rohrberg D, Pollen AA, Hauser MD (2002) Temporal cues in the antiphonal long-calling behaviour of cottontop tamarins. Anim Behav 64:427–438. https://doi.org/10.1006/anbe.2002.3074

Ghiglione R (1986) L'homme communiquant. (A. Colin, Ed.). Paris, France

Ginsburg GP, Kilbourne BK (1988) Emergence of vocal alternation in mother-infant interchanges. J Child Lang 15:221–235. https://doi.org/10.1017/S0305000900012344

Golinkoff RM (1986) 'I beg your pardon?': The preverbal negotiation of failed messages. J Child Lang 21:205–226. https://doi.org/10.1017/S0305000900006826

Golinkoff RM (1993) When is communication a "meeting of minds"? J Child Lang 20:199–207

Goll Y, Demartsev V, Koren L, Geffen E (2017) Male hyraxes increase countersinging as strangers become 'nasty neighbours'. Anim Behav 134:9–14. https://doi.org/10.1016/j.anbehav.2017.10.002

Gregory SW (1990) Analysis of fundamental frequency reveals covariation in interview partners' speech. J Nonverbal Behav 14(4):237–251. https://doi.org/10.1007/BF00989318

Griffin ZM, Bock K (2000) What the eyes say about speaking. Psychol Sci 11(4):274–279. https://doi.org/10.1111/1467-9280.00255

Grubb P, Butynski TM, Oates JF, Bearder SK, Disotell TR, Groves CP, Struhsaker TT (2003) Assessment of the diversity of African Primates. Int J Primatol 24(6):1301–1357

Hage SR, Nieder A (2013) Single neurons in monkey prefrontal cortex encode volitional initiation of vocalizations. Nat Commun 4:2409. https://doi.org/10.1038/ncomms3409

Haimoff EH (1986) Convergence in the duetting of monogamous Old World primates. J Hum Evol 15(1):51–59. https://doi.org/10.1016/S0047-2484(86)80065-3

Hall ET (1959) The silent language. Doubleday, New York

Hartbauer M, Kratzer S, Steiner K, Römer H (2005) Mechanisms for synchrony and alternation in song interactions of the bushcricket Mecopoda elongata (Tettigoniidae: *Orthoptera*). J Comp Physiol A Neuroethol Sens Neural Behav Physiol 191(2):175–188. https://doi.org/10.1007/s00359-004-0586-4

Hauser MD, Marler P (1993) Food-associated calls in rhesus macaques (*Macaca mulatta*): II. Costs and benefits of call production and suppression. Behav Ecol 4:206–212

Hedwig B (2006) Pulses, patterns and paths: neurobiology of acoustic behaviour in crickets. J Comp Physiol A Neuroethol Sens Neural Behav Physiol. https://doi.org/10.1007/s00359-006-0115-8

Heldner M, Edlund J (2010) Pauses, gaps and overlaps in conversations. J Phon 38(4):555–568. https://doi.org/10.1016/j.wocn.2010.08.002

Henry L, Craig AJFK, Lemasson A, Hausberger M (2015) Social coordination in animal vocal interactions. Is there any evidence of turn-taking? The starling as an animal model. Front Psychol 6:1416. https://doi.org/10.3389/fpsyg.2015.01924

Hilbrink EE, Gattis M, Levinson SC (2015) Early developmental changes in the timing of turntaking: a longitudinal study of mother-infant interaction. Front Psychol 6:1492. https://doi.org/10.3389/978-2-88919-825-2

Hyman J (2003) Countersinging as a signal of aggression in a territorial songbird. Anim Behav 65 (6):1179–1185. https://doi.org/10.1006/anbe.2003.2175

Indefrey P, Levelt WJM (2004) The spatial and temporal signatures of word production components. Cognition 92(1–2):101–144. https://doi.org/10.1016/j.cognition.2002.06.001

Jaffe J, Beebe B, Feldstein S, Crown CL, Jasnow MD (2001) Rhythms of dialogue in infancy: coordinated timing in development. Monogr Soc Res Child Dev 66(2): 1–132. Retrieved from http://www.ncbi.nlm.nih.gov/pubmed/11428150

Jiang J, Dai B, Peng D, Zhu C, Liu L, Lu C (2012) Neural synchronization during face-to-face communication. J Neurosci 32(45):16064–16069. https://doi.org/10.1523/jneurosci.2926-12.2012

Jordan K, Weiss D, Hauser M, Mcmurray B (2004) Antiphonal responses to loud contact calls produced by Saguinus oedipus. Int J Primatol 25(2):465–475

Jürgens U (1995) Neuronal control of vocal production in non-human and human Primates. In: Zimmermann JU, Newman JD (eds) Current topics in primate vocal communication. Springer, Boston, pp 199–206. https://doi.org/10.1007/978-1-4757-9930-9_10

Jürgens U, Ploog D (1970) Cerebral representation of vocalization in the squirrel monkey. Exp Brain Res 10(5):532–554. https://doi.org/10.1007/BF00234269

Jürgens U, Maurus M, Ploog D, Winter P (1967) Vocalization in the squirrel monkey (*Saimiri sciureus*) elicited by brain stimulation. Exp Brain Res 4:114–117. https://doi.org/10.1007/BF00240356

Jürgens U, Lu C-L, Quondamatteo F (1996) C-fos expression during vocal mobbing in the new world monkey *Saguinus fuscicollis*. Eur J Neurosci 8(1):2–10. https://doi.org/10.1111/j.1460-9568.1996.tb01162.x

Kappeler PM, Van Schaik CP (2002) Evolution of primate social systems. Int J Primatol 23 (4):707–740

Keller H, Lohaus A, Völker S, Cappenberg M, Chasiotis A (1999) Temporal contingency as an independent component of parenting behavior. Child Dev 70(2):474–485. https://doi.org/10.1111/1467-8624.00034

Kerbat-Orecchioni C (1990) Les interactions verbales. Armand Collin, Paris

Koda H (2004) Flexibility and context-sensitivity during the vocal exchange of coo calls in wild Japanese macaques (*Macaca fuscata yakui*). Behaviour 141(10):1279–1296. https://doi.org/10.1163/1568539042729685

Kohler E, Keysers C, Umiltà MA, Fogassi L, Gallese V, Rizzolatti G (2002) Hearing sounds, understanding actions: action representation in Mirror neurons. Science 297:846–848

Kurtić E, Brown GJ, Wells B (2013) Resources for turn competition in overlapping talk. Speech Comm 55:1–23. https://doi.org/10.1016/j.specom.2012.10.002

Laporte MNC, Zuberbühler K (2010) Vocal greeting behaviour in wild chimpanzee females. Anim Behav 80(3):467–473. https://doi.org/10.1016/j.anbehav.2010.06.005

Leavens DA, Hostetter AB, Wesley MJ, Hopkins WD (2004) Tactical use of unimodal and bimodal communication by chimpanzees, Pan troglodytes. Anim Behav 67(3):467–476. https://doi.org/10.1016/j.anbehav.2003.04.007

Leavens DA, Russell JL, Hopkins WD (2009) Multimodal communication by captive chimpanzees (Pan troglodytes). Anim Cogn 13(1):33–40. https://doi.org/10.1007/s10071-009-0242-z

Lemasson A (2011) What can forest guenons « tell » us about the origin of language? In: Vilain A, Schwartz JL, Vauclair JC (eds) Primate communication and human language: vocalisation, gestures, imitation and deixis in humans and non-humans. John Benjamins Publishing, Amsterdam, pp 39–70

Lemasson A, Hausberger M (2004) Patterns of vocal sharing and social dynamics in a captive group of Campbell's monkeys (Cercopithecus campbelli campbelli). J Comp Psychol 118(3):347–359. https://doi.org/10.1037/0735-7036.118.3.347

Lemasson A, Hausberger M (2011) Acoustic variability and social significance of calls in female Campbell's monkeys (*Cercopithecus campbelli campbelli*). J Acoust Soc Am 129(5):3341–3352. https://doi.org/10.1121/1.3569704

Lemasson A, Gautier JP, Hausberger M (2003) Vocal similarities and social bonds Campbell's monkey (*Cercopithecus campbelli*). C R Biol 326(12):1185–1193. https://doi.org/10.1016/j.crvi.2003.10.005

Lemasson A, Hausberger M, Zuberbühler K (2005) Socially meaningful vocal plasticity in adult Campbell's monkeys (Cercopithecus campbelli). J Comp Psychol 119:220–229. https://doi.org/10.1037/0735-7036.119.2.220

Lemasson A, Blois-Heulin C, Jubin R, Hausberger M (2006) Female social relationships in a captive group of Campbell's monkeys (*Cercopithecus campbelli campbelli*). Am J Primatol 68:1161–1170. https://doi.org/10.1002/ajp.20315

Lemasson A, Gandon E, Hausberger M (2010) Attention to elders' voice in non-human primates. Biol Lett 6(3):325–328. https://doi.org/10.1098/rsbl.2009.0875

Lemasson A, Glas L, Barbu S, Lacroix A, Guilloux M, Remeuf K, Koda H (2011a) Youngsters do not pay attention to conversational rules: is this so for nonhuman primates? Sci Rep 1:12–15. https://doi.org/10.1038/srep00022

Lemasson A, Ouattara K, Petit EJ, Zuberbühler K (2011b) Social learning of vocal structure in a nonhuman primate? BMC Evol Biol 11(362):1–7. https://doi.org/10.1186/1471-2148-11-362

Lemasson A, Guilloux M, Rizaldi, Barbu S, Lacroix A, Koda H (2013a) Age- and sex-dependent contact call usage in Japanese macaques. Primates 54:283–291. https://doi.org/10.1007/s10329-013-0347-5

Lemasson A, Ouattara K, Zuberbühler K (2013b) Exploring the gaps between primate calls and human language. In: Botha M, Everaert R (eds) The evolutionary emergence of language: evidence and inference. Oxford University Press, Utrecht, pp 181–203

Lemasson A, Jubin R, Masataka N, Arlet M (2016) Copying hierarchical leaders' voices? Acoustic plasticity in female Japanese macaques. Sci Rep 6(21289). https://doi.org/10.1038/srep21289

Lemasson A, Pereira H, Levréro F (2018) Social basis of vocal interactions in Western lowland gorillas (*Gorilla g. gorilla*). J Comp Psychol 132(2):141–151. https://doi.org/10.1037/com0000105

Levinson SC (1983) Pragmatics. Cambridge University Press, Cambridge

Levinson SC (2016) Turn-taking in human communication – origins and implications for language processing. Trends Cogn Sci 20(1):6–14. https://doi.org/10.1016/j.tics.2015.10.010
Levinson SC, Holler J (2014) The origin of human multi-modal communication. Philos Trans R Soc Lond B Biol Sci 369(1651). https://doi.org/10.1098/rstb.2013.0302
Levréro F, Touitou S, Frédet J, Nairaud B, Guéry J-P, Lemasson A (2019) Social bonding drives vocal exchanges in bonobos. Sci Rep 9:711. https://doi.org/10.1038/s41598-018-36024-9
Lindsay L, Gambi C, Rabagliati H (2019) Preschoolers optimize the timing of their conversational turns through flexible coordination of language comprehension and production. Psychol Sci 30 (4):504–515
MacKinnon J, MacKinnon K (1980) The behavior of wild spectral tarsiers. Int J Primatol 1 (4):361–379. https://doi.org/10.1007/BF02692280
MacLarnon A, Hewitt G (2004) Increased breathing control: another factor in the evolution of human language. Evol Anthropol 13:181–197. https://doi.org/10.1002/evan.20032
Masataka N (2003) The onset of language. In: The onset of language. Cambridge University Press. https://doi.org/10.1017/CBO9780511489754
Masataka N, Biben M (1987) Temporal rules regulating affiliative vocal exchanges of squirrel monkeys. Behaviour 101:311–319
McComb K, Semple S (2005) Coevolution of vocal communication and sociality in primates. Biol Lett 1(4):381–385. https://doi.org/10.1098/rsbl.2005.0366
McGarva AR, Warner RM (2003) Attraction and social coordination: mutual entrainment of vocal activity rhythms. J Psycholinguist Res 32(3):335–354. https://doi.org/10.1023/A:1023547703110
Mennill DJ, Boag PT, Ratcliffe LM (2003) The reproductive choices of eavesdropping female black-capped chickadees, Poecile atricapillus. Naturwissenschaften 90(12):577–582. https://doi.org/10.1007/s00114-003-0479-3
Miller CT, Wang X (2006) Sensory-motor interactions modulate a primate vocal behavior: antiphonal calling in common marmosets. J Comp Physiol A Neuroethol Sens Neural Behav Physiol 192:27–38. https://doi.org/10.1007/s00359-005-0043-z
Miller CT, DiMauro A, Pistorio A, Hendry S, Wang X (2010) Vocalization induced CFos expression in marmoset cortex. Front Integr Neurosci 4(128). https://doi.org/10.3389/fnint.2010.00128
Mitani JC, Nishida T (1993) Contexts and social correlates of long-distance calling by male chimpanzees. Anim Behav 45(4):735–746. https://doi.org/10.1006/anbe.1993.1088
Nguyen N, Delvaux V (2015) Role of imitation in the emergence of phonological systems. J Phon 53:46–54. https://doi.org/10.1016/j.wocn.2015.08.004
Nishitani N, Schürmann M, Amunts K, Hari R (2005) Broca's region: from action to language. Physiology 20(1):60–69. https://doi.org/10.1152/physiol.00043.2004
Nummela SU, Jovanovic V, de la Mothe L, Miller CT (2017) Social context-dependent activity in marmoset frontal cortex populations during natural conversations. J Neurosci 37 (29):7036–7047. https://doi.org/10.1523/JNEUROSCI.0702-17.2017
Okobi DEJ, Banerjee A, Matheson AMM, Phelps SM, Long MA (2019) Motor cortical control of vocal interaction in neotropical singing mice. Science 363(6430):979–983. https://doi.org/10.1126/science.aau1758
Ouattara K, Lemasson A, Zuberbuhler K (2009a) Campbell's monkeys concatenate vocalizations into context-specific call sequences. Proc Natl Acad Sci 106:22026–22031. https://doi.org/10.1073/pnas.0908118106
Ouattara K, Lemasson A, Zuberbühler K (2009b) Campbell's monkeys use affixation to alter call meaning. PLoS One 4:e7808. https://doi.org/10.1371/journal.pone.0007808
Papoušek M, Papoušek H (1989) Forms and functions of vocal matching in interactions between mothers and their precanonical infants. First Lang 9(6):137–157. https://doi.org/10.1177/014272378900900603
Pardo JS, Gibbons R, Suppes A, Krauss RM (2012) Phonetic convergence in college roommates. J Phon 40:190–197. https://doi.org/10.1016/j.wocn.2011.10.001

Pika S, Wilkinson R, Kendrick KH, Vernes SC (2018) Taking turns: bridging the gap between human and animal communication. Proc R Soc B 285(1880):1–9. https://doi.org/10.1098/rspb.2018.0598

Pollick AS, Gouzoules H, De Waal FBM (2005) Audience effects on food calls in captive brown capuchin monkeys, Cebus apella. Anim Behav 70(6):1273–1281. https://doi.org/10.1016/j.anbehav.2005.03.007

Pougnault L, Levréro F, Mulot B, Lemasson A (2020) Breaking conversational rules matters to captive gorillas: A playback experiment. Sci Rep in press

Reby D, McComb K (2003) Vocal communication and reproduction in deer. Adv Study Behav 33:231–264. https://doi.org/10.1016/S0065-3454(03)33005-0

Reissland N, Stephenson T (1999) Turn-taking in early vocal interaction: A comparison of premature and term infants' vocal interaction with their mothers. Child Care Health Dev 25 (6):447–456. https://doi.org/10.1046/j.1365-2214.1999.00109.x

Rizzolatti G, Arbib MA (1998) Language within our grasp. Trends Neurosci 21(5):188–194

Roberts F, Margutti P, Takano S (2011) Judgments concerning the valence of inter-turn silence across speakers of American English, Italian, and Japanese. Discourse Process 48:331–354. https://doi.org/10.1080/0163853X.2011.558002

Sacks H, Schegloff EA, Jefferson G (1974) A simplest systematics for the organization of turn-taking for conversation. Language 50:696–735. https://doi.org/10.2307/412243

Schaffer R (1977) Study on mother-infant interaction. Academic Press, New York

Schamberg I, Cheney DL, Clay Z, Hohmann G, Seyfarth RM (2016) Call combinations, vocal exchanges and interparty movement in wild bonobos. Anim Behav 122:109–116. https://doi.org/10.1016/j.anbehav.2016.10.003

Schegloff EA (2002) Overlapping talk and the organization of turn-taking for conversation. Lang Soc 29(01):1–63. https://doi.org/10.1017/s0047404500001019

Schwartz JJ (1987) The function of call alternation in anuran amphibians: a test of three hypotheses. Evolution 41(3):461. https://doi.org/10.2307/2409249

Sewall KB, Young AM, Wright TF (2016) Social calls provide novel insights into the evolution of vocal learning. Anim Behav 120:163–172. https://doi.org/10.1016/j.anbehav.2016.07.031

Silk JB (2002) Grunts, Girneys, and good intentions: the origins of strategic commitment in nonhuman primates. In: Nesse R (ed) Commitment: evolutionary perspectives. Russell Sage Press, New York, pp 138–157

Simões CS, Vianney PVR, de Moura MM, Freire MAM, Mello LE, Samesshima K, Araùjo J, Nicolelis M, Ribeiro S (2010) Activation of frontal neocortical areas by vocal production in marmosets. Front Integr Neurosci 4:1–12. https://doi.org/10.3389/fnint.2010.00123

Slocombe KE, Kaller T, Turman L, Townsend SW, Papworth S, Squibbs P, Zuberbühler K (2010) Production of food-associated calls in wild male chimpanzees is dependent on the composition of the audience. Behav Ecol Sociobiol 64(12):1959–1966. https://doi.org/10.1007/s00265-010-1006-0

Snowdon CT, Cleveland J (1984) "Conversations" among pygmy marmosets. Am J Primatol 7 (1):15–20. https://doi.org/10.1002/ajp.1350070104

Snowdon CT, Elowson AM (1999) Pygmy marmosets modify call structure when paired. Ethology 105(10):893–908. https://doi.org/10.1046/j.1439-0310.1999.00483.x

Soltis J, Bernhards D, Donkin H, Newman JD (2002) Squirrel monkey chuck call: vocal response to playback chucks based on acoustic structure and affiliative relationship with the caller. Am J Primatol 7:119–130. https://doi.org/10.1002/ajp.10039

Stephens GJ, Silbert LJ, Hasson U (2010) Speaker-listener neural coupling underlies successful communication. Proc Natl Acad Sci 107(32):14425–14430. https://doi.org/10.1073/pnas.1008662107

Stern DN, Gibbon J (1979) Temporal expectancies of social behaviours in mother-infant play. In: Thoman EB (ed) Origins of the infant's social responsiveness. Lawrence Erlbaum Associates, Hillsdale, pp 409–429

Stern DN, Jaffe J, Beebe B, Bennett SL (1975) Vocalizing in unison and in alternation: two modes of communication within the mother- infant dyad. Ann N Y Acad Sci 263(1):89–100. https://doi.org/10.1111/j.1749-6632.1975.tb41574.x

Stern DN, Beebe B, Jaffe J, Bennett SL (1977) The infant's stimulus world during social interaction: a study of caregiver behaviors with particular reference to repetition and timing. In: Schaffer HR (ed) Studies in mother–infant interaction. Academic Press, New York, pp 177–202

Stivers T, Enfield NJ, Brown P, Englert C, Hayashi M, Heinemann T, Hoymann G, Rossano F, de Ruiter J, Levinson SC (2009) Universals and cultural variation in turn-taking in conversation. Proc Natl Acad Sci 106(26):10587–10592. https://doi.org/10.1073/pnas.0903616106

Sugiura H (1993) Temporal and acoustic correlates in vocal exchange of coo calls in Japanese macaques. Behaviour 124:3–4. https://doi.org/10.1163/156853993X00588

Sugiura H (1998) Matching of acoustic features during the vocal exchange of coo calls by Japanese macaques. Anim Behav 55:673–687. https://doi.org/10.1006/anbe.1997.0602

Sugiura H (2007) Adjustment of temporal call usage during vocal exchange coo calls in Japanese macaques. Ethology 113:528–533. https://doi.org/10.1111/j.1439-0310.2007.01349.x

Sugiura H, Masataka N (1995) Temporal and acoustic flexibility in vocal exchanges of coo calls in Japanese macaques (*Macaca fuscata*). In: Zimmermann E, Newman JD, Jürgens U (eds) *Current topics in primate vocal communication*. Plenum Press, New York, pp 121–140. https://doi.org/10.1007/978-1-4757-9930-9

Sutton D, Larson C, Lindeman RC (1974) Neocortical and limbic lesion effects on primate phonation. Brain Res 71:61–75

Takahashi DY, Narayanan DZ, Ghazanfar AA (2013) Coupled oscillator dynamics of vocal turn-taking in monkeys. Curr Biol 23:2162–2168. https://doi.org/10.1016/j.cub.2013.09.005

Takahashi DY, Fenley AR, Ghazanfar AA (2016) Early development of turn-taking with parents shapes vocal acoustics in infant marmoset monkeys. Philos Trans R Soc B Biol Sci 371 (1693):20150370. https://doi.org/10.1098/rstb.2015.0370

Tanaka T, Sugiura H, Masataka N (2006) Cross-sectional and longitudinal studies of the development of group differences in acoustic features of coo calls in two groups of Japanese macaques. Ethology 112(1):7–21. https://doi.org/10.1111/j.1439-0310.2006.01103.x

Ter Maat M, Truong KP, Heylen D (2010) How turn-taking strategies influence users' impressions of an agent. In: Intelligent virtual agents: proceedings of the 10th international conference, IVA, pp 441–453. https://doi.org/10.1007/978-3-642-15892-6_48

Thierry B (2007) Unity in diversity: lessons from macaque societies. Evol Anthropol 16 (6):224–238. https://doi.org/10.1002/evan.20147

Thierry B, Iwaniuk N, Pellis SM (2000) The influence of phylogeny on the social behaviour of macaques (Primates: *Cercopithecidae*, genus *Macaca*). Ethology 106:713–728

Toarmino CR, Wong L, Miller CT (2017) Audience affects decision-making in a marmoset communication network. Biol Lett 13(1):20160934. https://doi.org/10.1098/rsbl.2016.0934

Trudgill P (2008) Colonial dialect contact in the history of European languages: on the irrelevance of identity to new-dialect formation. Lang Soc 37(2):241–254. https://doi.org/10.1017/S0047404508080287

Vehrencamp SL, Ellis JM, Cropp BF, Koltz JM (2014) Negotiation of territorial boundaries in a songbird. Behav Ecol 25(6):1436–1450. https://doi.org/10.1093/beheco/aru135

Wilson M, Wilson TP (2005) An oscillator model of the timing of turn-taking. Psychon Bull Rev 12 (6):957–968. https://doi.org/10.3758/BF03206432

Yamaguchi C, Izumi A, Nakamura K (2009) Temporal rules in vocal exchanges of phees and trills in common marmosets (Callithrix jacchus). Am J Primatol 71:617–622. https://doi.org/10.1002/ajp.20697

Yoshida S, Okanoya K (2005) Evolution of turntaking: a bio-cognitive perspective. Cogn Stud 12 (3):153–165

Zelick R, Narins PM (1985) Characterization of the advertisement call oscillator in the frog Eleutherodactylus coqui. J Comp Physiol A 156(2):223–229. https://doi.org/10.1007/BF00610865

# Chapter 5
# Language Evolution from a Perspective of Broca's Area

**Masumi Wakita**

**Abstract** Broca's area is one of language-related brain areas that typically locates in the posterior part of the inferior frontal region of the left hemisphere. Broca's area plays an important role in the perception and production of sequential/hierarchical structures of language. A contribution of two major frontotemporal fiber bundles connected to Broca's area also supports our language faculty: the dorsal and ventral pathways. The dorsal pathway supports syntactics, while the ventral pathway supports semantic processes and local phrase structure.

Homologues of these brain structures have been identified in nonhuman primates. However, Broca's area of humans exhibits disproportionate expansion compared to that of chimpanzees. In addition, the dorsal pathway connections have undergone considerable modification as evidenced by differences compared to both chimpanzees and macaques, while the ventral pathway in nonhuman primate is well developed as in human. These structural differences indicate that nonhuman primates can recognize meaning of individual calls or tones but cannot unify constituents to represent temporally structured sequences. Such assumption is consistent with the well-documented evidence from auditory memory task. Differences in the language-related brain structures between human and nonhuman primate species indicate that language faculty emerged in recent human evolution, since our lineage branched from a common ancestor with chimpanzees.

Difference between human and nonhuman primates is firstly found in vocal production. Monkey's call production is mediated by limbic and brain stem regions, not by Broca's area. Thus, their calls largely represent involuntary symptoms of specific emotional and arousal states like laugh and cry of humans. But this system may be sufficient for the species-specific vocalizations that does not typically require rapid articulation. Secondly, difference between human and nonhuman primates is also found in the sequential processing of vocal production and perception. In the vocalization of monkeys, monosyllabic calls are typically produced, or the same calls are repeated. Different types of calls are occasionally combined but in a

M. Wakita (✉)
Primate Research Institute, Kyoto University, Inuyama, Japan
e-mail: wakita.masumi.2e@kyoto-u.ac.jp



N. Masataka (ed.), *The Origins of Language Revisited*,
https://doi.org/10.1007/978-981-15-4250-3_5

stereotyped order. Thus, combined calls unlikely generate new meaning. Both laboratory and field studies of nonhuman primate call/tone perception indicate that they are sensitive to some aspects of call/tone streams. However, they may perceive call/tone sequences based on local acoustic feature or perceive them as holistic chunks. Recently, the involvement of the inferior frontal region along the ventral stream in the processing tonal sequential rule was found in both humans and monkeys.

Thus, it is reasonable to assume that modification of pre-existing brain structures, i.e., expansion of Broca's area and extension of the frontotemporal connections to this area, allowed humans language faculties. Consequently, humans can intentionally control speech production, create new words by combining several different syllables, and can learn new language. However, Broca's area may not be a structure designed exclusively for language processing. Studies have revealed that Broca's area contributes in the domains that require sequential/hierarchical processing, such as music and action domains along with language domain. Language is one of unique features of human. From a perspective of the function of Broca's area, however, language may be a byproduct of the evolution of sequential/hierarchical processing.

**Keywords** Broca's area · Language faculty · Vocal production · Movement

## 5.1 Introduction

Human language consists of a hierarchical structure in which phonemes are combined to form words, phrases, and sentences. This hierarchy is governed by several sets of rules and forms the basis for speech structure and discourse. Studies have investigated the brain regions involved in the structural analysis of language and have revealed a posterior-anterior functional gradient with substantial overlap in the left inferior frontal region: phonological sequence processing at the word level is localized in the Brodmann area (BA) 6/44, the processing of syntactic rules for words and sentences is localized in BA 44/45, and semantic processing at the sentence level is related to BA 45/47 activity (Friederici and Kotz 2003; Hagoort 2005; Uchiyama et al. 2008; Uddén and Bahlmann 2012). Thus, Broca's area (BA 44 and 45) is involved in syntactic (or structural) and semantic language analysis.

In this chapter, I provide a brief overview of the comparative anatomy of the Broca's area in humans and other primates, as well as the functional uniqueness of the human Broca's area as it relates to language. I then discuss the auditory behavior of nonhuman primates with a "Broca's area" that is constructed differently compared to that of humans. Additionally, by surveying the role of Broca's area in nonlanguage domains, I discuss how hierarchical (or temporal structural) analysis based on BA 44 as the posterior region of Broca's area and its dorsal connection provides a neurobiological basis for the unique human cognition related to language.

## 5.2 Broca's Area in Humans

The classical definition of Broca's area comprises the left posterior part of the inferior frontal region, i.e., BA 44 and 45. This region is essential for linguistic faculty, particularly sequencing phonemes and words, thereby enabling the detection of words from continuous speech and the analysis of grammatical structures (Price 2000; Damasio et al. 2004). Broca's area is a part of the language network; thus, its connectivity should be considered when seeking to understand its functions.

Broca's area receives signals from the temporal auditory regions via two main pathways: the dorsal and ventral pathways (Catani et al. 2005; Dick and Tremblay 2012; Friederici et al. 2017). The arcuate fasciculus of the dorsal pathway directly connects the posterior region of Broca's area (BA 44) to Wernicke's area in the posterior region of the superior temporal gyrus and the middle and inferior temporal gyri. The ventral pathway connects the anterior region of Broca's area (BA 45) with Wernicke's area and the posterior region of the inferior temporal gyri via the inferior fronto-occipital fasciculus. The ventral pathway also connects BA 45 with the anterior temporal pole via the uncinate fasciculus.

Thus, the dorsal pathway connects regions that are crucial for sequencing phonemes and words, thereby enabling the acquisition of new words and the analysis of grammatical structures. In contrast, the ventral pathway connects regions that are crucial for evaluating the meanings of words and sentences (Burton et al. 2000; Price 2000; Humphries et al. 2006, 2007; Hickok and Poeppel 2007; Saur et al. 2008; Specht et al. 2008; Cunillera et al. 2009; Rodd et al. 2010; Saur et al. 2010; Friederici 2011; Rolheiser et al. 2011).

In addition to being a language domain, Broca's area has also been known to be involved in music. The activation of BA 6/44 for rhythm discrimination and melody matching may reflect the processing of sequential sounds (Platel et al. 1997; Brown and Martinez 2007). BA 44/45 is involved in harmonic evaluation (Tillmann et al. 2003; Brown and Martinez 2007). The similarities between linguistic and musical analysis raise the possibility that hierarchical processing in these different domains is subserved by Broca's area (Maess et al. 2001; Koelsch et al. 2002; Patel 2003; Brown et al. 2006; Schön and François 2011). In the musical domain, the inferior frontal region is generally responsive bilaterally in the auditory perception of harmony violation (Maess et al. 2001; Koelsch et al. 2002; Tillmann et al. 2003) and in melody and harmony discrimination (Brown and Martinez 2007).

## 5.3 Homologues of Broca's Area in Nonhuman Primates

Given its contribution to language, the comparative anatomy and physiology of a Broca's area homologue in nonhuman primates is of obvious interest in studying the evolution of language. Due to cytoarchitectonic features (dysgranular cortex in BA 44 and granular cortex in BA 45), both areas were found in human, macaque, and ape brains at the inferior part of the frontal lobe (Schenker et al. 2010; Zilles and Amunts 2018). Human and nonhuman primates have comparable cortical structures in the inferior frontal region with the exception that galago has neither BA 44 nor 45 (Preuss and Goldman-Rakic 1991) and that BA 44 is missing in the common marmoset (Fukushima et al. 2019). Thus, the human Broca's region appears to be a phylogenetically new region in the history of primates. Particularly, BA 44 is proposed to have evolved from evolutionally old neighboring regions (Amunts and Zilles 2012), i.e., BA 6, BA 9/46, and the frontal opercular cortex, which share similar features with BA 44 in terms of action observation, working memory, and linguistic syntax, respectively.

The cytoarchitecture of BA 44 and 45 of nonhuman primates differs considerably from those in humans. Thus far, detailed quantitative data on the size of these areas have been reported for humans and chimpanzees. BA 44 in the left and right hemispheres are 6.6- and 4.1-fold larger, respectively, and left and right BA 45 are 6.0- and 5.0-fold larger, respectively, in humans compared to chimpanzees (Schenker et al. 2010); additionally, the overall brain size is roughly 3.6-fold larger in humans than chimpanzees. Notably, the superior temporal cortex (area Tpt) in the left and right hemispheres, which is homologous regions of Wernicke's area, are only 4.2- and 2.0-fold larger in humans than chimpanzees (Spocter et al. 2010). Thus, there have been disproportionate increases in Broca's area (particularly left BA 44) during human evolution.

Dissimilarity in Broca's area between humans and nonhuman primates has also been found in its microstructure. Notably, the spacing between cortical minicolumns caused by the neuropil-rich region is wider in human BA 44 and 45 than in nonhuman primates (Schenker et al. 2008; Rilling 2014). There is also higher neuronal density in the minicolumns in human compared to nonhuman primate brains (Palomero-Gallagher and Zilles 2019). Such cortical modifications may indicate the availability of more space for neural connections.

Recent comparative neuroanatomical studies have elucidated that the arcuate fasciculus of macaques and chimpanzees is rudimentary compared with that in humans, although other connections of the dorsal pathway are relatively preserved (Schmahmann et al. 2007; Perani et al. 2011; Amunts and Zilles 2012; Dick and Tremblay 2012; Petrides et al. 2012; Thiebaut de Schotten et al. 2012; Rilling 2014). The arcuate fasciculus is minimal in nonhuman primates, and its temporal terminations are predominantly in the parietal lobe. In the monkey brain, the connections from the posterior region of the superior temporal cortex that is homologous to Wernicke's area terminate anteriorly to the regions that are homologous to the posterior region of Broca's area (BA 44). Thus, BA 44 in nonhuman primates processes little auditory information (if any) via the dorsal pathways. The dorsal pathway potentially carries information about spatial location (Romanski et al. 1999), while the human dorsal pathway contributes to the structural analysis of sound sequences.

Unlike the dorsal pathway, nonhuman primates have relatively well-developed ventral pathways. The prefrontal region that is homologous to the anterior region of Broca's area (BA 45) and the anterior and middle regions of the superior temporal cortices are connected via the uncinate fasciculus and the extreme capsule of the ventral pathway. Thus, their ventral pathway is involved in auditory object identification (Romanski et al. 1999). Notably, hearing species-specific calls has been shown to activate homologues of human Broca's and Wernicke's areas in monkeys (Poremba et al. 2004; Gil-da-Costa et al. 2006; Petkov et al. 2008; Wilson et al. 2015). In chimpanzees, communicative signaling has activated the homolog of Broca's area (Taglialatela et al. 2008; Taglialatela et al. 2011).

Thus, the comparative neuroanatomical findings mentioned above indicate that the uniqueness of human language function may be linked to a disproportionate expansion of the Broca's area (particularly left BA 44) and an extension of the arcuate fasciculus for syntax and hierarchically ordered information.

## 5.4 Auditory Behavior in Wild Monkeys

Unlike the dorsal pathways, both human and nonhuman primates have well-developed ventral pathways that contribute to semantic processing of sound. However, since the dorsal pathway of nonhuman primates does not permit syntactic processing of sound, the capacity that their ventral pathway can process may be limited to a single tone, such as typical calls of monkey species. Notably, monkey species recognize the meaning of calls as a form of "vocabulary." Playback experiments have indicated that some monkey species can recognize the individual identity, including the kinship and social rank, of a caller (Cheney and Seyfarth 1980; Bergman et al. 2003) and have different alarm calls that can specify distinct predators (Seyfarth et al. 1980; Macedonia 1990; Zuberbühler et al. 1999; Arnold and Zuberbühler 2008; Ouattara et al. 2009; Price et al. 2015). The majority of their calls consist of single-element vocalizations but are often repeated; different elements are also occasionally combined (Itani 1963; Winter et al. 1966; Cheney and Seyfarth 1980; Cleveland and Snowdon 1981; Bezerra and Souto 2008).

Some nonhuman primate species have been reported to emit different calls sequentially (Mitani and Marler 1989; Arnold and Zuberbühler 2008; Schel et al. 2010; Wheeler 2010; Cäsar et al. 2012). Are they sensitive to the temporal order of call sequences? In classical playback experiments, Mitani and Marler (1989) tested gibbon behavior by broadcasting artificial call sequences in which the positions and/or transitions of call types were not naturally heard. They found that significantly more territorial calls were elicited when they heard arranged songs than when normal songs were broadcasted. Accordingly, the authors suggested that gibbons recognized the call sequences. However, such findings were only sufficient to show that gibbons can use local transitions to recognize familiar and unfamiliar songs. It was difficult to prove whether gibbons can analyze the entire call sequence because their natural call combinations typically occur in a stereotyped order.

In another example, Arnold and Zuberbühler (2012) tested monkeys' ability to understand the sequential structure of calls by systematically modifying call configurations. Male putty-nosed monkeys uttered alarm calls of "pyow" (P) and "hack"

(H) in the appearance of leopards and eagles, respectively. Additionally, combined alarm call sequences (PH) induced progression in other members of the group (Arnold and Zuberbühler 2008). Such findings may imply an idiomatic function of call combinations. However, when Arnold and Zuberbühler (2012) played back three variations of "P-H" sequences (e.g., "PPPPPH," "PPPHHH," and "PHHHHH"), equivalent responses were elicited from other members of the group. In their study, detailed observations revealed that the number of P-calls in the sequence and the sequence length, rather than an entire sequence pattern, were related to the travel distance of the group. Although different types of calls were sequentially emitted, due to the variable components and length as well as a long silent gap between calls within natural PH sequences, combined calls are unlikely to be interpreted as idioms (Schlenker et al. 2016). Moreover, monkeys lack an essential cortical motor network for volitional vocal control (Jürgens 2002; Simonyan 2014; Kumar et al. 2016), and motor planning of long purposeful utterances may not be possible. Therefore, analyzing the temporal structure of calls may not be part of monkeys' natural behavior.

## 5.5 Auditory Perception in Monkeys

In addition to field studies, laboratory studies have also tested the abilities of nonhuman primates to analyze the temporal structure of tone sequences. Studies have shown that nonhuman primates can learn tonal sequences. Izumi (2001) and Brosch et al. (2004) trained macaque monkeys to discriminated tonal sequences and demonstrated that they were able to abstract an ascending or descending direction of pitch change independent of their absolute pitch. However, Izumi (1999, 2003) showed that monkeys' ability to detect frequency changes was influenced by a silent gap inserted between two tones; similarly, monkeys' ability to detect intertone gaps declined when the frequencies of two tones were different. However, Brosch et al. (2006) revealed that such a categorical perception of the pitch contour was not affected by modifying the intertone intervals or the tone duration. These examples indicate that macaques memorize and identify tonal sequences according to their pitch contour, i.e., to the sequential up-and-down patterning between adjacent notes. However, they cannot integrate auditory pitch with a temporal change to perceive a tonal sequence such as a melody contour.

In another instance, Wright et al. (2000) found that, after melody discrimination training, the monkeys could generalize their responses to the +1 and +2 octave-transposed test melodies. Such results may imply an ability to recognize melody contours. However, such a generalization did not occur when test melodies were arranged in a different key by transposing by +1/2 or +2/3 octaves despite the preservation of the relative pitch relations between adjacent tones. This lack of generalization indicated that monkeys' discrimination of auditory sequences is largely controlled by tonality or local acoustic cues rather than by the temporal structure of the entire tone sequence (also see D'Amato and Salmon 1984; D'Amato and Colombo 1988).

Auditory sequence analysis of monkeys likely relies on the ventral pathway, particularly the frontal operculum cortex, which is one of frontal terminations of the ventral pathway and is responsive to the relationship of adjacent tones (Friederici et al. 2006a). Common marmosets, a New World monkey species that lack BA 44, could not discriminate ABAB and AABB patterns after a long training, although they could efficiently discriminate constituent sounds (Wakita 2019). Notably, they could detect a pattern change (e.g., a transition from ABAB to AABB patterns) (Wakita n.d.). Such results indicated that the ventral connection is not useful to the structural analysis of an auditory sequence but that it is sufficient for the online processing of an auditory sequence.

A poorly organized dorsal pathway likely influences not only auditory sequence perception but also auditory list memory. It has been shown that monkeys can hold sound information of a single sound for longer than 10 seconds (D'Amato and Colombo 1985; Kojima 1985; Colombo and D'Amato 1986; Fritz et al. 2005). However, monkeys find it difficult to retrieve a target sound from the sequentially presented sample list (Wright 2007). Scott and colleagues suggested that the memory trace of a preceding sound can be disrupted by incoming distractor stimuli that are presented during the retention period (Scott et al. 2012).

## 5.6 Brain Activity in Monkeys

Recent comparative studies elucidated similarity and dissimilarity between macaque and human auditory sequence processing. Such studies employed a habituation-dishabituation procedure: the monkeys and humans were passively exposed to sequences of sounds structured according to a given rule and then tested with acoustically new stimuli, which were arranged either congruently or incongruently with the original rule. Functional imaging studies revealed that, in both macaques and humans, the cortical regions that represented partial properties of auditory sequences were found in the ventral pathway (the anterior supratemporal sulcus and the anterior insular/frontal opercular) (Wang et al. 2015; Wilson et al. 2015). Furthermore, the cortical regions that represented global features of the sequence were found in the bilateral inferior frontal areas (mainly BA 44) and posterior superior temporal sulcus only in humans; no such region was found in macaque brain (Wang et al. 2015). Thus, semantic information or physical properties such as local transitions of sound or the number of items within a sequence may be available to macaques. However, humans abstract syntactical information or the structural regularity of auditory sequences. These findings indicate that monkeys may perceive tone sequences as unstructured pools of individual tones and thus cannot perceive global temporal patterns of sound.

Milne et al. (2016) compared macaque mismatch response patterns to the deviant test stimuli with those obtained from human infants and adults (Mueller et al. 2012).

Despite differences in experimental conditions, they found that macaque brain activation to auditory sequences was more similar to that in human infants than human adults. Such results imply a strong engagement of the ventral, rather than dorsal, pathways in both human infants and macaques for certain sequencing operations. An electrophysiological study (Milne et al. 2016) revealed that contrasts in event-related potential patterns evoked by congruent and incongruent sequences from monkeys are more similar to those from human infants than human adults. Thus, such findings indicate similar strategies for detecting rule deviations in sound sequences in both monkeys and human infants.

## 5.7 Ontogeny of Language Acquisition

Processing the temporal structure of auditory sequences is also the core of human language and is subserved specifically by BA 44 and its dorsal connection to the posterior temporal cortex, and thus the arcuate fasciculus. How can the language acquisition process be linked to the maturation of this fiber that is unique to humans?

Although the ventral fiber tract is clearly present at birth, the dorsal tracts, particularly the arcuate fasciculus, are very immature (Dubois et al. 2008; Perani et al. 2011). Prelinguistic infants are able to detect word-like segments within an auditory stream, although such chunks cannot be stored in a long-term memory store (Jusczyk 1999; Yoshida et al. 2010; Kudo et al. 2011; Shukla et al. 2011; Erickson and Thiessen 2015; Friedrich and Friederici 2015; Minagawa et al. 2017). Infants appear to perceive speech as holistic chunks dependent on pitch changes or prosodic information rather than phoneme sequences. Using such a strategy, they may learn the prosodic patterns of their target language before they can analyze the sequential structure of relevant language.

By the age of 4, children have recruited Broca's area and Wernicke's area during sentence understanding (Brauer and Friederici 2007; Dick et al. 2010; Moore-Parks et al. 2010). However, the segregation of syntactic and semantic information at the sentence level occurs slowly and gradually. In children aged from 5 to 7, the arcuate fasciculus is not well myelinized (Brauer et al. 2011), although Broca's area and Wernicke's area are functionally connected (Xiao et al. 2016). In addition, confusion between semantics and syntax relation is seen in the mid portions of the left superior temporal gyrus (Skeide et al. 2014). Consequently, it is difficult to understand sentences in which unfamiliar events are described or words are arranged in noncanonical order (Brauer et al. 2011; Knoll et al. 2012). The degree of myelination of the arcuate fasciculus determines the accuracy of understanding noncanonical sentences (Skeide and Friederici 2016).

By the age of 10, although children can understand noncanonical sentences, BA 45 and BA 44 are recruited during noncanonical sentence understanding. In adults, the noncanonicity of sentences selectively induces activation from BA 44 (Friederici et al. 2006b). Syntax and semantics are still not fully separated in 10-year-old children's Broca's area (Skeide and Friederici 2016). Thus, the maturation of this fiber tract is a neuroanatomical precondition for the acquisition of syntax.

Rüschemeyer et al. (2006) showed that BA 44 of an adult native speaker becomes active only when processing a structurally complex sentence; BA 44 is less active when processing a simple sentence. However, BA 44 of non-native speakers is consistently active, regardless of the degree of complexity of sentences (see also Rüschemeyer et al. 2005; Yokoyama et al. 2006). Moreover, Jeon and Friederici (2013) demonstrated that the syntactical analysis of complex non-native sentences recruited BA 47. Thus, substantial experience is needed until BA 44 represents the structure of language. However, the less-proficient syntactical processing of non-native speakers is compensated for by the semantic information in the sentences. Thus, in an early learning stage, as found during early ontogeny of Broca's area, the dorsal pathway is supported by the ventral pathway even in adults.

## 5.8 Broca's Areas in the Action Domain

Broca's area is not an exclusive region for the language and music domains. Broca's area is also a node within the action-perception and action-production network that consists of temporal-parietal-frontal dorsal and temporal-frontal ventral circuits (Nishitani et al. 2005; Borra and Luppino 2019). Hecht et al. (2013a) compared frontal, parietal, and occipitotemporal activations during the observation of object-directed grasping actions from macaques, chimpanzees, and humans and found differences underlying the action understanding mechanism among these species.

In macaques, the ventral connection, which supports object recognition and the physical results of observed actions, was greater than the dorsal connection, which supports spatial/kinematics, the detail of observed actions (Johnson-Frey et al. 2003; Beauchamp and Martin 2007; Goldenberg 2009). Notably, the macaque F5c that corresponds to human BA 44 primarily represents the consequence rather than the process of the observed action. Therefore, macaque mirror neurons respond to actions toward objects (transitive actions) but not to actions without goals (intransitive actions) (Rizzolatti et al. 1996; Nelissen et al. 2005).

Chimpanzees have a stronger dorsal connection between the superior temporal sulcus and inferior parietal cortex compared to that of macaques. BA 44 in humans and the FCBm cortex in chimpanzees (i.e., the homologue of human BA 44) map both transitive and intransitive actions onto one's own motor system with somatotopic specificity (Buccino et al. 2001; Hecht et al. 2013b). In macaques and chimpanzees, however, there was greater frontal activation and less parietal activation during action observation compared to humans (Denys et al. 2004; Peeters et al. 2009; Hecht et al. 2013b). Frontal and parietal activation were comparable in humans.

In humans but not in macaques or chimpanzees, however, the dorsal connection between Broca's area and parietal mirror regions extends into the superior parietal cortex, which supports spatial awareness and attention (Husain and Nachev 2007;

Hecht et al. 2013b). Thus, such a human-unique connection may transfer information from the spatial/kinematic trajectories of others' actions through space to the human Broca's area. These findings indicate that humans understand observed object-related actions in a different manner than nonhuman primates despite having similar brain regions and connections.

It has also been demonstrated that Broca's area and its homologue in the right hemisphere of humans support cognitive control processes such as rule-based action selection, information retrieval, and hierarchical control (Petrides 2005; Koechlin and Jubault 2006; Badre and D'Esposito 2009). Chimpanzees show structured actions in the wild, such as simple tool use for nut cracking. However, their ability to show a hierarchically structured action, as seen in a nesting-cup task, is limited (Hayashi 2007). The macaque action repertoire is limited to nonhierarchical object-directed actions, such as potato washing (Kawamura 1959) and stone handling (Huffman 1984). Thus, controlling a structurally complex action is a feature that is unique to humans and is likely linked to a recent modification of Broca's area and its right homologue and dorsal connection.

The effect of expertise is also found in the action domain (e.g., Bengtsson et al. 2005; Calvo-Merino et al. 2005; Cross et al. 2006; Wakita and Hiraishi 2011; Wakita 2014, 2016). For instance, in a near-infrared spectroscopy study by Wakita (2014, 2016), two groups of participants (well-trained vs. less-trained groups in music) observed silent movies of hierarchically organized hand movements playing familiar and unfamiliar melodies. The results revealed an increased activation in Broca's area under the unfamiliar melody condition only in the well-trained group. Thus, Broca's area is a center for structural analysis. However, whether this area represents such a relationship depends on the individual's experience.

## 5.9 What Do Stone Tools Tell About Language Evolution?

Evolution in tool making along with fossil cranial size may carry implications for studying language evolution. Stone tool making was refined during the *H. erectus* epoch. Oldowan tools (sharp-edged flakes) that are made by striking one stone with another stone were gradually replaced with Acheulean tools (hand axes) that are made by a multilevel process of shaping until a desired design is obtained. During the same time period, the hominin brain underwent a roughly threefold increase from 450 cc (*Australopithecus garhi*) to 1350 cc (*Homo heidelbergensis*). This indicates a link between the cognitive skill to make complex tools and increments in cranial capacity (Hillert 2015).

Stout et al. (2011) recently found that, in modern humans, the observation of Acheulean compared with Oldowan tool making induced more activations of Broca's area and the left parietal region irrespective of the participants' tool-making skill. Activation of the right parietal region was added depending on the degree of expertise. In addition, advanced stone tool making was supported by the dorsal pathway, which overlaps with that in the language domain. These findings suggest that neural substrates for tool making may be involved in more general human capacities, including language and music, which require hierarchical processing. Therefore, the invention of fundamental linguistic capacities may have occurred during the *H. erectus* epoch; the coevolution of language, music, and action can also be suggested.

## 5.10 Conclusion

Broca's area, which is one of brain regions essential for human language faculty, is evolutionally novel. In humans, Broca's area, particularly left BA 44, is expanded. The arcuate fasciculus of the dorsal pathway is also extended in humans compared with chimpanzees. Thus, the neurobiological substrates for syntactical analysis underwent a large modification during hominization. In early ontogeny, syntactic understanding of children is related to the degree of the segregation of syntactic and semantic processing regions and the degree of myelination of the arcuate fasciculus. Even in adults, representation of syntax in BA 44 is influenced by the degree of expertise. Broca's area is also involved in ordering or structural analysis in nonlanguage domains. The evolutionally young Broca's area and its dorsal connection network anatomically develops gradually during early ontogeny. The function of Broca's area appears functionally plastic even in adulthood such that the representation of structural information of general human capacities including language, music, and action, which require hierarchical processing, are represented and can be determined by individual experience.

**Acknowledgments** Studies supported by JSPS Grants-in-Aid JP26540066 and JP18K12011 are included in this chapter. The author thanks Mitsue Wakita for her comments on the earlier version of manuscript.

## References

Amunts K, Zilles K (2012) Architecture and organizational principles of Broca's region. Trends Cogn Sci 16:418–426

Arnold K, Zuberbühler K (2008) Meaningful call combinations in a non-human primate. Curr Biol 18:R202–R204

Arnold K, Zuberbühler K (2012) Call combinations in monkeys: compositional or idiomatic expressions? Brain Lang 120:303–309

Badre D, D'Esposito M (2009) Is the rostro-caudal axis of the frontal lobe hierarchical? Nat Rev Neurosci 10:659–669

Beauchamp MS, Martin A (2007) Grounding object concepts in perception and action: evidence from fMRI studies of tools. Cortex 43:461–468

Bengtsson S, Nagy Z, Skare S, Forsman L, Forssberg H, Ullén F (2005) Extensive piano practicing has regionally specific effects on white matter development. Nat Neurosci 8:1148–1150

Bergman TJ, Beehner JC, Cheney DL, Seyfarth RM (2003) Hierarchical classification by rank and kinship in baboons. Science 302:1234–1236
Bezerra BM, Souto A (2008) Structure and usage of the vocal repertoire of *Callithrix jacchus*. Int J Primatol 29:671–701
Borra E, Luppino G (2019) Large-scale temporo–parieto–frontal networks for motor and cognitive motor functions in the primate brain. Cortex 118:19–37
Brauer J, Friederici AD (2007) Functional neural networks of semantic and syntactic processes in the developing brain. J Cogn Neurosci 19:1609–1623
Brauer J, Anwander A, Friederici AD (2011) Neuroanatomical prerequisites for language functions in the maturing brain. Cereb Cortex 21:459–466
Brosch M, Selezneva E, Bucks C, Scheich H (2004) Macaque monkeys discriminate pitch relationships. Cognition 91:259–272
Brosch M, Oshurkova E, Bucks C, Scheich H (2006) Influence of tone duration and intertone interval on the discrimination of frequency contours in a macaque monkey. Neurosci Lett 406:97–101
Brown S, Martinez MJ (2007) Activation of premotor vocal areas during musical discrimination. Brain Cogn 63:59–69
Brown S, Martinez MJ, Parsons LM (2006) Music and language side by side in the brain: a PET study of the generation of melodies and sentences. Eur J Neurosci 23:2791–2803
Buccino G, Binkofski F, Fink GR, Fadiga L, Fogassi L, Gallese V, Seitz RJ, Zilles K, Rizzolatti G, Freund HJ (2001) Action observation activates premotor and parietal areas in a somatotopic manner: an fMRI study. Eur J Neurosci 13:400–404
Burton MW, Small SL, Blumstein SE (2000) The role of segmentation in phonological processing: an fMRI investigation. J Cogn Neurosci 12:679–690
Calvo-Merino B, Glaser DE, Grèzes J, Passingham RE, Haggard P (2005) Action observation and acquired motor skills: an fMRI study with expert dancers. Cereb Cortex 15:1243–1249
Cäsar C, Byrne R, Young RJ, Zuberbühler K (2012) The alarm call system of wild black-fronted titi monkeys, *Callicebus nigrifrons*. Behav Ecol Sociobiol 66:653–667
Catani M, Jones DK, Ffytche DH (2005) Perisylvian language networks of the human brain. Ann Neurol 57:8–16
Cheney DL, Seyfarth RM (1980) Vocal recognition in free-ranging vervet monkeys. Anim Behav 28:362–376
Cleveland J, Snowdon CT (1981) The complex vocal repertoire of the adult cotton-top tamarin *(Saguinus oedipus Oedipus)*. Zeit Tierpsycholgie 58:231–270
Colombo M, D'Amato MR (1986) A comparison of visual and auditory short-term memory in monkeys (*Cebus paella*). Q J Exp Psychol Sec B 38:425–448
Cross ES, Hamilton AF, Grafton ST (2006) Building a motor simulation de novo: observation of dance by dancers. NeuroImage 31:1257–1267
Cunillera T, Camara E, Toro JM, Marco-Pallares J, Sebastian-Galles N, Ortiz H, Pujol J, Rodriguez-Fornells A (2009) Time course and functional neuroanatomy of speech segmentation in adults. NeuroImage 48:541–553
D'Amato MR, Colombo M (1985) Auditory matching-to-sample in monkeys (*Cebus apella*). Anim Learn Behav 13:375–382
D'Amato MR, Colombo M (1988) On tonal pattern perception in monkeys (*Cebus paella*). Anim Learn Behav 16:417–424
D'Amato MR, Salmon DP (1984) Processing of complex auditory stimuli (tunes) by rats and monkeys (*Cebus apella*). Anim Learn Behav 12:184–194
Damasio H, Tranel D, Grabowski T, Adolphs R, Damasio A (2004) Neural systems behind word and concept retrieval. Cognition 92:179–229
Denys K, Vanduffel W, Fize D, Nelissen K, Sawamura H, Georgieva S, Vogels R, Van Essen D, Orban GA (2004) Visual activation in prefrontal cortex is stronger in monkeys than in humans. J Cogn Neurosci 16:1505–1516

Dick AS, Tremblay P (2012) Beyond the arcuate fasciculus: consensus and controversy in the connectional anatomy of language. Brain 135:3529–3550

Dick AS, Solodkin A, Small SL (2010) Neural development of networks for audiovisual speech comprehension. Brain Lang 114:101–114

Dubois J, Dehaene-Lambertz G, Perrin M, Mangin J, Cointepas Y, Duchesnay E, Le Bihan D, Hertz-Pannier L (2008) Asynchrony of the early maturation of white matter bundles in healthy infants: quantitative landmarks revealed noninvasively by diffusion tensor imaging. Hum Brain Mapp 29:14–27

Erickson LC, Thiessen ED (2015) Statistical learning of language: theory, validity, and predictions of a statistical learning account of language acquisition. Dev Rev 37:66–108

Friederici AD (2011) The brain bases of language processing: from structure to function. Physiol Rev 91:1357–1392

Friederici AD, Kotz SA (2003) The brain basis of syntactic processes: functional imaging and lesion studies. NeuroImage 20:S8–S17

Friederici AD, Bahlmann J, Heim S, Schubotz RI, Anwander A (2006a) The brain differentiates human and non-human grammars: functional localization and structural connectivity. Proc Natl Acad Sci USA 103:2458–2463

Friederici AD, Fiebach CJ, Schlesewsky M, Bornkessel ID, von Cramon DY (2006b) Processing linguistic complexity and grammaticality in the left frontal cortex. Cereb Cortex 16:1709–1717

Friederici AD, Chomsky N, Berwick RC, Moro A, Bolhuis JJ (2017) Language, mind and brain. Nat Hum Behav 1:713–722

Friedrich M, Friederici AD (2015) The origins of word learning: brain responses of 3-month-olds indicate their rapid association of objects and words. Dev Sci 20:1–13

Fritz J, Mishkin M, Saunders R (2005) In search of an auditory engram. Proc Natl Acad Sci USA 102:9359–9364

Fukushima M, Ichinohe N, Okano H (2019) Neuroanatomy of the marmoset. In: Marini R, Wachtman L, Tardif S, Mansfield K, Fox J (eds) The common marmoset in captivity and biomedical research. Academic Press, London, pp 43–62

Gil-da-Costa R, Martin A, Lopes M, Muñoz M, Fritz JB, Braun A (2006) Species-specific calls activate homologs of Broca's and Wernicke's areas in the macaque. Nat Neurosci 9:1064–1070

Goldenberg G (2009) Apraxia and the parietal lobes. Neuropsychologia 47:1449–1459

Hagoort P (2005) On Broca, brain, and binding: a new framework. Trends Cogn Sci 9:416–423

Hayashi M (2007) A new notation system of object manipulation in the nesting-cup task for chimpanzees and humans. Cortex 43:308–318

Hecht EE, Gutman DA, Preuss TM, Sanchez MM, Parr LA, Rilling JK (2013a) Process versus product in social learning: comparative diffusion tensor imaging of neural systems for action execution-observation matching in macaques, chimpanzees, and humans. Cereb Cortex 23:1014–1024

Hecht EE, Murphy LE, Gutman DA, Votaw JR, Schuster DM, Preuss TM, Orban GA, Stout D, Parr LA (2013b) Differences in neural activation for object-directed grasping in chimpanzees and humans. J Neurosci 33:14117–14134

Hickok G, Poeppel D (2007) The cortical organization of speech processing. Nat Rev Neurosci 8:393–402

Hillert DG (2015) On the evolving biology of language. Front Psychol 6:1796

Huffman M (1984) Stone play of *Macaca fuscata* in Arashiyama B troop: transmission of a non-adaptive behavior. J Hum Evol 13:725–735

Humphries C, Binder JR, Medler DA, Liebenthal E (2006) Syntactic and semantic modulation of neural activity during auditory sentence comprehension. J Cogn Neurosci 18:665–679

Humphries C, Binder JR, Medler DA, Liebenthal E (2007) Time course of semantic processes during sentence comprehension: an fMRI study. NeuroImage 36:924–932

Husain M, Nachev P (2007) Space and the parietal cortex. Trends Cogn Sci 11:30–36

Itani J (1963) Vocal communication of the wild Japanese monkey. Primates 4:11–66

Izumi A (1999) The effect of marker frequency disparity on the discrimination of gap duration in monkeys. Perception 28:437–444
Izumi A (2001) Relative pitch perception in Japanese monkeys (*Macaca fuscata*). J Comp Psychol 115:127–131
Izumi A (2003) Effect of temporal separation on tone-sequence discrimination in monkeys. Hear Res 175:75–81
Jeon HA, Friederici AD (2013) Two principles of organization in the prefrontal cortex are cognitive hierarchy and degree of automaticity. Nat Commun 4:2041
Johnson-Frey SH, Maloof FR, Newman-Norlund R, Farrer C, Inati S, Grafton ST (2003) Actions or hand-object interactions? Human inferior frontal cortex and action observation. Neuron 39:1053–1058
Jürgens U (2002) Neural pathways underlying vocal control. Neurosci Biobehav Rev 26:235–258
Jusczyk PW (1999) How infants begin to extract words from speech. Trends Cogn Sci 3:323–328
Kawamura S (1959) The process of sub-culture propagation among Japanese macaques. Primates 2:45–60
Knoll LJ, Obleser J, Schipke CS, Friederici AD, Brauer J (2012) Left prefrontal cortex activation during sentence comprehension covaries with grammatical knowledge in children. NeuroImage 62:207–216
Koechlin E, Jubault T (2006) Broca's area and the hierarchical organization of human behavior. Neuron 50:963–974
Koelsch S, Gunter TC, von Cramon DY, Zysset S, Lohmann G, Friederici AD (2002) Bach speaks: a cortical "language-network" serves the processing of music. NeuroImage 17:956–966
Kojima S (1985) Auditory short-term memory in the Japanese monkey. Int J Neurosci 25:255–262
Kudo N, Nonaka Y, Mizuno N, Mizuno K, Okanoya K (2011) On-line statistical segmentation of a non-speech auditory stream in neonates as demonstrated by event-related brain potentials. Dev Sci 14:1100–1106
Kumar V, Croxson PL, Simonyan K (2016) Structural organization of the laryngeal motor cortical network and its implication for evolution of speech production. J Neurosci 36:4170–4181
Macedonia JM (1990) What is communicated in the antipredator calls of lemurs: evidence from playback experiments with ring-tailed and ruffed lemurs? Ethology 86:177–190
Maess B, Koelsch S, Gunter TC, Friederici AD (2001) Musical syntax is processed in Broca's area: an MEG study. Nat Neurosci 4:540–545
Milne AE, Mueller JL, Männel C, Attaheri A, Friederici AD, Petkov CI (2016) Evolutionary origins of non-adjacent sequence processing in primate brain potentials. Sci Rep 6:36259
Minagawa Y, Hakuno Y, Kobayashi A, Naoi N, Kojima S (2017) Infant word segmentation recruits the cerebral network of phonological short-term memory. Brain Lang 170:39–49
Mitani JC, Marler P (1989) A phonological analysis of male gibbon singing behavior. Behaviour 109:20–45
Moore-Parks EN, Burns EL, Bazzill R, Levy S, Posada V, Mueller R-A (2010) An fMRI study of sentence-embedded lexical-semantic decision in children and adults. Brain Lang 114:90–100
Mueller JL, Friedericim AD, Männel C (2012) Auditory perception at the root of language learning. Proc Natl Acad Sci USA 109:15953–15958
Nelissen K, Luppino G, Vanduffel W, Rizzolatti G, Orban GA (2005) Observing others: multiple action representation in the frontal lobe. Science 310:332–336
Nishitani N, Schürmann M, Amunts K, Hari R (2005) Broca's region: from action to language. Physiology 20:60–69
Ouattara K, Lemasson A, Zuberbühler K (2009) Campbell's monkeys concatenate vocalizations into context-specific call sequences. Proc Natl Acad Sci USA 106:22026–22031
Palomero-Gallagher N, Zilles K (2019) Differences in cytoarchitecture of Broca's region between human, ape and macaque brains. Cortex 118:132–153
Patel AD (2003) Language, music, syntax and the brain. Nat Neurosci 6:674–681

Peeters R, Simone L, Nelissen K, Fabbri-Destro M, Vanduffel W, Rizzolatti G, Orban GA (2009) The representation of tool use in humans and monkeys: common and uniquely human features. J Neurosci 29:11523–11539
Perani D, Saccumann MC, Scifo P, Anwander A, Spada D, Baldoli C, Poloniato A, Lohmann G, Friederici AD (2011) Neural language networks at birth. Proc Natl Acad Sci USA 108:16056–16061
Petkov C, Kayser C, Steudel T, Whittingstall K, Augath M, Logothetis NK (2008) A voice region in the monkey brain. Nat Neurosci 11:367–374
Petrides M (2005) Lateral prefrontal cortex: architectonic and functional organization. Philos Trans R Soc B Biol Sci 360:781–795
Petrides M, Tomaiuolo F, Yeterian EH, Pandya DN (2012) The prefrontal cortex: comparative architectonic organization in the human and the macaque monkey brains. Cortex 48:46–57
Platel H, Price C, Baron JC, Wise R, Lambert J, Frackowiak RS, Lechevalier B, Eustache F (1997) The structural components of music perception. A functional anatomical study. Brain 120:229–243
Poremba A, Malloy M, Saunders R, Carson RE, Herscovitch P, Mishkin M (2004) Species-specific calls evoke asymmetric activity in the monkey's temporal poles. Nature 427:448–451
Preuss TM, Goldman-Rakic PS (1991) Myelo- and cytoarchitecture of the granular frontal cortex and surrounding regions in the strepsirhine primate *Galago* and the anthropoid primate *Macaca*. J Comp Neurol 310:429–474
Price CJ (2000) The anatomy of language: contributions from functional neuroimaging. J Anat 197:335–359
Price T, Wadewitz P, Cheney D, Seyfarth R, Hammerschmidt K, Fischer J (2015) Vervets revisited: a quantitative analysis of alarm call structure and context specificity. Sci Rep 5:13220
Rilling JK (2014) Comparative primate neurobiology and the evolution of brain language systems. Curr Opin Neurobiol 28:10–14
Rizzolatti G, Fadiga L, Gallese V, Fogassi L (1996) Premotor cortex and the recognition of motor actions. Brain Res Cogn Brain Res 3:131–141
Rodd JM, Longe OA, Randall B, Tyler LK (2010) The functional organisation of the fronto-temporal language system: evidence from syntactic and semantic ambiguity. Neuropsychologia 48:1324–1335
Rolheiser T, Stamatakis EA, Tyler LK (2011) Dynamic processing in the human language system: synergy between the Arcuate fascicle and extreme capsule. J Neurosci 31:1649–1657
Romanski LM, Tian B, Fritz J, Mishkin M, Goldman-Rakic PS, Rauschecker JP (1999) Dual streams of auditory afferents target multiple domains in the primate prefrontal cortex. Nat Neurosci 2:1131–1136
Rüschemeyer SA, Fiebach CJ, Kempe V, Friederici AD (2005) Processing lexical semantic and syntactic information in first and second language: fMRI evidence from German and Russian. Hum Brain Mapp 25:266–286
Rüschemeyer SA, Zysset S, Friederici AD (2006) Native and non- native reading of sentences: an fMRI experiment. NeuroImage 31:354–365
Saur D, Kreher BW, Schnell S, Kummerer D, Kellmeyer P, Vry MS, Umarova R, Musso M, Glauche V, Abel S, Huber W, Rijntjes M, Hennig J, Weiller C (2008) Ventral and dorsal pathways for language. Proc Natl Acad Sci USA 105:18035–18040
Saur D, Schelter B, Schnell S, Kratochvil D, Kupper H, Kellmeyer P, Kummerer D, Kloppel S, Glauche V, Lange R, Mader W, Feess D, Timmer J, Weiller C (2010) Combining functional and anatomical connectivity reveals brain networks for auditory language comprehension. NeuroImage 49:3187–3197
Schel AM, Candiotti A, Zuberbühler K (2010) Predator-deterring alarm call sequences in Guereza colobus monkeys are meaningful to conspecifics. Anim Behav 80:799–808
Schenker NM, Buxhoeveden DP, Blackmon WL, Amunts K, Zilles K, Semendeferi K (2008) A comparative quantitative analysis of cytoarchitecture and minicolumnar organization in Broca's area in humans and great apes. J Comp Neurol 510:117–128

Schenker NM, Hopkins WD, Spocter MA, Garrison A, Stimpson CD, Erwin JM, Hof PR, Sherwood CC (2010) Broca area homologue in chimpanzees (*Pan troglodytes*): probabilistic mapping, asymmetry and comparison to humans. Cereb Cortex 20:730–742

Schlenker P, Chemla E, Zuberbühler K (2016) What do monkey calls mean? Trends Cogn Sci 12:894–904

Schmahmann JD, Pandya DN, Wang R, Dai G, D'Arceuil HE, de Crespigny AJ, Wedeen VJ (2007) Association fibre pathways of the brain: parallel observations from diffusion spectrum imaging and autoradiography. Brain 130:630–653

Schön D, François C (2011) Musical expertise and statistical learning of musical and linguistic structures. Front Psychol 2:167

Scott BH, Mishkin M, Yin P (2012) Monkeys have a limited form of short-term memory in audition. Proc Natl Acad Sci USA 109:12237–12241

Seyfarth RM, Cheney DL, Marler P (1980) Monkey responses to three different alarm calls: evidence for predator classification and semantic communication. Science 210:801–803

Shukla M, White KS, Aslin RN (2011) Prosody guides the rapid mapping of auditory word forms onto visual objects in 6-mo-old infants. Proc Natl Acad Sci USA 108:6038–6043

Simonyan K (2014) The laryngeal motor cortex: its organization and connectivity. Curr Opin Neurobiol 28:15–21

Skeide MA, Friederici AD (2016) The ontogeny of the cortical language network. Nat Rev Neurosci 17:323–332

Skeide MA, Brauer J, Friederici AD (2014) Syntax gradually segregates from semantics in the developing brain. NeuroImage 100:106–111

Specht K, Huber W, Willmes K, Shah NJ, Jäncke L (2008) Tracing the ventral stream for auditory speech processing in the temporal lobe by using a combined time series and independent component analysis. Neurosci Lett 442:180–185

Spocter MA, Hopkins WD, Garrison AR, Bauernfeind AL, Stimpson CD, Hof P, Sherwood CC (2010) Wernicke's area homologue in chimpanzees (Pan troglodytes) and its relation to the appearance of modern human language. Philos Trans R Soc B Biol Sci 277:2165–2174

Stout D, Passingham R, Frith C, Apel J, Chaminade T (2011) Technology, expertise and social cognition in human evolution. Eur J Neurosci 33:1328–1338

Taglialatela JP, Russell JL, Schaeffer JA, Hopkins WD (2008) Communicative Signaling activates 'Broca's' homolog in chimpanzees. Curr Biol 18:343–348

Taglialatela JP, Russell JL, Schaeffer JA, Hopkins WD (2011) Chimpanzee vocal signaling points to a multimodal origin of human language. PLoS One 6:e18852

Thiebaut de Schotten M, Dell'Acqua F, Valabregue R, Catani M (2012) Monkey to human comparative anatomy of the frontal lobe association tracts. Cortex 48:82–96

Tillmann B, Janata P, Bharucha JJ (2003) Activation of the inferior frontal cortex in musical priming. Brain Res Cogn Brain Res 16:145–161

Uchiyama Y, Toyoda H, Honda M, Yoshida H, Kochiyama T, Ebe K, Sadato N (2008) Functional segregation of the inferior frontal gyrus for syntactic processes: a functional magnetic-resonance imaging study. Neurosci Res 61:309–318

Uddén J, Bahlmann J (2012) A rostro-caudal gradient of structured sequence processing in the left inferior frontal gyrus. Philos Trans R Soc B Biol Sci 19:2023–2032

Wakita M (2014) Broca's area processes the hierarchical organization of observed action. Front Hum Neurosci 7:937

Wakita M (2016) Interaction between perceived action and music sequences in the left prefrontal area. Front Hum Neurosci 10:656

Wakita M (2019) Auditory sequence perception in common marmosets (*Callithrix jacchus*). Behav Process 162:55–63

Wakita M (n.d.) Common marmosets (*Callithrix jacchus*) cannot recognize global configurations of sound patterns but can recognize adjacent relations of sounds (submitted manuscript)

Wakita M, Hiraishi H (2011) Effects of handedness and viewing perspective on Broca's area activity. Neuroreport 22:331–336

Wang L, Uhrig L, Jarraya B, Dehaene S (2015) Representation of numerical and sequential patterns in macaque and human brains. Curr Biol 25:1966–1974

Wheeler B (2010) Production and perception of situationally variable alarm calls in wild tufted caputin monkeys (*Cebus paella nigritus*). Behav Ecol Sociobiol 64:989–1000

Wilson B, Kikuchi Y, Sun L, Hunter D, Dick F, Smith K, Thiele A, Griffiths TD, Marslen-Wilson WD, Petkov CI (2015) Auditory sequence processing engages evolutionarily conserved regions of frontal cortex in macaques and humans. Nat Commun 6:8901

Winter P, Ploog D, Latta J (1966) Vocal repertoire of the squirrel monkey *(Saimiri sciureus)*, its analysis and significance. Exp Brain Res 1:359–384

Wright AA (2007) An experimental analysis of memory processing. J Exp Anal Behav 88:405–433

Wright AA, Rivera JJ, Hulse SH, Shyan M, Neiworth JJ (2000) Music perception and octave generalization in rhesus monkeys. J Exp Psychol Gen 129:291–307

Xiao Y, Friederici AD, Margulies DS, Brauer J (2016) Development of a selective left-hemispheric fronto-temporal network for processing syntactic complexity in language comprehension. Neuropsychologia 83:274–282

Yokoyama S, Okamoto H, Miyamoto T, Yoshimoto K, Kim J, Iwata K, Jeong H, Uchida S, Ikuta N, Sassa Y, Nakamura W, Horie K, Sato S, Kawashima R (2006) Cortical activation in the processing of passive sentences in L1 and L2: an fMRI study. NeuroImage 30:570–579

Yoshida KA, Iversen JR, Patel AD, Mazuka R, Nito H, Gervain J, Werker JF (2010) The development of perceptual grouping biases in infancy: a Japanese-English cross-linguistic study. Cognition 115:356–361

Zilles K, Amunts K (2018) Cytoarchitectonic and receptorarchitectonic organization in Broca's region and surrounding cortex. Curr Opin Behav Sci 21:93–105

Zuberbühler K, Cheney DL, Seyfarth RM (1999) Conceptual semantics in a nonhuman primate. J Comp Psychol 113:33–42

# Chapter 6
# Social Scaffolding of Vocal and Language Development

**Hirokazu Doi**

**Abstract** Human infants achieve remarkable feats in language acquisition within the first 2 years of life. Social inputs from other individuals, who primarily tend to be caregivers, facilitate their language development. Across several non-human species, social inputs shape vocal learning through social reinforcement, sensory enhancement, and vocal imitation. Similar mechanisms appear to contribute to the development of vocal production and phonetic perception in humans. However, it remains unclear whether such similarities stem from shared evolutionary roots or convergent evolution. In addition to such shared mechanisms, humans use other types of social learning strategies to acquire the ability to analyse and produce more complex forms of linguistic information. The findings of basic researches will surely contribute to the development of behavioural and pharmacological intervention programmes that provide more enriched social scaffolding of language development to at-risk children.

**Keywords** Vocal communication · Language acquisition · Imitation · Multimodal information · Social reinforcement

## 6.1 Contribution of Environmental Factors to Language Development

The ability to send and encode complex messages that are embedded within a sound stream is undoubtedly a great feat for mankind. Thus, the fact that typically developing infants acquire the basic sets of cognitive and motor skills that are necessary for linguistic communication before the age of 2 years has intrigued many parents and researchers alike. This is especially surprising when one considers 'poverty of the stimulus'; in other words, the amount of linguistic input to which infants are

H. Doi (✉)
School of Science and Engineering, Kokushikan University, Tokyo, Japan
e-mail: hdoi@kokushikan.ac.jp



N. Masataka (ed.), *The Origins of Language Revisited*,
https://doi.org/10.1007/978-981-15-4250-3_6

exposed during early development appears to be too minimal to allow them to grasp the underlying structures and many rules of their mother tongues (Pullum and Scholz 2002). Understandably, researchers have been tempted to argue that an innate machinery or language module is embedded within an infant's brain and remains dormant until its fullest potential is unleashed during the appropriate developmental stage. An influential theory of universal grammar (Chomsky 1965) is the widely accepted nativist perspective, and it has succeeded in theorising about the mechanism of language development, especially the acquisition of the grammatical structure of one's mother tongue to some extent. Similarly, the development of precise motor control, which is required for the production of linguistic sounds (Fitch 2000, for a review), had largely been attributed to the innately programmed maturation of the vocal apparatus (Fitch and Giedd 1999) and neural systems that are responsible for motor control (Smith et al. 1995). Behavioural genetic research has facilitated the identification of a component of the genetic predisposition that enables linguistic communication, thereby offering further support to the nativist view of linguistic ability. It is well-known that a rare mutation in the FOXP2 gene leads to a severe deficiency in or complete lack of linguistic function (Vargha-Khadem et al. 2005; Lai et al. 2001). Further, a series of twin studies have shown that substantial variances in linguistic ability are attributable to hereditary factors (Stromswold 2001).

The nativist view succinctly explains the process of language development but does not account for individual differences in linguistic ability. Widely cited reports underscored intra-generational gaps in language development (Anderson and Freebody 1981; Farkas and Beron 2004; Hart and Risley 1995; Snow et al. 1998). According to the reports, there is a prominent delay in linguistic acquisition among children who belong to households with a low socio-economic status (SES) than those who hail from relatively affluent communities. This gap widens as they grow older and is considered to be one of the leading factors that determine educational achievement and economic success.

In an attempt to address this problem, researchers have investigated the environmental factors that are linked to language acquisition in ecologically valid environments (Fernald et al. 2013; Weisleder and Fernald 2013). Many studies typically focused on the association between the amount of linguistic input (e.g. number of caregiver utterances that are directed towards infants) and later language development. As expected, they found that exposure to large amounts of linguistic material during infancy and toddlerhood leads to faster acquisition of more sophisticated language. In addition to the quantity of language input, studies have also underscored the importance of the quality of linguistic input by undertaking fine-grained analysis of the interaction between infants and their caregivers (Cartmill et al. 2013; Hirsh-Pasek et al. 2015). For example, Cartmill et al. (2013) investigated the quality of linguistic input and later development of vocabulary by video-recording mother-infant interactions. The quality of linguistic input was defined as the accuracy with which adults who were naïve to the research purpose guessed the word that the caregiver had uttered from a muted video of a mother-infant interaction. Their main finding was that the input quality of interactions between mothers and their 1.5–2-year-old infants predicted vocabulary size 3 years later.

Past studies have shown that it is not only quantity but also the quality of linguistic input that plays a facilitative role in infant language development (Cartmill et al. 2013; Hirsh-Pasek et al. 2015). The contribution of environmental factors (i.e. quality and quantity of linguistic input) is a promising observation that has implications for the development of intervention programmes that promote language acquisition among children who live in adverse environments. At the same time, in many of these studies, the quality of linguistic input was loosely defined. Consequently, the precise mechanism through which environmental factors, especially caregiver interactions, influence infant language development remains poorly understood.

In real-life situations, remarks and linguistic sounds are not uttered in a vacuum. Instead, caregivers make these utterances and sounds within the context of social interactions with infants. Linguistic inputs that are directed towards infants are usually embedded within a flow of social exchanges and accompanied by various kinds of multimodal information. Consider the example of a caregiver who changes an infant's diaper. The caregiver repeatedly utters the infant's name and makes short remarks using a high-pitched voice. Simultaneously, the caregiver may make exaggerated faces and provide tactile stimulation to gain the infant's attention. In response, the infant may laugh and smile, which presumably entrains the caregiver to expend greater effort to amuse the infant. In such a context, the following questions emerge: which components of such a social interaction contribute to infant language development, and what are their unique contributions?

## 6.2 Social Context Plays a Role in the Vocal Learning of Animals

Since the focus has shifted from the innateness of language functions to the environmental determinants of language development, research findings on how social interaction influences different aspects of language development in humans have accumulated (Kuhl 2007; Syal and Finlay 2010). At the same time, a substantially larger number of articles on the influences of social context on the vocal communication and development of animals have been published. Thus, a review of the main observations of non-human species may serve as a good starting point from which one can explore the means by which social interaction promotes language acquisition in humans.

Many studies on vocal learning have been conducted using songbirds. The birdsongs of young birds are immature; there is substantial variance in the timing of motor control of their vocal instruments, and the birdsongs of young birds contain many acoustical features that are not observed in the birdsongs of their adult counterparts (Syal and Finlay 2010; Goldstein et al. 2003; Petkov and Jarvis

2012). Young songbirds refine their motor control and the acoustic structure of their birdsongs through post-natal experiences. As a result, the learning process that the production of birdsongs entails is often considered to be a model of speech development in humans (Syal and Finlay 2010; Goldstein et al. 2003).

The ability to engage in vocal learning has been observed in many non-human species other than songbirds, including mammals such as dolphins (Janik 2000), whales (Noad et al. 2000), bats (Wilkinson and Boughman 1998; Boughman 1998), elephants (Stoeger and Manger 2014), and some species of primates (Sugiura 1998; Masataka and Fujita 1989). An illuminating example is the series of studies that Boughman (1998) conducted using spear-nosed bats. Spear-nosed bats form social groups, and members of different groups produce group-distinctive calls. When female spear-nosed bats forage and find a food-rich location, they emit group-distinctive calls to gather other members of their social group so that they can collectively defend the foraging site (Wilkinson and Boughman 1998). In another study that was conducted by the same group (Boughman 1998), young captive bats were transferred from one cage to another one in which bats of another social group were housed. At first, the acoustic features of the calls of young bats were slightly different from those of their roommates. However, after several months, the young bats that had been transferred to another cage had begun to produce calls that were very similar to and ultimately indistinguishable from those of the adult bats. Thus, through social interaction with other group members, the spear-nosed bats had learned to produce group-distinctive calls and use them effectively to gain survival benefits both as a social group and as individuals.

Bottleneck dolphins are another well-known example of mammals that have an excellent ability to engage in vocal learning. They can mimic novel sounds with very high accuracy in the first attempt, and they have been known to demonstrate this ability throughout their lifespans. Once dolphins form a social group or alliance, they begin to produce shared whistles through vocal learning (Watwood et al. 2004). The production of shared whistles is considered to strengthen and promote group cohesion (Sewall et al. 2016; Sewall 2012; Tyack 2008).

As the aforementioned examples illustrate, the usage of a learned vocal repertoire as a ‘password’ to ascertain group membership or transmit affiliative or confrontational messages has been widely observed (Sewall et al. 2016; Sewall 2012; Tyack 2008). At the same time, little is known about the mechanisms, especially the neural one, through which vocal learning is achieved in avian and mammalian species through social experience. In this regard, the existing literature delineates at least three contributions of social interactions: social reinforcement, sensory enhancement, and vocal imitation.

### *6.2.1 Social Reinforcement*

Reinforcement learning is one of the most widely studied mechanisms of behavioural learning. The recent surge in interdisciplinary research has revealed that the neural process that underlies reinforcement learning can be modelled using the classic algorithms of machine learning (Glimcher 2011). Thus, the nature of reinforcement learning not only is well understood but can also be described as a set of algorithmic steps, which is quite rare in neuroscientific and behavioural studies.

Given the success of reinforcement learning in explaining a wide range of learning processes, it is logical to postulate that vocal learning may also be shaped by reward-based reinforcement learning. Direct support for such a contention is offered by the findings of a laboratory study that was conducted by Manabe and Dooling (1997). They trained budgerigars to emit calls that were similar to a 'template call' using food reward. When production of the template call was reinforced, all the budgerigars learned to emit the template call in response to a cue. Conversely, when the researchers trained the budgerigars to produce non-template calls, they learned to avoid the production of the template call, which in turn resulted in an increase in the diversity of emitted calls.

The findings of Manabe and Dooling's study (1997) suggest that reinforcement learning can function as a mechanism that underlies vocal learning. However, the kind of signal that serves as a reinforcer in natural environments outside the laboratory is yet to be identified. Nevertheless, recent studies on infant marmosets have revealed that the temporally contingent calls of adult marmosets serve as social reinforcers that facilitate call maturation (Takahashi et al. 2016, 2017). In this study, infant marmosets were reared in two different environments. In one environment, infant calls with mature characteristics (e.g. low entropy) were contingently rewarded with adults' calls. This social contingency was experimentally diminished in the other environment. Among infants who had been reared in the high-contingency environment, the entropy of their calls had steadily decreased, and this was indicative of a smooth maturation of their calls. This was not the case for the infants that had been raised in the low-contingency environment. Thus, social exchanges with adults, or more specifically temporally contingent vocal exchanges, facilitated vocal maturation in marmosets. This finding is especially surprising given that the social contingency was experimentally manipulated for only 40 min per day. This underscores the possibility that small differences in the social environment may have a large impact on vocal and possibly language development.

### *6.2.2 Sensory Enhancement*

As has been briefly described with regard to interactions between human infants and their caregivers, the utterances that adults direct towards infants are often accompanied by abundant multimodal information. Chen et al. (2016) investigated how multimodal information influences vocal learning in songbirds. They compared the song learning of young zebra finches that were allowed to multimodally interact with a tutor bird and those that were passively exposed to the vocalisations of a tutor bird. As expected, the song structure of pupil birds that had socially interacted with a tutor bird was more similar to that of the tutor, when compared to the song structure of their counterparts that had only been passively exposed to the tutor's songs. Interestingly, social interaction only in auditory modality did not facilitate pupil's song learning compared to passive exposure to tutor's song. Thus, the multimodality of social interactions is of primary importance to zebra finches' song learning.

What role does visual information play in guiding the vocal learning of zebra finches? This question was addressed in the analysis of attentiveness of pupil birds. The researchers found that more attentive songbirds emitted songs that were more structurally similar to those of the tutor birds, irrespective of the quality of social interaction. Importantly, social interaction increased the attention that young birds paid to their tutor's songs. Further, the positive outcomes of attentiveness were more pronounced among pupils that had multimodally interacted with their tutor. These findings underscore two points. First, social interaction motivates young birds to pay more attention to adults' songs. Visual information appears to be quite efficient in gaining the attention of young birds; indeed, pupil birds were more attentive to the tutors' songs when the tutor birds directed their attention towards their pupils. Second, and more interestingly, the songs that tutors had produced during their multimodal interaction contained some characteristics that facilitated song learning. Indeed, when tutors sang to their pupils during the multimodal interaction, their songs entailed more repetitions of introductory notes, longer intervals between motifs, and a lower mean frequency and spectral entropy; these acoustical features are homologous to those that characterise infant-directed speech or 'motherese' among humans (Golinkoff et al. 2015). Taken together, social interaction influences the behaviours of both pupil and tutor birds. The visual attention of tutors motivates pupils to pay attention to the tutor's songs, and in turn, the presence of pupils motivates tutors to modify their songs to facilitate song learning.

### 6.2.3 *Vocal Imitation*

Since the ground-breaking discovery of the mirror neuron in the premotor region of macaque monkey, the mirror system has been linked to a variety of social behaviours (Iacoboni et al. 1999; Rizzolatti and Fabbri-Destro 2008). Accordingly, the mirror system is proposed to be involved in linguistic processing (Rizzolatti and Arbibb 1998; Levy 2011), but very few studies have examined this mechanism. Mirror neurons fire during both the observation of another individual's action and the execution of the same movement, thereby representing the sensorimotor correspondence between a visual image of an action and the motor command that generates it. The activation of the mirror system was first identified as a response to visual information. However, it was later found to also occur in response to auditory information that accompany articulatory movement among humans. For example, Fadiga et al. (2002) reported that listening to speech sounds triggers preparation of speech-related tongue movement. Thus, it is logical to contend that the mirror system also plays a role in vocal learning. This conjecture is supported by findings in previous lesion studies. Forebrain high vocal centre (HVC) is roughly analogous to human Broca's area in birdsongs' brain and sends projections to neural pathways that play essential roles in singing and song learning (Mooney 2014). Lesions to HVC impairs songbirds' ability to process conspecific's songs (Brenowitz 1991; Gentner et al. 2000), which indicates the possibility that HVC neurons function as auditory mirror system in songbirds. Accordingly, researchers have revealed that HVC neurons in swamp sparrows and Bengalese finches demonstrate a pattern of 'auditory-vocal mirroring' (Prather et al. 2008; Mooney 2014). A subset of HVC neurons respond to certain sequences in the own and conspecific's birdsongs and demonstrate neural firing when the songbirds sing the same sequence themselves. These subsets of HVC mirror neurons in swamp sparrows categorically represent the perceptual boundaries of the note durations of a primary song, which is a birdsong homologue of phoneme boundaries in humans. There is a geographical difference (a kind of dialect) in the categorical boundaries of note durations. Interestingly, HVC mirror neurons represent such regional differences in categorical boundaries (Prather et al. 2009; Mooney 2014). This supports the claim that auditory-vocal mirroring plays a role in post-natal vocal learning through sensory exposure.

## 6.3 How Social Interaction Shapes Human Language Development

Human infants undergo remarkable development of their ability for sensory analysis during the first year of life. Traditional perspectives posit that infants are born as a 'blank slate' with no meaningful representations of the outside world. As peripheral sensory organs mature and infants make many observations of events, they accumulate experiences and knowledge about how the world around them works (James 1890). However, evidences from the field of developmental psychology drastically changed this view by demonstrating that infants as young as only a few days or even hours old exhibit signs of their ability to represent basic knowledge about the world (Spelke et al. 1992; Spelke 2000). Similar observations have been made in the domain of social cognition as well. Early studies showed that neonates were able to recognize schematic faces that were composed of three black rectangles, which were configured as an inverse triangle (Slater 2002, for a review). Similarly, human neonates demonstrate sensitivity to point-light displays of biological motion (Simion et al. 2008). Another well-known example is Meltzoff and Moore (1977)'s observation that neonates mimic the facial movements of adults. These and other observations indicate that human infants are innately endowed with representations of and predispositions to pay attention to socially meaningful information. As their social brain matures, they acquire more refined abilities to exchange social messages as multimodal sensory information.

As noted earlier, past studies have shown that it is not only quantity but also the quality of linguistic input that determines the course of language development. The quality of linguistic input is determined by how linguistic material is embedded within a stream of multimodal social information. At the same time, any effort on the part of the message sender, which is typically the caregiver, to enhance the quality of linguistic input will be futile if the receiver, namely, an infant, cannot analyse and respond to the multimodal information that accompanies the linguistic material. Thus, it appears that the type of social information that caregivers use to enhance the quality of linguistic input varies depending on the developmental stage of their infant's socio-cognitive functions. This in turn indicates that the timing of emergence of linguistic ability is at least partly constrained by the development of social cognition (see, Tomasello 2000, for similar view). In this regard, an understanding of human social cognition is indispensable to the elucidation of the course of language development in humans.

Some mammalian and avian species achieve vocal learning by recruiting at least three mechanisms: social reinforcement, sensory enhancement, and vocal imitation (which have been described in the preceding section). The number of studies on the role of social experience in the development of auditory processing in non-human species is relatively small. However, a few studies reported that pattern of neuronal responses in zebra finches, whose auditory processing shares many characteristics with human speech processing (Ohms et al. 2010), is tuned by exposure to conspecific's songs (Chen et al. 2017; Prather et al. 2008; Mooney 2014). Thus, similar processes to those observed in non-human species such as songbirds may play a role in the development of vocal control and phonetic perception in human infants as well. However, it remains an elusive enterprise to delineate whether such similarities, if any, are the consequences of phylogenetic linkage or convergent evolution.

The complexity and flexibility of human language is unmatched by the vocal communication of non-human species. In human vocal communication, a word is rarely uttered in isolation. Instead, multisyllabic words are concatenated in accordance with grammatical rules and successively uttered as a sound stream. Thus, the first hurdle that hinders human infants from analysing linguistic inputs after acquiring the ability for phonetic perception is the segmentation of the acoustic stream of linguistic sounds into the unit of words. Another important task that infants are required to engage in is the comprehension of the semantics of linguistic input. As the classic Gavagai problem (Quine 1960) illustrates, language learners are confronted with a great ambiguity in understanding the correspondence between an unknown word and its referent. Several previous studies have reported signs of the precursors of semantics in non-human species (Fedurek and Slocombe 2011). For example, it has been reported that dogs possess the ability for fast word-to-object mapping and long-term retention of correspondence (Kaminski et al. 2004). However, the number of learned words and semantic flexibility of human language is unparalleled.

The need to segment a sound stream into words and comprehend the semantics of them poses a unique problem to the full development of the linguistic ability of human infants. The question of if and how social information helps infants tackle these tasks is important because the corresponding answers are likely to reveal the uniqueness of human language acquisition.

### 6.3.1 Vocal Learning

During the first year of life, an infant's vocal ability undergoes drastic transition (i.e. repeated vowels to babbling and well-formed syllables), and the first words are uttered approximately 10 to 15 months after birth (Owens 2005). The timeline of such vocal development is partly constrained by the anatomical and physiological maturation of the vocal instrument and neuronal maturation of the centres that are responsible for vocal control (Fitch and Giedd 1999; Smith et al. 1995). At the same time, many researchers have proposed that multimodal inputs from others and an infant's self-generated input contribute to the refinement of the ability to produce syllabic sounds. This claim has been supported by the findings of studies that have been conducted among infants with hearing loss. For example, Oller and Eilers (1988) found that infants with hearing loss start producing well-formed syllables much later than those without hearing loss.

How do social inputs from others contribute to vocal development? Many studies that have examined this question have found that the temporal contingency of caregiver behaviours influences the syllabic nature of infant vocalisation (Bloom 1988; Bloom et al. 1987; Hsu and Fogel 2001; Goldstein et al. 2003; Goldstein and Schwade 2008). Across a series of early studies, Bloom (1988) manipulated the social contingency of mothers' vocalisations during their social interactions with their 3-month-old infants. In one condition, the mothers were encouraged to smile and provide tactile stimulation and vocalisation after each infant vocalisation, thereby complying with the 'turn-taking rule' of communication. In the other condition, the turn-taking rule was violated; specifically, the same kinds of stimulation were provided to infants at pre-cued timings. Bloom's (1988) main finding was that the proportion of syllabic vocalisation had increased from baseline when the turn-taking rule was maintained but not when it was violated. Bloom (1988) replicated the same experiment using non-verbal sounds as mothers' vocal stimulation. Interestingly, the proportional increase in infants' syllabic vocalisation was not observed when mothers provided non-verbal stimulation after infant vocalisation. Goldstein and Schwade (2008) made similar observations of the babbling of 9.5-month-old infants. Specifically, infants mimicked the acoustic features of their mothers' vocalisations (e.g. produce a consonant-vowel sound in response to one's mother's consonant-vowel vocalisation) when their mothers responded to them contingently, but no such trend was observed when mothers responded non-contingently.

Taken together, these findings indicate that temporally contingent caregiver vocalisation facilitates more speech-like infant vocalisation. One obvious mechanism that underlies this phenomenon is vocal imitation. As mentioned earlier, even neonates possess the basic ability to represent sensorimotor correspondence between the visual image of bodily motion and one's own motor program (Meltzoff and Moore 1977; Rizzolatti and Fabbri-Destro 2008). Thus, it is not surprising that the viewing of articulatory movement triggers the production of a more mature form of infant vocalisation, especially when infants interact with caregivers in an *en face* position; this was the case in the laboratory studies that have been discussed in the preceding paragraphs. Consistent with this view, Masataka (1992) delineated the possibility that infants who are as young as 5 months of age can modify the acoustic features of their vocalisations to render them more similar to those of their mothers' vocalisations (see also Ko et al. 2016, for similar findings among 12–30-month-olds). Similarly, Kuhl and Meltzoff (1996) found that 3–5-month-old infants can mimic the vowel sounds of adults.

The question of how the temporal contingency of caregiver responses to infant vocalisation influences the tendency for infants to mimic the acoustic features of an adult's sounds and consequently produce more mature vocalisations remains elusive. Mutual entrainment that is founded on temporally contingent social interaction might increase the perceptual salience of a social partner against the environment. Infants who have reached 5 months of age have strong expectations that their partner in a social interaction should respond to their behaviours in a temporally contingent manner (Rochat 2001). Hence, infants may volitionally change the allocation of their attentional resources based on this expectation and consequently end up paying more attention to meaningful information, which in turn augments vocal imitation when a social contingency is maintained. Such an increase in attentiveness towards the multimodal information that caregiver vocalisation entails and accompanying articulatory movement may underlie the increase in the frequency of more mature vocalisation, which is similar to what has been observed in the case of song learning in zebra finches (Chen et al. 2016).

Interestingly, several studies have shown that caregivers also modify their vocalisations in response to the acoustic features of infant vocalisation. For example, Gros-Louis et al. (2006) found that mothers provided differential verbal feedback in response to infant vocalisation based on its acoustic features. More specifically, when compared to vowel vocalisation, an infant's consonant-vowel vocalisation was more likely to beget mother's interactive response that was accompanied by vocal sounds. The reason that underlies such a modification in maternal response is unclear, but it seems likely that mothers intuitively try to facilitate the production of mature infant vocalisation by socially rewarding advanced forms of infant vocalisation and affording infants the opportunity to observe more templates for vocal imitation.

### *6.3.2 Phonetic Perception*

An infant's auditory system is tuned to discriminate the phonemes of their mother tongue until the age of 9–12 months (Best et al. 1995). Interestingly, before this stage, infants possess the capability to discriminate phonemes that are not categorically differentiated in their mother tongue. Thus, during this developmental stage, there is a decline in the ability to discriminate between phonemes that do not belong to one's mother tongue. Can this decline be reversed by exposure to a foreign language? Kuhl et al. (2011) tackled this question by investigating the effects of short-term exposure to a foreign language. As the researchers had hypothesised, the decline was attenuated among infants who had participated in sessions in which they were exposed to a foreign language. Specifically, infants who were residing in the United States and had been exposed to Mandarin showed lower levels of decline in the ability to categorise phonemes that are unique to Mandarin than their counterparts who had not been exposed to a foreign language. Interestingly, however, the effect of short-term exposure was evident only when the infants had socially interacted with live tutors.

Off-line analysis of infant behaviours during exposure sessions revealed that infants paid more attention to a foreign language speaker or the objects that he or she was referring to, when they were interacting with a live tutor than when they were passively exposed to foreign language materials and a speaker through a television. Thus, attentiveness to information that is relevant to linguistic input plays a key role in the development of receptive language as well as vocal learning. Given the contribution of temporal contingency to vocal learning (Bloom 1988; Goldstein and Schwade 2008), it is logical to postulate that real-world social interaction also facilitates phonetic learning by enhancing the attentiveness of infants.

Yet another strategy that mothers use to direct the attention of infants towards linguistic materials is infant-directed speech (IDS) or motherese. When compared to adult-directed speech (ADS), IDS has several peculiar characteristics such as a higher pitch, a slower tempo, and an exaggerated prosodic contour and intonation (Kuhl 2007; Golinkoff et al. 2015; Sulpizio et al. 2018). The benefit that IDS confers on language learning is well-documented (Kuhl 2007; Golinkoff et al. 2015). A slow tempo and an exaggerated intonation make it easier for infants to perceive phonemic units, segment sound stream into words, and consequently grasp the grammatical structure of language. From the perspective of attentional effects, the positive valence of IDS (i.e. happiness) makes it attentionally salient to infants (Singh et al. 2009).

During the period in which phonetic discrimination abilities are immature, visual information that is gained from the face, especially the mouth region, facilitates the disambiguation of a linguistic sound. Lewkowicz and Hansen-Tift (2012) found that infants strategically use facial information depending on their developmental stage of speech processing ability. In their study, fixation patterns for speaking faces, which were presented using an audio-visual movie format, were measured among 4-, 6-, 8-, 10-, and 12-month-old infants (Fig. 6.1). Four-month-old infants viewed the eye region for a longer duration than the mouth region, but 8–10-month-old infants shifted their attention from the eyes to the mouth region; these infants represented the developmental period during which one learns to discriminate between the phonemes of one's mother tongue. When infants reach 12 months of age, and phonetic learning of their mother tongue is almost complete, they shift their attention back from the mouth to the eye region when they view faces that speak native but not non-native languages. This pattern of development offers strong support to the view that the multimodal information that is gained from facial regions plays an indispensable role in phonetic learning.

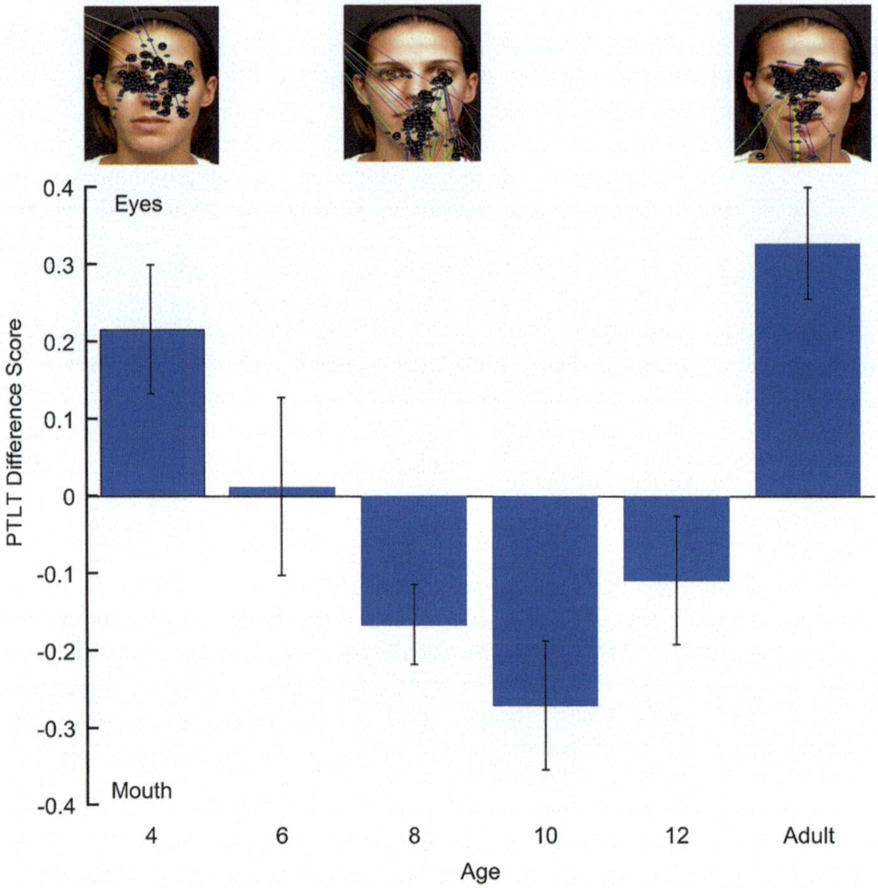


**Fig. 6.1** Developmental change in proportion of total looking time (PTLT) to eye and mouth regions. This figure was taken from Lewkowicz and Hansen-Tift (2012). The PTLT difference scores in the vertical axis were calculated by subtracting PTLT to the mouth region from that to the eye region

### *6.3.3 Word Acquisition*

The most fundamental question that is relevant to language acquisition may pertain to why infants map referents to linguistic sounds but not to other types of auditory stimuli in the environment (Tomasello 2000; Woodward and Hoyne 1999). One key to answer this question seems to be 'communicative intent' of speaker of linguistic materials (Tomasello 2000). When an adult speaks to a toddler, the toddler understands 'that the adult is making these funny noises in an attempt to communicate with her' (Tomasello 2000). This insight motivates children to explore what the adult is trying to communicate through a particular sound, thereby helping them link a linguistic sound to its referent.

Consider a situation in which you are surrounded by people who are speaking a language that is completely foreign to you. One of them utters a word, and you have to find out what the novel word means. The word could refer to a single object, collection of objects, particular attribute of an object, or an abstract concept. As this example illustrates, the referent of a novel word cannot be determined using a bottom-up approach, and this problem is called 'referential indeterminacy' (Nimtz 2005; Tomasello 2000). Nevertheless, human infants somehow manage to acquire the ability to correctly map words to their referents. Past investigations into word acquisition have centred around how human infants solve this problem.

The constraints theory of word learning (Markman 1990, 1992) attempts to answer this question by introducing several mapping rules or 'constraints' as follows: 'words can be extended to basic-level categories'. In contrast, the social-pragmatic theory of word learning, which has been championed by Tomasello (2000), emphasises the important role that communicative intent plays in helping children navigate towards the resolution of referential indeterminacy.

Communicative intent can be transmitted through several different forms, one of the most important of which is joint attention (Moore and Dunham 1995). In a joint attention episode, two individuals simultaneously pay attention to the same object, thereby forming a triadic relationship. When a word is spoken during a joint attention episode that involves an adult, a child can intuitively grasp what an adult is referring to by the linguistic sound. Following another's gaze not only helps children disambiguate the referent but also facilitates the processing of the referred object, which in turn seemingly aids the long-term retention of word-object mapping (Okumura et al. 2007). The link between joint attention and later vocabulary development has been observed in several longitudinal studies (Morales et al. 2000; Brooks and Meltzoff 2008; Mundy et al. 2003). Interestingly, the effect of joint attention on lexical acquisition is amplified when joint attention episodes are initiated by children (Tomasello and Farrar 1986; Baumwell et al. 1997). The findings of Tomasello and Farrar (1986) indicated the possibility that children with mothers who tended to follow their attentional focus rather than intrusively initiate a joint attention episode had a larger vocabulary size at later developmental stage. Thus, children are no longer passive learners at least at this developmental stage; instead, they actively search for information that they are curious about (Begus et al. 2014).

Joint attention and gaze-following are not solely sufficient to scaffold word-to-referent mapping. To fully grasp the communicative intent of another, the ability to make inferences about his or her mental state is indispensable. Early studies have argued that children acquire representations of others' mental states typically between the ages of 3 and 4 years (Wimmer and Perner 1983). Given that vocabulary size is said to rapidly increase during the second year of life (Ganger and Brent 2004), information about complex mental states such as beliefs may only make limited contributions to word learning during the stage of 'vocabulary explosion'. However, more recent studies have shown that children as young as 24 months old possess the ability to represent others' beliefs (Onishi and Baillargeon 2005;

Southgate et al. 2007). Therefore, communicative intent may bear greater significance to word acquisition than what has previously been believed.

Another perspective offers a different explanation to the process of referent disambiguation: caregivers present the referents of a word in such a manner that it bears the greatest perceptual salience against the environment (Yu and Smith 2012). This makes it almost unnecessary for children to disambiguate the meaning of a word either by deciphering communicative intent (Tomasello 2000; Tomasello and Farrar 1986) or adapting constraints (Markman 1990, 1992). In their ingenious study, Pereira et al. (2014) video-recorded the first-person perspective view of 15–20-month-old infants during sessions in which they played with their caregivers using unfamiliar objects with novel names. They found that children tended to remember object names better when caregivers uttered the name of the object while moving the referent object towards the centre of the visual field of the child. Since placing an object in the centre of one's visual field demands high attentional priority, all that the children were required to achieve in this situation was to associate the auditory signal with the most perceptually salient information in the environment. The findings of these lines of studies imply that caregivers can assist the word learning of toddlers by intentionally, or intuitively, increasing the perceptual salience of a referent object when they introduce object names to their children. This can be deemed to be another type of social scaffolding that caregivers provide to children to help them increase their vocabulary size.

## 6.4 Practical Implications and Directions for Future Research

Investigations into environmental influences on human language development have partly been motivated by the objective to bridge intra-generational gaps in linguistic ability by facilitating the language development of children who live in adverse environments. This section summarises the specific means by which the findings of basic research studies can be used to develop and implement interventions that promote the language development of at-risk children.

### *6.4.1 Implementation of Parent-Focused Interventions*

Past studies have found that social variables such as the temporal contingency of a social response (Bloom 1988; Goldstein and Schwade 2008), use of IDS (Kuhl 2007; Golinkoff et al. 2015), and simultaneous presentation of vocal and facial information (Lewkowicz and Hansen-Tift 2012; Hillairet de Boisferon et al. 2018) contribute to the quality of linguistic input and consequently influence later development of linguistic abilities. Thus, the most straightforward strategy to facilitate the language development of children who live in adverse environments is to disseminate empirical findings and offer educational courses to caregivers, particularly expectant mothers and fathers. Researchers have already begun to implement such parent-focused language intervention programmes. For example, Buschmann et al. (2008) demonstrated the efficacy of a short-term intervention for parents of children with expressive language delays using a randomised control design. Their intervention programme provided training on child-oriented interactions, and it fostered the parental skill of social assistance, which facilitates children's language development. Parents of children with problems in receptive/expressive language tend to be more aware of the need for interventions than their counterparts with typically developing children. In contrast, adults who belong to a low SES community are generally difficult to reach, and this makes it difficult to provide them with parent-focused interventions that facilitate the linguistic development of children who hail from such communities. Thus, the identification of incentives that will effectively motivate parents who belong to a low SES community to participate in such intervention programmes is an important topic that applied researchers must address to resolve the intra-generational gap in language development.

### 6.4.2 Interventions for Children with Depressive Mothers

With regard to parental problems, the children of depressive mothers constitute another group that is vulnerable to delays in language development. The prevalence of post-partum depression is relatively high (Gavin et al. 2005), and maternal depression, even subclinical one, poses a great risk to the language development of children with varying SESs. There have been many reports about the detrimental effects of maternal depression on mother-infant interactions. Specifically, mother-infant dyads with depressive mothers are less efficient in managing socially contingent interaction (Skotheim et al. 2013; Weinberg et al. 2006). Moreover, mothers with depressive state are not good at producing IDS (Kaplan et al. 2001). Indeed, infants with depressive mothers fail to learn from their own mother's IDS but are capable of doing so, when presented with an unfamiliar healthy mother's IDS (Kaplan et al. 2002).

Currently, the most commonly prescribed medication to depression is selective serotonin reuptake inhibitors (SSRIs). Although relatively little information is available, a systematic review underscored the effectiveness of SSRIs in ameliorating the symptoms of post-partum depression (Hoffbrand et al. 2001). In addition, Kaplan et al. (2001) reported that IDS of depressive mothers taking SSRI had comparable quality to IDS of healthy mothers. Thus, pharmacological inventions that are provided to depressive mothers may be beneficial in promoting the language development of their children. However, the effects of SSRIs should also be cautiously examined from the perspective of its influences on the quality of mother-infant interactions and children's language development, because a recent study raised the possibility that SSRI medication to mothers alters the trajectory of speech perception development of children during pre- and post-natal period (Weikum et al. 2012).

### 6.4.3 *Interventions for Children with Autism Spectrum Disorders and Their Parents*

Even in the case of parents who are able to provide their children with rich social environments, interventions may be needed to help their children maximally benefit from their social environments. Children with autism spectrum disorder (ASD) appear to be vulnerable to delays in language development (De Fossé et al. 2004). ASD is characterised by a collection of core symptoms (Lenroot and Yeung 2013), one of which is atypicality in social communication. People with ASD are not good at decoding emotion faces and voices (Doi et al. 2013), often fail to follow the gaze of another (Gillespie-Lynch et al. 2013), and do not spontaneously make inferences about the mental states of others (Senju et al. 2009). Typically, a diagnosis is made when a child is approximately 3 years old. Nevertheless, infants who are later diagnosed with ASD manifest atypical patterns of social development (e.g. less frequent fixation on another person's face, reduced cortical activation in response to gaze information) when they are as young as 12 months old (Osterling et al. 2002; Elsabbagh et al. 2012). Further, several researchers have argued that some of the core symptoms of ASD stem from atypical development of the mirror system (Rizzolatti and Fabbri-Destro 2010). All these findings and theorisations suggest that infants with ASD may receive limited assistance to enhance their linguistic ability from their social environments.

Studies have found many aspects of linguistic function to be atypical in people with ASD. A peculiar symptom of the now-classic Kanner's autism is prominent delays in language development. This has led to the contention that atypicality in neural function in ASD is left-lateralised (McCann 1982), though empirical studies have yielded only mixed support for this postulation (Doi and Shinohara 2017). More recent studies have found that phoneme perception and the audio-visual representation of speech are atypical even in people with high-functioning ASD (Saalasti et al. 2012; Smith and Bennetto 2007). In an event-related potential study, Kuhl et al. (2005) measured cortical activation in response to syllables among 2.5–6-year-old children with and without ASD. They focused on an event-related potential component, namely, mismatched negativity (MMN), which is elicited by a deviant syllable that is embedded within a stream of standard syllables. Their main finding was that children with ASD but not the control children failed to demonstrate MMN in response to deviant syllables, thereby underscoring the less efficient speech perception of children with ASD.

Researchers have developed several training programs to promote the communicative functions, including linguistic abilities, of children with ASD (Paul 2008, for a review). A majority of these programs are child-focused interventions that are conducted within natural environments. Parent-focused interventions are relatively few in number, but the findings of empirical studies offer hints about how caregiver behaviours can be modified to promote the language development of children with ASD. In one such study, Dimitrova et al. (2016) investigated the association between parents' translations of children's gestures and vocabulary development. They found that children with ASD exhibited fewer gestures to point at or request for a specific object than typically developing children. However, when parents translated the gestures of children (e.g. by uttering the names of the referent), the referent names were more likely to be incorporated into the children's vocabulary than the referents of untranslated gestures among both typically developing children and their counterparts with ASD. This finding indicates that increasing the frequency of gesture translation can help expand the vocabulary of children with ASD. In another study, Warlaumont et al. (2014) found that caregiver responses to children's vocalisation are less dependent on whether the vocalisation is speech-like or not among children with ASD than among their typically developing counterparts. Such a lack of discriminability in caregiver responses may weaken the positive effects of social reinforcement on the vocal learning of children with ASD. Thus, parent-focused interventions that teach them to provide quality-contingent responses to their child's vocalisation are likely to promote the vocal learning of children with ASD.

In addition to behavioural interventions, pharmacological interventions that utilise oxytocin administration as a remedy for ASD symptoms have also recently received substantial attention. The findings of past clinical trials have observed significant variability in response to oxytocin administration among people with ASD and yielded mixed support for the efficacy of this type of intervention (Parker et al. 2017 for a brief review). Nevertheless, if oxytocin administration does indeed modulate central nervous system function in the expected manner, it may be effective in ameliorating the language problems of children with ASD in several ways. First, the findings of several oxytocin administration studies have indicated that the nasal administration of oxytocin can enhance attentiveness to socially meaningful information such as the eye region (Auyeung et al. 2015). Thus, it is only logical to expect that increased attentiveness to social information induced by oxytocin administration can promote the language development of children with ASD. Second, it is possible that oxytocin administration facilitates human vocal learning. Animal studies have found that the neural substrates that are recruited in vocal learning and production are influenced by the oxytocinergic neural system (Theofanopoulou et al. 2017, for a review). Songbird homologue of dopaminergic circuit of the basal ganglia plays primary roles in motor control of vocal apparatus and song learning (Gale and Perkel 2010). Further, Tanaka et al. (2018) revealed that dopaminergic projection from periaqueductal grey to HVC drives social learning of birdsong from tutor birds. An emerging view suggests that the function of the dopaminergic system is modulated by oxytocin in humans (Gregory et al. 2015; Doi et al. 2017; Scheele et al. 2013). Taking all these into consideration, it is conceivable that oxytocin administration facilitates vocal development in human children through its influences on oxytocinergic and dopaminergic neural systems. However, caution should be exercised in implementing such intervention, because little is known about the efficacy and safety of oxytocin administration on young children.

## 6.5 Conclusion

Human language development is indeed a great feat. Given the conundrum of the poverty of stimulation, several researchers have claimed that an innate machinery is largely responsible for human language development. However, research findings in the field of developmental psychology have underscored the importance of social inputs from other individuals, mainly caregivers, in navigating infant language learning. This view has gained further support from the findings of behavioural and neuroscientific animal studies, which indicated the possibility that similar social mechanisms are involved in vocal and phonetic learning, when compared to human language development.

Taken together, the findings that have been reviewed in this chapter underscore the effectiveness of interventions that modify social environments in promoting language development during childhood. These interventions can take many forms. For example, they may be parent-focused interventions that improve the manner in which caregivers interact with their infants or pharmacological treatments that are directly provided to children. However, few empirical studies have examined the efficacy of such intervention strategies. Further, clinical trials have largely failed to examine the effects of pharmacological interventions on language development. The study of language development is interdisciplinary in nature. Accordingly, researchers from various scientific disciplines should be invited to make a concerted effort towards the creation of practical changes that promote the language development of children who live in adverse environments or amidst pathological conditions.

## References

Anderson RC, Freebody P (1981) Vocabulary knowledge. In: Guthrie JT (ed) Reading comprehension and education. International Reading Association, Newark, DE, pp 77–117

Auyeung B, Lombardo MV, Heinrichs M, Chakrabarti B, Sule A, Deakin JB, Bethlehem RA, Dickens L, Mooney N, Sipple JA, Thiemann P, Baron-Cohen S (2015) Oxytocin increases eye contact during a real-time, naturalistic social interaction in males with and without autism. Transl Psychiatry 5:e507. https://doi.org/10.1038/tp.2014.146

Baumwell L, Tamis-LeMonda CS, Bornstein MH (1997) Maternal verbal sensitivity and child language comprehension. Infant Behav Dev 20(2):247–258

Begus K, Gliga T, Southgate V (2014) Infants learn what they want to learn: responding to infant pointing leads to superior learning. PLoS One 9(10):e108817

Best CT, McRoberts GW, LaFleur R, Silver-Isenstadt J (1995) Divergent developmental patterns for infants' perception of two nonnative consonant contrasts. Infant Behav Dev 18(3):339–350

Bloom K (1988) Quality of adult vocalizations affects the quality of infant vocalizations. J Child Lang 15(3):469–480

Bloom K, Russell A, Wassenberg K (1987) Turn taking affects the quality of infant vocalizations. J Child Lang 14(2):211–227

Boughman JW (1998) Vocal learning by greater spear-nosed bats. Proc R Soc B Biol Sci 265 (1392):227–233

Brenowitz EA (1991) Altered perception of species-specific song by female birds after lesions of a forebrain nucleus. Science 251(4991):303–305

Brooks R, Meltzoff AN (2008) Infant gaze following and pointing predict accelerated vocabulary growth through two years of age: a longitudinal, growth curve modeling study. J Child Lang 35 (1):207–220

Buschmann A, Jooss B, Rupp A, Feldhusen F, Pietz J, Philippi H (2008) Parent based language intervention for 2-year-old children with specific expressive language delay: a randomised controlled trial. Arch Dis Child 94(2):110–116

Cartmill EA, Armstrong BF III, Gleitman LA, Goldin-Meadow S, Medina TN, Trueswell JC (2013) Quality of early parent input predicts child vocabulary 3 years later. Proc Natl Acad Sci USA 110(28):11278–11283

Chen Y, Matheson LE, Sakata JT (2016) Mechanisms underlying the social enhancement of vocal learning in songbirds. Proc Natl Acad Sci USA 113(24):6641–6646

Chen Y, Clark O, Woolley SC (2017) Courtship song preferences in female zebra finches are shaped by developmental auditory experience. Proc R Soc B Biol Sci 284(1855):pii: 20170054. https://doi.org/10.1098/rspb.2017.0054

Chomsky N (1965) Aspects of the theory of syntax. MIT Press, Cambridge, MA. isbn:0-262-53007-4

De Fossé L, Hodge SM, Makris N, Kennedy DN, Caviness VS Jr, McGrath L et al (2004) Language-association cortex asymmetry in autism and specific language impairment. Ann Neurol 56:757–766. https://doi.org/10.1002/ana.20275

Dimitrova N, Özçalışkan Ş, Adamson LB (2016) Parents' translations of child gesture facilitate word learning in children with autism, down syndrome and typical development. J Autism Dev Disord 46(1):221–231

Doi H, Shinohara K (2017) fNIRS studies on hemispheric asymmetry in atypical neural function in developmental disorders. Front Hum Neurosci 11:137. https://doi.org/10.3389/fnhum.2017.00137. eCollection 2017

Doi H, Fujisawa TX, Kanai C, Ohta H, Yokoi H, Iwanami A, Kato N, Shinohara K (2013) Recognition of facial expressions and prosodic cues with graded emotional intensities in adults with Asperger syndrome. J Autism Dev Disord 43(9):2099–2113

Doi H, Morikawa M, Inadomi N, Aikawa K, Uetani M, Shinohara K (2017) Neural correlates of babyish adult face processing in men. Neuropsychologia 97:9–17

Elsabbagh M, Mercure E, Hudry K, Chandler S, Pasco G, Charman T, Pickles A, Baron-Cohen S, Bolton P, Johnson MH, Team BASIS (2012) Infant neural sensitivity to dynamic eye gaze is associated with later emerging autism. Curr Biol 22(4):338–342

Fadiga L, Craighero L, Buccino G, Rizzolatti G (2002) Speech listening specifically modulates the excitability of tongue muscles: a TMS study. Eur J Neurosci 15(2):399–402

Farkas G, Beron K (2004) The detailed age trajectory of oral vocabulary knowledge: differences by class and race. Soc Sci Res 33(3):464–497

Fedurek P, Slocombe KE (2011) Primate vocal communication: a useful tool for understanding human speech and language evolution? Hum Biol 83(2):153–173

Fernald A, Marchman VA, Weisleder A (2013) SES differences in language processing skill and vocabulary are evident at 18 months. Dev Sci 16(2):234–248

Fitch WT (2000) The evolution of speech: a comparative review. Trends Cogn Sci 4(7):258–267

Fitch WT, Giedd J (1999) Morphology and development of the human vocal tract: a study using magnetic resonance imaging. J Acoust Soc Am 106(3 Pt 1):1511–1522

Gale SD, Perkel DJ (2010) Anatomy of a songbird basal ganglia circuit essential for vocal learning and plasticity. J Chem Neuroanat 39(2):124–131

Ganger J, Brent MR (2004) Reexamining the vocabulary spurt. Dev Psychol 40(4):621–632

Gavin NI, Gaynes BN, Lohr KN, Meltzer-Brody S, Gartlehner G, Swinson T (2005) Perinatal depression: a systematic review of prevalence and incidence. Obstet Gynecol 106 (5 Pt 1):1071–1083

Gentner TQ, Hulse SH, Bentley GE, Ball GF (2000) Individual vocal recognition and the effect of partial lesions to HVC on discrimination, learning, and categorization of conspecific song in adult songbirds. J Neurobiol 42:117–133

Gillespie-Lynch K, Elias R, Escudero P, Hutman T, Johnson SP (2013) Atypical gaze following in autism: a comparison of three potential mechanisms. J Autism Dev Disord 43(12):2779–2792

Glimcher PW (2011) Understanding dopamine and reinforcement learning: the dopamine reward prediction error hypothesis. Proc Natl Acad Sci USA 108(Suppl 3):15647–15654

Goldstein MH, Schwade JA (2008) Social feedback to infants' babbling facilitates rapid phonological learning. Psychol Sci 19(5):515–523

Goldstein MH, King AP, West MJ (2003) Social interaction shapes babbling: testing parallels between birdsong and speech. Proc Natl Acad Sci USA 100(13):8030–8035

Golinkoff RM, Can DD, Soderstrom M, Hirsh-Pasek K (2015) (Baby)talk to me: the social context of infant-directed speech and its effects on early language acquisition. Curr Dir Psychol Sci 24 (5):339–344

Gregory R, Cheng H, Rupp HA, Sengelaub DR, Heiman JR (2015) Oxytocin increases VTA activation to infant and sexual stimuli in nulliparous and postpartum women. Horm Behav 69:82–88

Gros-Louis J, West MJ, Goldstein MH, King AP (2006) Mothers provide differential feedback to infants' prelinguistic sounds. Int J Behav Dev 30(6):509–516

Hart B, Risley T (1995) Meaningful differences in the everyday experience of young American children. Brookes, Baltimore, MD, p 1995

Hillairet de Boisferon A, Tift AH, Minar NJ, Lewkowicz DJ (2018) The redeployment of attention to the mouth of a talking face during the second year of life. J Exp Child Psychol 172:189–200

Hirsh-Pasek K, Adamson LB, Bakeman R, Owen MT, Golinkoff RM, Pace A, Yust PK, Suma K (2015) The contribution of early communication quality to low-income Children's language success. Psychol Sci 26(7):1071–1083

Hoffbrand S, Howard L, Crawley H (2001) Antidepressant drug treatment for postnatal depression. Cochrane Database Syst Rev. 2001(2):CD002018

Hsu HC, Fogel A (2001) Infant vocal development in a dynamic mother-infant communication system. Infancy 2(1):87–109

Iacoboni M, Woods RP, Brass M, Bekkering H, Mazziotta JC, Rizzolatti G (1999) Cortical mechanisms of human imitation. Science 286(5449):2526–2528

Janik VM (2000) Whistle matching in wild bottlenose dolphins (*Tursiops truncatus*). Science 289 (5483):1355–1357

Kaminski J, Call J, Fischer J (2004) Word learning in a domestic dog: evidence for "fast mapping". Science 304(5677):1682–1683

Kaplan PS, Bachorowski J-A, Smoski MJ, Zinser M (2001) Role of clinical diagnosis and medication use in effects of maternal depression on infant-directed speech. Infancy 2 (4):537–548

Kaplan PS, Bachorowski JA, Smoski MJ, Hudenko WJ (2002) Infants of depressed mothers, although competent learners, fail to learn in response to their own mothers' infant-directed speech. Psychol Sci 13(3):268–271

Ko ES, Seidl A, Cristia A, Reimchen M, Soderstrom M (2016) Entrainment of prosody in the interaction of mothers with their young children. J Child Lang 43(2):284–309

Kuhl PK (2007) Is speech learning 'gated' by the social brain? Dev Sci 10(1):110–120

Kuhl PK, Meltzoff AN (1996) Infant vocalizations in response to speech: vocal imitation and developmental change. J Acoust Soc Am 100(4 Pt 1):2425–2438

Kuhl PK, Coffey-Corina S, Padden D, Dawson G (2005) Links between social and linguistic processing of speech in preschool children with autism: behavioral and electrophysiological measures. Dev Sci 8(1):F1–F12

Kuhl PK, Tsao F-M, Liu H-M (2011) Foreign-language experience in infancy: effects of short-term exposure and social interaction on phonetic learning. Proc Natl Acad Sci 100(15):9096–9101

Lai CS, Fisher SE, Hurst JA, Vargha-Khadem F, Monaco AP (2001) A forkhead-domain gene is mutated in a severe speech and language disorder. Nature 413(6855):519–523
Lenroot RK, Yeung PK (2013) Heterogeneity within autism spectrum disorders: what have we learned from neuroimaging studies? Front Hum Neurosci 7:733. https://doi.org/10.3389/fnhum.2013.00733
Levy F (2011) Mirror neurons, birdsong, and human language: a hypothesis. Front Psych 2:78. https://doi.org/10.3389/fpsyt.2011.00078
Lewkowicz DJ, Hansen-Tift AM (2012) Infants deploy selective attention to the mouth of a talking face when learning speech. Proc Natl Acad Sci USA 109(5):1431–1436
Manabe K, Dooling RJ (1997) Control of vocal production in budgerigars (*Melopsittacus undulatus*): selective reinforcement, call differentiation, and stimulus control. Behav Process 41(2):117–132
Markman EM (1990) Constraints children place on word meanings. Cogn Sci 14:57–77
Markman EM (1992) Constraints on word learning: speculations about their nature, origins and domain specificity. In: Modularity and constraints in language and cognition: the Minnesota symposium on child psychology. Psychology Press, Hillsdale, pp 59–101
Masataka N (1992) Early ontogeny of vocal behavior of Japanese infants in response to maternal speech. Child Dev 63(5):1177–1185
Masataka N, Fujita K (1989) Vocal learning of Japanese and rhesus monkeys. Behaviour 109 (3/4):191–199
McCann BS (1982) Hemispheric asymmetries and early infantile autism. J Autism Dev Disord 11 (4):401–411
Meltzoff AN, Moore MK (1977) Imitation of facial and manual gestures by human neonates. Science 198:75–78
Mooney R (2014) Auditory–vocal mirroring in songbirds. Philos Trans R Soc B 369 (1644):20130179. https://doi.org/10.1098/rstb.2013.0179
Moore C, Dunham P (1995) Joint attention: its origins and role in development. Lawrence Erlbaum Associates, Hillsdale
Morales M, Mundy P, Delgado CEF, Yale M, Messinger D, Neal R, Schwartz HK (2000) Responding to joint attention across the 6- through 24-month age period and early language acquisition. J Appl Dev Psychol 21(3):283–298
Mundy P, Fox N, Card J (2003) EEG coherence, joint attention and language development in the second year. Dev Sci 6(1):48–54
Nimtz C (2005) Reassessing referential indeterminacy. Erkenntnis 62(1):1–28
Noad MJ, Cato DH, Bryden MM, Jenner M-N, Jenner KCS (2000) Cultural revolution in whale songs. Nature 408(6812):537
Ohms VR, Gill A, Van Heijningen CA, Beckers GJ, ten Cate C (2010) Zebra finches exhibit speaker-independent phonetic perception of human speech. Proc R Soc B. Biol Sci 277 (1684):1003–1009
Okumura Y, Kanakogi Y, Kobayashi T, Itakura S (2007) Individual differences in object-processing explain the relationship between early gaze-following and later language development. Cognition 166:418–424
Oller DK, Eilers RE (1988) The role of audition in infant babbling. Child Dev 59(2):441
Onishi KH, Baillargeon R (2005) Do 15-month-old infants understand false beliefs? Science 308 (5719):255–258
Osterling JA, Dawson G, Munson JA (2002) Early recognition of 1 year-old infants with autism spectrum disorder versus mental retardation. Dev Psychopathol 14(2):239–251
Owens RE (2005) Language development: an introduction. Pearson, Boston, pp 125–136
Parker KJ, Oztan O, Libove RA, Sumiyoshi RD, Jackson LP, Karhson DS, Summers JE, Hinman KE, Motonaga KS, Phillips JM, Carson DS, Garner JP, Hardan AY (2017) Intranasal oxytocin treatment for social deficits and biomarkers of response in children with autism. Proc Natl Acad Sci USA 114(30):8119–8124
Paul R (2008) Interventions to improve communication in autism. Child Adolesc Psychiatr Clin N Am 17(4):835–856

Pereira AF, Smith LB, Yu C (2014) A bottom-up view of toddler word learning. Psychon Bull Rev 21(1):178–185
Petkov CI, Jarvis ED (2012) Birds, primates, and spoken language origins: behavioral phenotypes and neurobiological substrates. Front Evol Neurosci 4:12. https://doi.org/10.3389/fnevo.2012.00012
Prather JF, Peters S, Nowicki S, Mooney R (2008) Precise auditory–vocal mirroring in neurons for learned vocal communication. Nature 451:305–310
Prather JF, Nowicki S, Anderson RC, Peters S, Mooney R (2009) Neural correlates of categorical perception in learned vocal communication. Nat Neurosci 12(2):221–228
Pullum GK, Scholz BC (2002) Empirical assessment of stimulus poverty arguments. Linguis Rev 19(1/2):9–50
Quine W (1960) Chapter 2: Translation and meaning. In: Word and object (New ed.). MIT Press, Cambridge, pp 23–72
Rizzolatti G, Arbibb MA (1998) Language within our grasp. Trends Neurosci 21(5):188–194
Rizzolatti G, Fabbri-Destro M (2008) The mirror system and its role in social cognition. Curr Opin Neurobiol 18(2):179–184
Rizzolatti G, Fabbri-Destro M (2010) Mirror neurons: from discovery to autism. Exp Brain Res 200 (3–4):223–237
Rochat P (2001) Social contingency detection and infant development. Bull Menn Clin 65 (3):347–360
Saalasti S, Kätsyri J, Tiippana K, Laine-Hernandez M, von Wendt L, Sams M (2012) Audiovisual speech perception and eye gaze behavior of adults with asperger syndrome. J Autism Dev Disord 42(8):1606–1615
Scheele D, Wille A, Kendrick KM, Stoffel-Wagner B, Becker B, Güntürkün O, Maier W, Hurlemann R (2013) Oxytocin enhances brain reward system responses in men viewing the face of their female partner. Proc Natl Acad Sci USA 110(50):20308–20313
Senju A, Southgate V, White S, Frith U (2009) Mindblind eyes: an absence of spontaneous theory of mind in Asperger syndrome. Science 325:883–885
Sewall KB (2012) Vocal matching in animals imitating the calls of group members and mates is a reliable signal of social bonds in some animal species. Am Sci 100(4):306–315
Sewall KB, Young AM, Wright TF (2016) Social calls provide novel insights into the evolution of vocal learning. Anim Behav 120:163–172
Simion F, Regolin L, Bulf H (2008) A predisposition for biological motion in the newborn baby. Proc Natl Acad Sci USA 105(2):809–813
Singh L, Morgan JL, Best CT (2009) Infants' listening preferences: baby talk or happy talk? Infancy 3(3):365–394
Skotheim S, Braarud HC, Høie K, Markhus MW, Malde MK, Graff IE, Berle JØ, Stormark KM (2013) Subclinical levels of maternal depression and infant sensitivity to social contingency. Infant Behav Dev 36(3):419–426
Slater A (2002) Visual perception in the newborn infant: issues and debates. Intellectica 34:57–76
Smith EG, Bennetto L (2007) Audiovisual speech integration and lipreading in autism. J Child Psychol Psychiatry 48(8):813–821
Smith A, Goffman L, Stark RE (1995) Speech motor development. Semin Speech Lang 16:87–99
Snow CE, Burns S, Griffin P (1998) Preventing reading difficulties in young children. National Academy Press, Washington, DC
Southgate V, Senju A, Csibra G (2007) Action anticipation through attribution of false belief by 2-year-olds. Psychol Sci 18(7):587–592
Spelke E (2000) Core knowledge. Am Psychol 55(11):1233–1243
Spelke ES, Breinlinger K, Macomber J, Jacobson K (1992) Origins of knowledge. Psychol Rev 99 (4):605–632
Stoeger AS, Manger P (2014) Vocal learning in elephants: neural bases and adaptive context. Curr Opin Neurobiol 28:101–107

Stromswold K (2001) The heritability of language: a review and Metaanalysis of twin, adoption, and linkage studies. Language 77(4):647–723
Sugiura H (1998) Matching of acoustic features during the vocal exchange of coo calls by Japanese macaques. Anim Behav 55(3):673–687
Sulpizio S, Kuroda K, Dalsasso M, Asakawa T, Bornstein MH, Doi H, Esposito G, Shinohara K (2018) Discriminating between mothers' infant- and adult-directed speech: cross-linguistic generalizability from Japanese to Italian and German. Neurosci Res 133:21–27
Syal S, Finlay BL (2010) Thinking outside the cortex: social motivation in the evolution and development of language. Dev Sci 14:417. https://doi.org/10.1111/j.1467-7687.2010.00997.x
Takahashi DY, Fenley AR, Ghazanfar AA (2016) Early development of turn-taking with parents shapes vocal acoustics in infant marmoset monkeys. Philos Trans R Soc B 371(1693):20150370
Takahashi DY, Liao DA, Ghazanfar AA (2017) Vocal learning via social reinforcement by infant marmoset monkeys. Curr Biol 27(12):1844–1852.e6
Tanaka M, Sun F, Li Y, Mooney R (2018) A mesocortical dopamine circuit enables the cultural transmission of vocal behavior. Nature 563:117–120
Theofanopoulou C, Boeckx C, Jarvis ED (2017) A hypothesis on a role of oxytocin in the social mechanisms of speech and vocal learning. Proc R Soc B Biol Sci 284(1861):pii: 20170988. https://doi.org/10.1098/rspb.2017.0988
Tomasello M (2000) The social-pragmatic theory of word learning. Pragmatics 10(4):401–413
Tomasello M, Farrar MJ (1986) Joint attention and early language. Child Dev 57(6):1454–1463
Tyack PL (2008) Convergence of calls as animals form social bonds, active compensation for noisy communication channels, and the evolution of vocal learning in mammals. J Comp Psychol 122 (3):319–331
Vargha-Khadem F, Gadian DG, Copp A, Mishkin M (2005) FOXP2 and the neuroanatomy of speech and language. Nat Rev Neurosci 6(2):131–138
Warlaumont AS, Richards JA, Gilkerson J, Oller DK (2014) A social feedback loop for speech development and its reduction in autism. Psychol Sci 25(7):1314–1324
Watwood SL, Tyack PL, Wells RS (2004) Whistle sharing in paired male bottlenose dolphins, *Tursiops truncatus*. Behav Ecol Sociobiol 55(6):531–543
Weikum WM, Oberlander TF, Hensch TK, Werker JF (2012) Prenatal exposure to antidepressants and depressed maternal mood alter trajectory of infant speech perception. Proc Natl Acad Sci USA 109(Suppl 2):17221–17227
Weinberg MK, Olson KL, Beeghly M, Tronick EZ (2006) Making up is hard to do, especially for mothers with high levels of depressive symptoms and their infant sons. J Child Psychol Psychiatry 47(7):670–683
Weisleder A, Fernald A (2013) Talking to children matters: early language experience strengthens processing and builds vocabulary. Psychol Sci 24(11):2143–2152
Wilkinson GS, Boughman JW (1998) Social calls coordinate foraging in greater spear-nosed bats. Anim Behav 55(2):337–350
Wimmer H, Perner J (1983) Beliefs about beliefs: representation and constraining function of wrong beliefs in young children's understanding of deception. Cognition 13(1):103–128
Woodward AL, Hoyne KL (1999) Infants' learning about words and sounds in relation to objects. Child Dev 70(1):65–77
Yu C, Smith LB (2012) Embodied attention and word learning by toddlers. Cognition 125 (2):244–262

# Chapter 7
# Emergence of the Distinction Between "Verbal" and "Musical" in Early Childhood Development

**Aleksey Nikolsky**

**Abstract** The pattern of acquisition of speech- and music-related skills during early stages of human infancy provides insight into the origins of language and music. Indiscriminate until shortly after birth, babies start gradually developing acoustic features in their vocalizations, as well as accompanying behaviors that make it possible to distinguish their attempts to speak from their attempts to sing. Comparative analysis of tonal organization of children's original (nonimitative) vocalizations in their developmental succession throughout the first 3 years of life casts light on several important acoustic features. These features play an important role in the separation of music skills from verbal skills and shaping the primordial music system the infant uses to address his/her musical needs.

Much of the existing scholarship makes a fundamental error by interpreting the earliest forms of human speech and music in terms of "adult" state of their mastery, regarding children's communication as a sort of "defective" imitation of adults' models. Moreover, such models are significantly biased toward Western classical music and Indo-European languages, which despite their cultural importance in the modern world, nevertheless, constitute only a small fraction of typology of tonal musical and phonological verbal organization. A much more comprehensive approach toward children's music and speech has been developed by Lev Vygotsky and his circle: Alexander Luria, Aleksei Leontyev, Alexander Zaporozhets, Peter Galperin, as well as Boris Teplov. They and their followers regarded children's speech and music as reflecting a child's own peculiar method of thinking. The Vygotskian approach shares much in common with that of Piaget and the neo-Piagetians, but offers an alternative framework for the explanation of the dichotomy between language and music—based on the methodology of intonation theory by Boleslav Yavorsky and Boris Asafyev. This theory was implemented in the State program of obligatory education within all territories of the former USSR; it had passed deep scrutiny throughout many years of administration over a massive population, which resulted in the creation of a special discipline of ear development that theoretically and practically dealt with the development of "musical hearing" as distinguished from "verbal hearing" throughout childhood. Unfortunately, much of

A. Nikolsky (✉)
Independent Researcher, Austin, TX, USA



N. Masataka (ed.), *The Origins of Language Revisited*,
https://doi.org/10.1007/978-981-15-4250-3_7

this literature is unknown to Western developmental specialists. This chapter covers this gap, familiarizing English-speaking scholars with a unique perspective on early musical and verbal development by Soviet and modern Russian ear-training specialists, with special attention to the issue of absolute pitch.

Advances in the methodology of intonational analysis have made it possible to adequately describe and more deeply understand the principles that govern the tonal organization of non-Western types of music—including those that are based on timbre rather than pitch. This approach can be effectively applied to the analysis of both ethnological and developmental data—to identify common patterns of ontogenetic and phylogenetic development.

**Keywords** Timbre-based music · Frequency-based music · Tonal organization · Musical mode · Modality · Tonality · Vygotskian school · Absolute pitch

## 7.1 Ontogeny and Phylogeny in a Historic Perspective on Science

The idea that studying the emergence of speech and music faculties in early childhood can reveal important patterns in the emergence of language and music in prehistoric societies is not new. The supposed "biogenetic law" holding that "ontogeny recapitulates phylogeny" was formulated in the very beginning of the nineteenth century by Johann Friedrich Meckel, formalized by Etienne Serres in the 1820s (Rehbock 1990) and elaborated into a theory of recapitulation by Ernst Haeckel in the 1860s (Breidbach 2002). Although Haeckel's mistakes were revealed already in the early twentieth century, as the embryology passed its embryonic state leading to the rejection of his theory, in the 1980s, a few embryologists voiced their concern that, after all, on a fundamental level, Haeckel was correct in assuming that all vertebrates develop a similar body plan (Sander 2002). In light of the collected evidence that certain genes control the ontogenetic development over a range of organisms, perhaps "recapitulation" indeed should be downgraded in its status from a "law" to just a "rule" while still be considered at least partially valid for the earliest developmental stages (Blackwell 2007). Alternatively, this can be considered to be fundamentally "imperfect" (Arthur 2011).

The recapitulation theory was comprehensively evaluated and refuted in Gould's classic *Ontogeny and Phylogeny* (Gould 1977). And prior to the establishment of evo-devo, Gould's arguments against ontophyletics appeared overwhelming (Müller 2013). However, it was often overlooked that Gould's resurrection of the heterochrony theory exposed a mechanism for evolutionary change capable of resurrecting the recapitulation theory—the modification of development through heterochrony (Müller 2008). Gould's disproof of Haeckel's "pure recapitulation" did not discard the general idea that developmental processes had been subjected to natural selection—the modern evo-devo synthesis only made the conceptual relationship between evolution and development much more complex (Coall et al. 2015).

In fact, the diversity of factors contributing to individual development and its open-endedness might justify the reversal of Haeckel's formula into: *phylogeny recapitulates ontogeny* (Zeleny 1981). Since social systems can be maintained in no other way but by individuals, the way in which individuals perceive the surrounding world and process the perceived data determines the social conventions that govern the emergence and the collapse of cultural phenomena (Hallpike 2014, 190). In this sense, the cultural phylogeny is constructed from the statistically dominant ontogenetic patterns. The proof for this can be found in the unprecedented massive social "experiments" in "upgrading" the social organization of the rural Uzbeks and the nomadic Kyrgyz to the industrial lifestyle, undertaken by Bolsheviks in the Middle Asia (Luria 1976, 161). Once the majority of the indigenous youngsters were brought up in a Westernized schooled way, an indigenous culture that was based on irrational beliefs dating back to animistic ideology became unsustainable and gave way to a rational worldview (Nell 1999). The entire line of human cultural evolution can be envisaged as a chain of similar "experiments"—changes in upbringing of children due to important environmental changes that required invention and mastering of new means for survival. The efficacy of such innovations resulted in the replacement of one culture with another, very much like the replacement of one cognitive scheme with a more effective one as an infant learns how to handle tasks that are common in his/her environment. A crucial role here belongs to the genetic component responsible for the succession of developmental stages.

The idea that biological evolution and the child's cognitive development, which both "follow much the same progression of evolutionary stages as that suggested in the archaeological record," has earned its place in the field of cultural anthropology (Foster 1994). In developmental epistemology, this idea has found even greater support. The entire research of Piaget was inspired by the possibility of reconstructing the prehistory of human thinking based on the ontogenesis of cognitive notions throughout childhood (Piaget 1970, 13). This approach met severe criticism from Lévi-Strauss, who held such parallels to be the manifestation of a Europocentric bias, ignited by the supposed aversion of scholars who were brought up within the framework of Western academia, to the "primitivism" of childish thinking in general, as well as of adult's thinking in preindustrial societies (Lévi-Strauss 1969, 88). It should be noted, however, that unlike such assumption by Lévi-Strauss, Piaget's theory did pass extensive testing in hundreds of studies all over the world during the 1960–1980s (Cole et al. 1971; Dasen 1972; Berry and Dasen 1974; Piaget 1976; Modgil and Modgil 1976; Dasen 1977; Poortinga 1977; Eckensberger et al. 1979; Dasen and Heron 1981; Mishra 1997; Segall et al. 1999)—revealing cultural differences neither in cognitive capacity nor in cognitive processes (Dasen and de Ribaupierre 1987). Pierre Dasen, one of the experts on cross-cultural developmental psychology, sounds unequivocal in his statement that he knows of no research that contradicts Piagetian claim of universality of the succession of sensorimotor, preoperational and operational stages at the structural level, with the only restriction that the stage of formal operations is reached only in societies that cultivate formal schooling (Dasen 2012). And this restriction is by no means "hardbound"—non-Western indigenous peoples were experimentally found to match or even outperform the Western urban population in certain cognitive tasks after having been exposed to formal training (Munroe and Munro 1997). If the succession of developmental stages stays the same across the currently existing cultures, there is no reason to believe that it was less universal in cultures of the past.

There must be some kind of a feedback loop between the individual cognitive development of each member of a sociocultural formation and the typology of representation of reality adopted within that formation, which in return directs the development of each youngster via the formative influence of his/her older siblings. This ongoing feedback is essential in the sociogenetic model of cognitive development advocated by Vygotsky and his school (Rene van der Veer and Valsiner 1991). According to this model, infants' inborn capacities are nearly congruent across different cultures, diverging farther and farther apart with each successive stage of development under cultural influences. In their longitudinal experiments with hundreds of twins, Vygotsky and Luria tried to establish the age at which a child's development becomes more driven by personal experience rather than by "genetic control" (i.e., driven by culture rather than by nature) and identified the age 7–12 as being critical in this transition (313–314).

Along the same line, Ian Cross, an authority on the evolution of music, holds that "the clearest traces of the impact of evolutionary processes on the mind are evident not in encultured adult behaviors but in the capacities of the infant mind" (Cross 2001a). He points out that very young infants acquire certain competences too rapidly to be explained by learning alone: they seem to innately know which sounds constitute speech, which—music, and which—non-conspecific sounds (e.g., bird's warbling or dog's barking). Such "priming" of infants for language and music comprises the influence of evolution in shaping the infant's predispositions, while culture shapes the expression of these predispositions into specific and distinct forms. Cross stresses that the need to process certain types of information rapidly and expertly without being taught to do so determines *the power of phylogeny over ontogeny in language and music communication*. Retaining the juvenile characteristics of our ancestors is part of that delicate biological balance between the capacity to effectively respond to events that are most common in one's immediate environment and critical for survival, on the one hand, and the ability to optimize or even create anew a necessary skill, on the other hand. Recapitulation and neoteny, respectively, support these hardwired yet flexible adaptive responses. And the former receives much greater importance at the very beginning of one's life, when the person is the most vulnerable and dependent on caretaker's help.

## 7.2 Toward Structural Distinction Between Music and Language

Language and music constitute the primary means for human infants to maintain effective communication with their caretakers, with whose help infants can acquire a set of skills optimal for their natural and cultural environment. Much has been written on the convoluted distinction between language and music, as well as their uneasy relation to animal communication. In short, music can be opposed to language across the pragmatics of communication: music *focuses* on "affective meaning," whereas language only *accounts* for it (Gussenhoven 2002). Of course, this distinction is not always clear-cut. Music can be used for conveying a referential meaning, as in military signals to gather or to go to bed. Speech, on the other hand, especially colloquial, is often used to express the affective state of a speaker.

For the lack of clear semiotic distinctions, scholars often prefer to rely on structural features to delineate between music and language. The most common is to distinguish music from speech by defining music as the arrangement of musical sounds in discrete units of pitch and rhythm—in contrast to continuous and somewhat arbitrary modulations in pitch levels and duration for the sounds of speech (Reybrouck and Podlipniak 2019). However, such view is nothing but the manifestation of the Europocentric bias of researchers who impose rules of such music system that is native to them on all other forms of music, disregarding their actual organization. For absolute majority of Western scholars, this is the framework of Western **tonality**—the system of *hierarchic subordination* of musical tones to one dominant pitch ("tonic") whose permanence enables fluctuations in harmonic tension (instability) and relaxation (stability) in the flow of music (Krumhansl 1990). But this is not the only way to arrange musical tones. Not any less common is the framework of **modality**—the system of *heterarchic coordination* of pitches based on permanence of patterns of their melodic relation ("musical mode") (Nikolsky 2015a), where tension and relaxation (Tsougras 2010) are generated not by referencing musical tones to the "tonic" but by altering the degrees of the musical mode (stability/instability of the entire mode) (Nikolsky 2016, Appendix-3). Yet, a third type of music systems is based on arranging musical tones in **timbre** rather than pitch (Nikolsky et al. 2019). In this kind of music, the changes in tension seem to originate mainly from contrasts between harmonicity and inharmonicity of sounds, as well as changes in their roughness[1] and a few other spectral parameters, such as spectral spread and spectral roll-off (Nikolsky 2017).[2]

[1]At present, research of which exact parameters of timbre are responsible for the perception of tension and relaxation in music remains in virginal state (Granot and Eitan 2011). The salience of specific harmonics can be perceived as increase or decrease in tension of sounds (Nazaikinsky and Rags 1964). Acoustic roughness is shown to have direct relation to the perceived timbral tension (Pressnitzer et al. 2000). The impression of timbral tension might originate in the conflicting interaction between the frequency- and time-based parameters of timbre (Volodin 1972). The dialectics of timbral tension and relaxation was laid into the foundation of the theory of phonism, elaborated by Nazaikinsky from the notion of phonism proposed by Tiulin for explanation of modernistic compositional styles of classical music (Tiulin 1937). According to Nazaikinsky, timbral tension constitutes a special form of harmonic interrelations between "dissonant" simultaneous combinations of musical tones, different from tonal and modal stability/instability by the peculiar integration of the holistic inseparability of a timbre with distinctions between different frequency components of a spectrum (Nazaikinsky 1988).

[2]There is some experimental evidence that listeners perceive the inflections of musical tension and relaxation due to timbral changes in music (Paraskeva and McAdams 1997). Granot and Eitan

In reality, by far, not every existing form of music is based on "pitch classes" and "rhythmic values." In fact, some ethnicities do not make music based on discretization of pitch at all. For instance, in Tuva, indigenous musicians perceive music in terms of changing the timbral colors rather than pitches, which is set in early childhood when children entertain themselves not by combining different pitch levels (as Western children usually do) but by reproducing natural sounds from their environment and creatively modifying and recombining them (Levin and Suzukei 2006). Many ethnicities, such as Inuits or Australian Aborigines do not even have words in their languages to distinguish between "melody" and "vocal timbre," which suggests that music makers in these cultures think musically in terms of timbre rather than frequency (Walker 1997). Evidently, such music is constructed not of discrete "pitch classes" but of "timbre classes" (Nikolsky et al. 2020). However, those scholars who consider discrete pitch to be the structural criterion of music qualify any music that is based on alternative principles as being nonmusic (Brown 2017)—even if such qualification goes against the conviction of its users. Sometimes, musicological conventions side with musicians' conventions in defining some difficult cases of what is considered to be music: e.g., rap songs (Adams 2015).[3] More consistent appears to be Patel's approach (Patel 2010b) in recognizing

---

uncovered the evidence for the combination of loudness and register as being capable of inducing the sensation of changes in tension (Granot and Eitan 2011). Clarke also maintains that timbral and dynamic aspects partake in the generation of the impression of musical movement (E. F. Clarke 2001). Volodin established the presence of the parameter of tension in listeners' perception of timbral differences in comparison of isolated tones (Volodin 1972). Mazepus identified the systemic use of timbral tension in many forms of indigenous music across Siberia—which he traced to the sensation of tension in the vocal folds and vocal articulations impeded by that tension (Mazepus 2009). Lerdahl recognized that timbre can be organized hierarchically into multidimensional timbral arrays, which would involve not only coordinational but also subordinational relations, thereby setting expectancies of resolutions in the manner of dissonance-consonance (Lerdahl 1987). An example of such hierarchical relationship between the timbral reductions of differing dimensionalities was demonstrated in the analysis of a traditional *shakuhachi* melody (Bolger and Griffith 2005). The gradations of tension in timbral modulations constitute the primary form of musical expression in deep throat singing of numerous traditions of Ural, Siberia, Mongolia, Far East, Kamchatka, Chukotka, Alaska, Canada, and Greenland (Sheikin 1996). This is also the only source of musical syntax in authentic traditions of Jew's harp music and musical bow across the world. For Jew's harps, the presence of syntactic/semantic organization by grouping of specific syllables in combination with a limited number of special effects is simply indisputable—most obvious in thorough method-books on Jew's harp playing, such as by Robert Zagretdinov (Zagretdinov 1997).

[3]The same problem occurs even within the decidedly "frequency-oriented" music systems, including Western classical music. Thus, *recitative semplice* in French and English operas could be scored or spoken: e.g., Bizet and Balfe have it written out as notes in some operas, and not notated at all in some others. In Russian Imperial opera productions, recitative was always sung—even in those operas such as Italian or French that originally used *recitative secco*, as a rule performed parlando. Operas by Russian composers rarely, if ever, use parlando in recitatives. So, according to the approach by such scholars as Brown, Reybrouck, and Podlipniak, such recitative would constitute nonmusic inserted into music (opera), which however appears inconsistent, since the very same set of sounds is pronounced "music" in French production of a particular opera and "nonmusic" in its Russian production. Such metamorphosis is hardly acceptable for a scientific study of music.

similar intermediate cases of music/language usage (e.g., recitative) as music—provided that such uses are designed to invoke emotional response in the listener.

This problem reveals the necessity to revise the definition of music and to base it not on purely structural but on functional semiotic criteria—one that reflects on the convention of the kind of sound humans conceive and perceive to be music. Such approach would require a scholar to evaluate the **tonal organization** (**TO**) of the sounds in question in order to establish whether or not they constitute music. The notion of tonal organization was conceptualized by François-Joseph Fétis in 1840 as *the method of joining musical tones together according to the sensibility of music-users* (Fétis 1994, XXV). In this sense, TO encapsulates tonality, modality, and "timbral music." TO can also be found in verbal speech, where it takes the form of "tonemes" (Pike 1948). Musical TO can be distinguished from tonemes by its complex multifactorial functional relations between musical tones that involve their arrangement according to simultaneous estimation of changes in rhythm, meter, pitch/register, dynamics, and articulation style (e.g., staccato/legato). These changes are usually categorized and rasterized into some kind of "classes" (most commonly, pitch or timbre) throughout the production or perception of music, so that every new sound is comparatively related to the previous sound in order to group them into a meaningful entity (Walker 1997, 322–3). Parsing of a music flow into such groups and gradient-based estimation of their relative tension is as crucial for music as the identification of syllables, and words based on phonemic contrasts are crucial for language.

Pragmatically, music can be defined as *such tonal organization that* ***entrains*** *listeners and performers and* ***transposes*** *performers' intentions in order to* ***emotionally*** *stir the listeners by vocal and/or instrumental performance*. This definition, elaborated after Cross (2001a, b), emphasizes three criteria that are imperative for music:

1. Entraining of psychophysiological reactions of listeners/performers to acoustic changes in the music flow secures the effective encoding and decoding of the information conveyed by specific idiomatic patterns of music—e.g., succession of a *fanfare*-like motif (a melodic motion whose sound outlines a major triad) and a *lamento*-like intonation (a descending step that is initiated on a strong beat) (Monelle 1991) would generate the synchronous change from happy to sad in the listeners (even if the performance is solitary).
2. Emotional contagion between the affective state assigned to a particular musical idiom and the actual affective state of the listener—e.g., detecting a fanfare motif in music causes the listener to experience joy, happiness, or some other kind of uplifted emotional state (ideally, this includes both the performer and the listener but can include the emotional change for the performer alone, if he/she happens to perform a fanfare motif while being in a sad state).
3. Transposability of intention to convey specific information secures the semantic integrity of a conveyed message over numerous instances of the use and reuse of exactly the same idiomatic pattern—e.g., whenever the musician makes the melody move by the sounds of a major triad, he/she intends to express the

same affective state of happiness (disregarding the differences in the circumstances of the performance).

None of these three criteria is found in the sounds of speech. Entrainment exclusively characterizes the perception of music as part of the social bonding effect typically involved in the interpersonal synchrony between performers and listeners even in situations of passive listening (Tarr et al. 2014). In speech, only some isolated utterances (e.g., laughter) are capable of triggering this effect. The specialization of music on the emotional contagion is most evident in the phenomenon of chills and chill-like reactions to music intended to express strong emotions, which is unique to music (Altenmüller et al. 2013). Transposability of intention is most obvious in music genres: e.g., most lullabies, across different cultures, engage the same or exceedingly similar set of structural features (Trehub et al. 1993). It seems that this similarity has to do with the natural sonic preferences of infants who generally prefer motherly singing to talking and have greater calming reaction to it (Nakata and Trehub 2004). In situations when mothers want their infants to rest, they most likely intuitively select the prosodic features that secure the strongest calming effect—which in many different cultures happens to coincide (Fernald 1992).

In terms of structure, TO of music usually can be identified by examining the pitch contour (Deutsch 2013b), rhythmic patterning, metric grid (Large 2008), and dynamic contour (Graves et al. 2013) which all interact (Granot and Eitan 2011) and contribute to the grouping of successive tones (Deutsch 2013b) into melodic motifs and phrases. The acoustic notion of "musicality" implies the complementarity of the pitch, rhythmo-metric and dynamic grouping, implemented in a systemic ongoing manner: unlike prosody of a tone language, musical prosody requires that every successive sound be evaluated in its pitch and dynamics, rhythmic value, and its position in the metric grid in relation to the previous sound. *It is this combined and continuous comparison of sounds in regard to their pitch, rhythm, meter, and dynamics that structurally distinguishes music from speech*. Any of these parameters taken in isolation does not suffice to distinguish music: a rap song or a dance performed on percussive instruments can have no pitches, a recitative or a religious chant might have no clear metric and rhythmic patterns, and a whisper song might have no dynamic changes. On the other hand, speech in tone languages uses pitch contrasts, poetic speech uses rhythmic values and metric feet, and sentence prosody often uses dynamics to distinguish one sentence type from another. Poetic speech comes exceedingly close to music, exposing the common roots of music and language, making the reliance on referential meaning over the entirety of speech the most salient distinction between verbal and musical prosodies (Lerdahl 2001). Just as much as the TO of speech prioritizes the intelligibility of words and their constituent syllables in order to support the retrieval of lexic meaning, the TO of music prioritizes the combined intelligibility of pitch, rhythmic, and dynamic groups within some metric grid (rigid or flexible) in order to support the intended emotional impact.

## 7.3 Tonal Organization (TO) of Music in Infancy

Music and language are considered structurally indistinguishable from each other in the earliest stages of verbal and musical development in infancy (Chen-Hafteck 1997). The reason for that seems to be biological. The same neural process of audio-vocal learning governs the acquisition of singing and speaking skills based on the common foundation of instinctive emotional communication—the same is true for both ontogenetic and phylogenetic developments (Gruhn 2009).

Semantically, reliance of preverbal infants on the emotional cues, found in prosodic features of infant-directed vocalizations of their caretakers, points in the direction of **music** as the closest prototype for infant's communication (Malloch and Trevarthen 2009). Animal calls, generally, also stay closer to human music than to human language, since animal vocal communication tends to specialize in displaying the information about the caller's affective state (Fitch 2006). Human infants start on the same page as primate cubs—by producing reflex-like utterances specific to a particular kind of stimulus (e.g., shrieking for pain) (Jürgens 1995). This semiotic similarity is accompanied by the anatomic similarity of the supralaryngeal vocal tract of infants and of nonhuman primates (Lieberman 1985). Animals express their affective states volitionally and do actually "feel" the perceived signal as long as it remains based on their innate "vocabulary" (Fitch 2010, 179–81). But if for animals such signaling remains hardwired disregarding their age and experience (Hauser 1996, 315), children quickly learn to modify the characteristic timbre and pitch contour of their cries according to what they need from their caretakers during the first year of their life (Wermke and Mende 2009). This opens doors to a long-lasting process of learning during which a prelinguistic infant builds and optimizes a repertory of calls and infers the rules of their connection in its own vocalizations as well as in response to the adult's vocalizations and behaviors in an attempt to keep them informed about his/her state. This stage is likely to be universal in human development (Wermke et al. 2007).

Not any less universal must be the *call/cry stage* in cultural evolution. One of the leading specialists in perception of music Klaus Scherer believes that the development of emotion in ontogeny mirrors that of the phylogeny, so that in both cases the new emotions emerge following the same pattern, are changed by age, and are determined by the degree of cognitive capacity for appraisal of events (Scherer 2013). This is important, because emotional communication lies at the heart of music semiosis (Nikolsky 2015b). Prehistoric musilanguage that preceded music and language (Brown 2000) most probably at first generated the conventions for affective music-like prosody and thereafter the intonational prosody, similar to language (Brown 2017). This must have occurred in response to the rising need to transfer and accumulate technological knowledge, triggered by the population growth and the growing demand in sustenance (Richerson et al. 2009). The emergence of language answered the necessity of coordinating the collective efforts between the participants in complex tasks and of passing the technological information to the next generation (Johansson 2015). Therefore, the investigation of the earliest stages in musical development can cast light on the evolutionary origin of music and language (Trainor and Hannon 2013). Especially the onset of musical development prior to birth and immediately after it provides insight into human prehistory (Parncutt 2016).

Phylogenetically, in all likelihood, music emerged prior to language, since the acoustic shape of primate vocalizations is mainly determined by music-like features which provide the basis for ontogenetic acquisition of verbal skills for humans (Koelsch 2009). An interesting argument for this was delivered by Vygotsky (2001, 58–61).[4] Starting from 1929, together with Luria, he ran the longitudinal study of speech and musical development in 150 pairs of mono- and dizygoric twins of 6–14 years of age with the purpose to estimate the impact of genetic versus environmental factors on different psychological functions (Luria 1962). Dizygotic twins were found to have the coefficient of 0.67 of musical abilities and of 0.89 of speech abilities versus, respectively, 0.93 and 0.96 for monozygotic twins (p. 59). This was interpreted as primarily the manifestation of their genetic differences, since the uteral conditions of all twins were identical and their early childhood experience remained very similar. Vygotsky and Luria decided that the greater similarity between the ratings of the monozygotic twins as opposed to dizygotic twins in respect to a particular function, the more that function was determined by heredity (p.60). Since the difference of the music coefficients was almost four times greater than the speech coefficients, Vygotsky was confident that music evolved earlier than speech.

It is regrettable that for the lack of methodology that could adequately capture the most important aspects of TO and support the comparative analysis of various vocalizations, the study of animal communication and of early human vocalizations so far has not delivered as much as they could potentially have done so. Both acoustic analysis of animal calls and human infants' vocalizations suffer from the same problem as takes of Western cognitive scientists on comparative ethnomusicology—they heavily lean toward the tonality model, thereby introducing a Western bias.

## 7.4 Pitch Reductionism and Its Detrimental Effects on the Study of Pretonal Forms of Music

The problem is that neither infants nor animals can be interviewed in order to supply scholars with information about what exactly they communicate and how they do it. At the same time, the acoustic features of their communication provide the most

[4]Unfortunately, this opus magnum, "Lectures on Pedology," is still unavailable in English. Its first part (seven lectures) was published in 1935 by the State Institute named after Gertzen in Leningrad, whereas its second part was stenographed by Serapion Korotayev, Vygotsky's assistant in the Institute, upon listening to these lectures. Korotayev prepared the materials for publication during his work at the Udmurt State University, while Vygotsky's legacy was banned by the Communist authorities. Korotayev's daughter completed the edition after Korotayev's death. Only the fourth lecture has been translated in English (Vygotsky 1994).

direct and valuable data for statistical analysis that could potentially establish the typology of the principal patterns in communication. Of course, researchers did develop techniques for conducting experimental studies of animal's and human infants' capacities to process auditory signals; however, such techniques remain quite limited in gathering information that could possibly enable comparative analysis of different types and typological variations of signals to the extent of reliably inferring the repertories of signals, their semantic values, and the rules of stitching signals together. As a result, absolute majority of studies that try to identify, categorize, and interpret such signals resort to the framework of the most verifiable model of TO—that of Western tonality. They typically estimate the auditory signals in terms of discrete pitch classes, pitch class sets (usually theorized as scales), arithmetically proportional rhythmic values, pitch intervals expressible in simple arithmetic ratios, and harmonic concepts of stability and instability that make sense only within the tonality framework. Application of all of these notions in the analysis of TO, even within the *modal* framework of Western "frequency-oriented" music (e.g., the so-called "Mixolydian" major typical for folk Slavic traditions), is prone to skew the results of the analysis (Pashina et al. 2005).[5] A convincing case of such "overinterpretation" of chromatic alterations in folk modal music demonstrate Ambrazevičius and Wiśniewska (2008). Even greater discrepancies are to be expected in the application of tonality-based analysis onto the music originally conceived as "indefinite in pitch" (Alekseyev 1976).[6]

[5]Von Hornbostel provides an example of such musicological misreading of the notated composition for shen composed within the traditional Chinese anhemitonic pentatonic system, giving the impression of being written in "E-flat major" (Hornbostel 1919). In general, modal music can be easily misinterpreted by false identification of "tonic." The same set of pitch classes G-A-B-C-D-E can be interpreted as a hexatonic G Major or A minor, which can be figured out only upon examining the melodic functions of these pitch classes in a particular piece of music (Hornbostel 1913). This goes to illustrate the absolute necessity of distinguishing between a "scale" and a "mode." Unfortunately, not all researchers of music make such distinction: e.g., Steven Brown and Richard Parncutt in their publications hold that "scale" and "mode" are synonymous. Such misunderstanding of music theory often leads to false generalizations. Thus, Doğantan-Dack assumes that resolution of unstable degrees of a pitch set into a stable degree at the end of music constitutes a universal rule in evolution of music (Doğantan-Dack 2013). In reality, a piece of modal music can finish on any degree of the mode, or even on a tone of "indefinite pitch," whose glissando presents a contrast to the usage of "definite pitch" in the rest of the music and therefore performs its function of marking the end of the music (Kholopov 2005).

[6]Alekseyev describes a situation common for traditional Yakut music explaining that when a singer is asked to repeat his/her song, they reproduce only the melodic contour and the rhythm—the exact intervals between the adjacent tones of the same tune change (1976, 148). If asked about the pitch differences, performers usually become surprised and deny any difference, reaffirming that the music is exactly "the same." At the same time, if tested in the manner of ear-training tasks, such singers have no trouble reproducing the pitch of an isolated tone perfectly in tune. When the singer is asked to sing the same song higher, he compresses its intervals to a smaller compass (Alekseyev 2013). Evidently, Yakut traditional music is based not on "pitch classes," but on "pitch contours." Similar to Yakut isomorphism was reported by List (1987) in relation to 11 performances of supposedly the same traditional lullaby *Black Bug* by Hopi Indians.

The extent of such distortions can be estimated by relating the findings of such tonality-based studies to the information gathered from indigenous adult users of "timbre-based music." Thus, Nenets rasping songs appear as "out of tune" and too noisy to Western listeners whose attempts to faithfully reproduce such music are not recognized as acceptable by Nenets—the latter of whom are totally unfamiliar with the notion of "wrong pitch" and instead focus primarily on the spectral features of sound, lyrics, and the shape of the melodic contour (Ojamaa and Ross 2011). Therefore, Nenets listeners find Western reproduction of rasping as timbrally "out of tune," while simply ignoring the exact pitch values of sounds of rasping (Ojamaa 2005). According to them, any pitch produced by a performer is necessarily "right" for his expression.

Yet another way of uncovering misattribution of TO is by comparing renditions of the same tune by different performers who belong to different music cultures that subscribe to different music systems. This is a rare musicological topic, but Eduard Alekseyev makes a convincing demonstration of how a Russian/Ukrainian popular song *Provody*, created within the framework of tonality (minor key), becomes anhemitonic pentatonic in a Buryat rendition and whole-tone tetratonic in a Yakut rendition—in both cases almost completely substituting the original pitch intervals with new ones (Alekseyev 1986, 148–55). Many Western cognitive scientists and even musicologists are likely to qualify such transformations as "mistakes" on part of indigenous performers and put them on the account of their supposedly deficient musical ear rather than to recognize them as the deliberate result of a systemic "conversion" of a tune from one music system that appears foreign and strange to a performer to another music system that is native and comfortable.

Albrecht Schneider (2001) describes how common such "overinterpretation" is in the acoustic analysis of tuning standards in African xylophone and East Asian gong/bell indigenous traditions which rely on timbre-oriented comparison of instrumental tones. Hence, Schneider qualifies the expression of tuning standards in terms of frequency as "pitch reductionism." Avoiding such reductionism is an imperative condition for finding out the real TO intended by the indigenous music makers and a great challenge for Western researchers (Schneider 2013). There is a rather convincing empirical evidence that the practice of tuning by pitch reference originates from earlier practices of tuning by various perceptual attributes of timbre.[7] A glimpse of timbre-oriented tuning is still observable in the gamelan tradition—formative for the entire region of Eastern Asia from the antiquity times onward (Blench 2007) and influential even for the Western classical music (Wen-Chung 1971).

[7]Timbral contribution to tuning is significant. As William Sethares demonstrates in the chapter "The Octave Is Dead" in his treatise *Tuning, Timbre, Spectrum, Scale*: "Introducing a dissonant octave—almost any interval can be made consonant or dissonant by proper choice of timbre" (Sethares 2005). And, indeed, in practice of gamelan tuning, tones about an octave apart are often deliberately mismatched to produce a special timbral effect (Hood 1971). Timbral contribution to tuning is very "real"—and too bad that it is less measurable than fundamental frequency. However, this difficulty should not be reason for dismissing it altogether. Good science explains the unknown and does not dismiss it.

The devil is in the tricky relationship between timbre and pitch in discretization of the spectral content. Discrimination of intervals is different for pure tones and for complex timbres (Vos 1988). In real-life conditions, a pitch often sounds "dirty," such as in auditioning of intervals on gongs, bells, cymbals, or xylophones. In such cases, ambiguity of spectral content affords multiple approaches to categorization of pitches, where harmonicity/inharmonicity and periodicity/non-periodicity (the prime acoustic sources for pitch evaluation) constitute only one of the possible axes of evaluation. Thus, gamelan performers estimate tuning of the intervals not in terms of frequency, but what they describe as "comfort," apparently referring to the interplay between the harmonic and inharmonic spectra that makes one tone perceptually stand out in relation to another (Li 2006). However, recently, gamelan tuners have started to follow the pitch standard set by their government's radio station and cater not to the preferences of instrumental players but of singers, which directed the tuning toward frequency evaluation (Vetter 1989). This development very likely reflects the cultural influence of the Western classical and popular musics—indigenous Balinese gamelan tradition does not incorporate vocal performance (Kartomi et al. 2008).

Similar situation is described by Gerhardt Kubik in regard to African traditions. Indigenous cultures rely on "deep listening" into the timbre of sound objects and selecting one of its attributes as a reference for tuning musical instruments (Kubik 1985)—in contrast to cultures influenced by Western music where tuning is done according to the reproduction of a specific harmony comprised by fixed frequency values, such as the chord generated by striking the open guitar strings (Kubik 1979).

The ubiquity of pitch reductionism in the modern globalized world is illustrated by the phenomenon of "musicalization"—intuitive assigning of pitch values to environmental sounds or signals that clearly have no relation to music (e.g., the squeaking of car breaks or an emergency alarm)—which is wide spread in societies that cultivate frequency-oriented music. The reason for this must be the difference between "musical" and "verbal" modes of listening: if speech perception relies on componential dynamic analysis of sound spectra (detecting the lowest formants), musical perception [for frequency-based music] relies on holistic analysis of periodicity, called forth to identify all natural series synchronously present in the entirety of spectrum (Walker 1997). Perception of timbral music comes closer to perception of speech in its focus on specific salient spectral components, breaking the entire spectrum into "foreground" objects to be accounted for and "background" objects to be ignored and thereafter tracking the contrasts between the "foreground" components in the flow of sounds (Nikolsky et al. 2019). The perception of timbre is rooted in environmental soundstage analysis based on the clues for sound-source identification and consistent delineation between an acoustic signal and the percept it provokes within the framework of structurally unclear and continuously changing tonal attributes of a sound (Fales 2002). The principal difference of this mode of listening from listening to speech is the *abstract, purely symbolic connection of the constituent verbal phonemes to lexic meaning as opposed to iconic or indexical connection of timbre classes of timbral music to their semantic values* (e.g., onomatopoeic-like sounds are usually iconic, while non-onomatopoeic indexical).

- Orientation of music toward frequency tends to correspond to the urban industrial environment that features lots of parallel and perpendicular lines, whereas orientation of music toward timbre produces indefinite-in-pitch intervals and degrees which correspond to life in open spaces like tundra or step where scarcity of landmarks promotes orientation not by increments between the visible objects within the three-dimensional frame of reference, but by reliance on natural phenomena like the direction of wind, sunlight, moonlight, the disposition of stars, relative time of travel, etc. (Nikolsky 2016, Appendix-7).

In effect, exposure to Western culture establishes the preference for frequency as the medium of TO, and its spatial representation in terms of vertical height, as confirmed by a study of pitch perception by urban Canadian, Indian, and Intuit children (Walker 1987). English musicians represent pitch/time as vertical/horizontal axes; Japanese musicians raised on Western music share the same representation, while Japanese musicians raised on Japanese traditional music encode pitch/time as horizontal/vertical axes; and Papua musicians are not receptive to any of these axes, but instead, they estimate the melody in terms of hue and loudness (Athanasopoulos and Moran 2013), which appears to constitute the timbral mode of music hearing. All in all, Western cultural influence corresponds to the adoption of standard of mapping vertical dimension to incremental differences in pitch (Ashley 2004). Children raised in a Western urban culture normally perceive pitch and rhythm in mathematical terms as "how much," "how many," and "what ratio" by the age of 11 years (Bamberger and DiSessa 2003). It is very possible that regular singing and listening to tonal music contribute to the establishment of a quantitative attitude toward the world. The way children are taught and learn music is shown to determine the developmental route and structure of their mental representations (Gruhn 2004).

The research conducted in the Soviet Union by the Vygotskian school led to the consensus that the emergence of frequency reference in early childhood reflects the child's experience in three-dimensional orientation and the projection of spatial characteristics onto the arrangement of musical sounds. Alexander Zaporozhets, pupil and colleague of Vygotsky, was the director of the Institute of the Preschool Education at the USSR Academy of Sciences and the head of the research lab which for a few decades conducted experiments on pitch perception in infancy. A number of studies, considered "classic" in Russia, were carried under his direction (Endovitskaya 1959; Endovitskaya 1963; Repina 1964; Lisina 1966; Mukhina and Lisina 1966; Repina 1966a; Repina 1966b). In sum, their findings demonstrated that frequency-oriented listening represented the acquisition of conventional sensory standards used to investigate perceived objects by means of materialistic models of space-like representation of reality in the auditory domain (Zaporozhets 2003, 26–7).

Another Vygotskian, Petr Galperin, regarded "objectivization" of sensory experience, "materialized" through various spatial properties and relations of objects as *the principal path for formation and mastering of cognitive operations by means of symbolic representation of reality* (Galperin 1999, 253–314). Here, music acted as an aid in abstracting the objective properties of the surrounding world—musical intervals, changes in the direction of the melody, the number of the simultaneously sounding pitches become the manifestations of the ways in which real things can be moved and how they relate to one another. And the main tool of such representation is the notion of the "sound object," constructed by the analogy to physical objects and supported by the rule of object permanence (Nazaikinsky 1973). As a result, *object permanence* becomes the attribute of the vertical axis of musical texture (the number of simultaneously sounding tones), while *space permanence* becomes the attribute of the axis of depth (the number of voices/parts used in music) (Nazaikinsky 1982, 82).

The iconic primordial nature of "frequency music" (in contrast to the symbolic nature of speech) manifests itself in the capacity of "sound objects" (perceived as pitches) to generate actions (perceived as melodic intervals that can leap or step) and form events (perceived as changes in musical movement, e.g., interruption, climax, and collapse)—similar to physical objects of the real world. A number of experimental studies conducted by Zaporozhets and his associates have demonstrated that children's realization of cross-modal correspondences between melodic tones and spatial arrangement of physical objects, or proportions in size between various creatures, or characteristics of a particular form of locomotion—all such cross-references have the power to accelerate the acquisition of fine-frequency categorization in childhood. Recent studies on early frequency discrimination that claim the opposite that neonates from birth attain the skill levels comparable to adults (Stefanics et al. 2009) should be taken cautiously and be thoroughly examined in their design. There has been a trend of "generous interpretation of the infant findings" among researchers who are looking for the earliest onset of various cognitive abilities (Trehub 2013). Trehub provides an example of such study that concludes that newborns are capable of detecting traits of TO related to musical keys just hours after birth (Perani et al. 2010). Such interpretation contradicts the known timeline of acquisition of harmonic skills, according to which children develop sensitivity to key structure not until 4–5 years of age (Corrigall and Trainor 2010).

The propensity of some researchers to overinterpret their findings and/or design their experiments with the purpose to discover the presence of patterns of TO that characterize adult music in vocalizations of young infants is a form of "pitch reductionism," based on the assumption that children somehow already possess germs of adult skills and therefore follow the same methods of encoding and decoding musical information as adults. For Western developmental psychologist, it is even more enticing than for a Western[8] musicologist to resort to this circular

[8]Here and earlier, I have emphasized that Western researchers are more prone than Eastern European and Asiatic researchers to commit to circular reasoning and discover traces of tonality in music that was created on timbral or modal rather than tonal principles. This is because in majority of Eastern European and Asiatic countries there are strong indigenous traditions of modal music which can significantly depart from the heptatonic diatonic Western stereotype by using pentatonic, hexatonic, hemiolic, chromatic, micro-chromatic, non-octave and/or symmetric modes (Nikolsky 2016, Appendix-8). Researchers familiar with these and timbral forms of music (especially when they are native to researchers) are likely to be more open-minded in evalutation of infants' musicking attempts than their "orthodox" Western colleagues.

logic. After all, children do follow a genetic script in their maturation that brings them up to attain a set of skills on par with adults. This can justify the evaluation of a melody performed by a young child in terms of its compliance to the Western tonality and attribute all deviations from "properly" tuned degrees of Western keys as children's "mistakes" due to immaturity of their technical skills. And majority of the studies dedicated to the analysis of children-made music indeed approach it this way.

Thus, one of the leading specialists on early children music, Lyle Davidson (1994), whose "contour theory" has become the mainstream theory of TO in early childhood, describes the earliest stage of children music-making during the first 3 years of life as revealing the ability to sing distinct pitches but lacking interval stability and tonal coherence between multiple musical phrases. Such description implies that infants are aware of pitch classes, pitch class sets, and tonicity, as well as the typology of pitch intervals, and only have technical problems in producing music in full accordance with them. However, there is no proof that it is common for 0–3-year-old infants to process music in terms of pitch classes, pitch class sets, interval classes, and interval class sets in reference to tonic. In fact, there is abundant evidence to the contrary: absolute majority of children of that age cannot identify different harmonic and melodic intervals upon hearing them, cannot keep track of "tonic" throughout a musical piece, and cannot reliably distinguish different major and minor keys—any music instructor who has had experience working with this age category in nurseries and preschools would testify to this. The best testimony to this is the State program of early musical education in Russia, which specifies the goals of music activities for children of different ages and includes none of the above-listed music skills for 3-year-olds (Kostina 2004).

The Soviet Union dedicated an unprecedented amount of resources (specialists, institutions, programs, etc.) to forge an effective system of musical education from the most basic to the most advanced levels and make this education available to the widest population. In 1921–1925, during his directorship of Glavprofobr (Chief Office of Professional Education) at Narkompros (People's Commissariat of Enlightenment), Boleslav Yavorsky included ear training in the State program of education (Fedorovich 2014, 89). Ever since ear-training programs have been systemically developed, administered to general population, and perfected by qualified specialists (Blium 1977). A full scope of professional musical ear development in Russia takes 14 years, with a wealth of standardized tests and exercises (Karasyova 2010). Not any less thorough and comprehensive is the methodological framework for music education in preschools and nurseries (Anisimov 2004). In all the countries of the former Soviet Union, after the Second World War, music preschool education was administered on a massive scale unparalleled by any Western country. Early music education was ideologically esteemed as the bastion in the upbringing of creative and socially adaptive members of society, reflected by the introduction of a new position of a "head of music" in the kindergarten personnel by the provision of the Council of Ministers of the USSR in November 1, 1947 (Kiilu 2011). The chief responsibility of this "head of music" was to carry out music education for the general population in strict accordance with the program adopted by the Ministry of Education. This program was created and improved by the mutual efforts of many specialists in music education across the country. According to this program, before the age of 3, teaching was limited mostly to listening to models and exercises involving unrestricted changing of voice in pitch, and at 3, children for the first time were given specially composed songs based on 2–4 tones within the range of a fourth with the syllables imitating animal calls and contrasting melodic shapes (ascending/descending) and interval sizes (steps/leaps) (Radynova et al. 1994, 101–104).

Nikolai Metlov, the pioneer and one of the main proponents of obligatory music education, during the 1920s, laid out a program of singing, rhythmic motion, performance on basic musical instruments, and music appreciation course (Metlov and Mikhailova 1935). It laid the foundation for the 1934 Program of Musical Development by Narkompros that featured the “expanding” model of gradual introduction of, at first, melodic and then harmonic intervals, starting from a step (Metlov 1985, 5). Nearly 90 years of ongoing testing and optimization of this program did not result in accelerating the time frame of mastery of pitch in TO, which indicates that children younger than 3 generally tend to process music not as conglomeration of pitch changes but in some other alternative method.

## 7.5 The Model of Early Developmental “Timbral Hearing”

Soviet researchers spent much effort trying to identify this earliest method of TO. The leading role here was played by Boris Teplov, the head of the psychology department in the Moscow State University and the author of the theory of timbral music hearing. Teplov was able to experimentally demonstrate that pitch contour can be comfortably followed without paying any attention to pitch—the proof for which is speech prosody: undifferentiated perception of spectral content is sufficient for noticing changes in direction in pitch contours (Teplov 1947, 88). Such “phonic,” broadband noise-like, holistic perception of tonal quality is exactly what very young children hear when they track melodies. The ability to resolve spectral change into frequency change has to be developed and is acquired only when perception of pitch becomes culturally important (89). Timbral hearing can lead to drastic discrepancies when it substitutes pitch hearing in harmonic analysis. Sofia Beliayeva-Ekzempliarskaya initiated a very unusual line of research in Russia right after the Bolshevik revolution. She asked kindergarten-age children without any formal music education to rate the pleasantness of two arrangement of a well-known folk tune, one in a “euphonic” manner and the other in a “cacophonic” manner: e.g., the melody and accompaniment in the same key versus the melody and accompaniment in two different keys (Beliayeva-Ekzempliarskaya 1925). Only one girl found the mismatched keys to sound worse than the matched keys. The majority of subjects did not hear any difference, and a few found the mismatched version even more pleasing. Evidently, their learning was occurring in the “timbral hearing” domain—which was their “natural” manner of listening in the absence of dedicated ear training. Similar results were found by Costa-Giomi who used a sudden abrupt modulation in the accompaniment of a song to see if the children would be able to adjust their singing to a new key (Costa-Giomi 2000). Less than half of the kindergarteners could stay in tune with the accompaniment. The percentage of older children (fourth graders) was higher, and they improved their performance after instruction. Evidently, commitment to timbral hearing is a matter of age and absence of dedicated ear training.

Teplov (1947, 192) commented on findings by Carl Stumpf (Stumpf 1890, 2:84) who conducted experiments with the preschool-age children, asking them to identify the number of simultaneously played tones. Children often mistook a harmonic octave for one tone, whereas they mistook major and minor third as well as major second—for three tones. Some children even recognized a major second as four tones. The same problem occurred with some older children without musical training. Teplov explains this numerical anomaly to be the result of children's commitment to timbral hearing. Their judgment of hearing too few (for octave) or too many (second and third) tones must have been caused by centering on the relative fullness of spectra in a way similar to appreciating the fullness of a particular timbre (i.e., oboe versus flute). Such manner of listening does not allow for resolving spectra into pitch, hence their failure to analyze the harmonic interval into two constituent tones, and instead their guessing a greater number of tones.

Perhaps the strongest support for Teplov's theory was delivered by Aleksei Leontyev (Leontyev 2009, 115–136). He started his research in 1958 by discovering the phenomenon of interference between categorizations of timbre and of musical pitch in perception of sung out verbal syllables. Leontyev measured the ability of 93 Russian-speaking non-musicians to discriminate pitch in dyadic combinations of the front vowel "Y" and the back vowel "U," each tuned to a different fundamental frequency within the range from 200 to 400 Hz. Only 13% of the subjects were consistent in recognizing the changes in frequency across all melodic dyads of Y-Y, U-U, U-Y, and Y-U, demonstrating a reliable skill to focus on the pitch parameter, ignoring the spectral parameter of vowel contrast. About a third of the subjects turned out to be completely "tone deaf"—they considered "U" as always lower in pitch and "Y" always higher, no matter if the actual F0 was higher or lower, even if the F0 difference neared an octave.

Then Leontyev ran the same test on 20 Vietnamese speakers and found out that none of them were "tone deaf." Since Vietnamese is a tonal language, "tone deafness," observed by him in Russian speakers, must have related to habituation to nontonal language, where timbral differentiation played the primary part and changes in pitch were unimportant.[9] To investigate the capacity of "tonally

[9]Another confirmation comes from a cross-cultural comparison between the Chinese- and English-speaking first graders: Chinese speakers show better singing performance than English speakers, indicating that speaking a tonal language enhances the perception and production of musical pitch (Rutkowski and Chen-Hafteck 2001). Even more specific is the finding of the experimental study by Mang who investigated the effects of age, gender, and language on the singing competency of 7–9-year-old Cantonese-speaking children (Mang 2006). She discovered that exclusive learners of tonal language acquired singing voice earlier than those children who learned English in parallel.

impaired" listeners to discriminate pitches, Leontyev measured their discrimination thresholds in detecting the difference between two tones of different frequencies in five attempts, where after the initial attempt they were asked to loudly sing the auditioned tones. Subjects showed dramatic reduction in their discrimination thresholds as a result of singing (e.g., a subject who could not judge the relative pitch in Y-U dyads, even with a difference of 1200 cents between them, achieved the threshold of 135 cents). In a further test, those "tone-deaf" subjects who could not sing in tune were trained to match their voice to the sound of an electric generator in two to six 30-minute-long training sessions. This reduced the discrimination threshold in pitch matching tests that these subjects had initially failed five times before. Evidently, during the study, they had developed pitch hearing. To test the contribution of vocalization, experimenters conducted a task of training in pitch discrimination in a control group with no singing and found only insignificant improvement in pitch sensitivity.

The conclusion Leontyev drew was that without the development of proper vocal activity and its inclusion in the receptor system, the musical ear would not become trained. This suggested that the use of speech and music was each associated with its own dedicated type of hearing. Since most phonemes in world languages are nontonal (only certain vowels constitute lexic tones), verbal ear in majority of individuals is based on fine *timbral* discrimination between consonants and vowels. If an infant learns to focus on the timbral aspect of verbal vocalizations and verbal development is most intense during infancy, a timbral mode of auditory analysis is likely to form the foundation for the development of musical ear. The child will use his/her "verbal ear" to process musical information and accommodate his/her perceptory schemes as needed for song production. Hence, the "pitch ear" develops with a considerable delay after the verbal spurt.

Both language and music are products of social development, crucial for a child's ability to manage his/her environment. Language and music, each, call for a distinct and autonomous method of processing information in order to avoid confusion between the two. Therefore, "pitch ear" becomes differentiated as a special faculty that *isolates the parameter of pitch* from timbre—as opposed to "verbal ear" in nontonal languages, which abstracts the *timbral properties of sounds*, isolating them from pitch (Leontyev 2009, 245–94). The proof of such specialization is the phenomenon of *whispering speech*, which is completely deprived of the tonal aspect by the exclusion of the vocal folds, yet it does not obstruct its understanding, provided whispering is sufficiently loud (Leontyev 2001, 213). The prevalence of the timbral aspect in the perception of verbal articulation is evident even in tonal languages, where whispering speech in long sequences supports up to 85% accuracy in the identification of the tonal components of the language, even in such tonally sensitive language as Thai (Abramson 1972). Recent advances in computer technology have even allowed for the development of software in automatic recognition of whispered speech in Mandarin (Lee et al. 2014). Quite similar to whisper in disclosing the self-sufficiency of timbral organization for speech is speaking in a monotone—for nontonal languages a monotone speech remains highly intelligible—

in sharp contrast to completely unintelligible monotimbral speech (where all phonemes are replaced with just one phoneme) (Patel 2010b, 51).

The reverse applies to music where a strictly monotone melody is hard to memorize and distinguish from another monotone melody—in contrary to a monotimbral but multitonal melody. The examples of strictly monotonic music are scarce and fragmentary—constituting only a section of a larger music form, such as Tibetan Tantric chants that can feature extended repetition of the same frequency, but usually contain other frequencies as well, and are framed by performance of multitonal vocal and instrumental music. Noteworthy is the affiliation of monotonic music with religion (Buddhist or Christian Orthodox), where the unusualness of monotonic limitation serves to illustrate austerity and focuses on purely spiritual matters, abstaining from a "normal" secular impulse to appreciate diversity in the flow of music. Even more rare is the employment of whispering in production of music. Thus, consistent whispering characterizes the genre of whispered inanga in Burundi, but its musical aspect is comprised entirely of its instrumental accompaniment, whereas vocalization in itself is realized by indigenous people as primarily verbal rather than musical expression (Fales 2002).

Leontyev considered "pitch ear" a dedicated musical *functional physiological organ of the nervous system* (Leontyev 2009, 286)—following the model of a "functional organ" by Aleksey Ukhtomsky (1978), who argued that "organ" was not necessarily a morphologically distinct anatomical part, but could constitute an organization of a specific function with *dynamic* rather than static attributes. Such organ would be comprised of some complex linking of numerous reflexes into an integral system possessing a highly generalized and qualitatively special function.

This view is shared by another pupil and colleague of Vygotsky, Luria, who considered "musical ear" and "verbal ear" to constitute different organs that not only differ functionally but also depended on *different locations in the brain* (Luria 2003, 85). He observed that in 800 cases of disruption of phonematic hearing, right-handed individuals with lesions in their left temporal lobes lost the ability to discriminate opposite phonemes, while maintaining the ability to discriminate pitches, whereas patients with lesions in their right temporal lobe retained the speech function but demonstrated symptoms of sensory amusia (154). Leontyev and Luria's conclusions seem to generally agree with the model of modularity of music processing, developed after Jerry Fodor (Fodor 2001), by Peretz and Coltheart (Peretz and Coltheart 2003). Their idea of tonal encoding of pitch as a distinct music-specific component inside the dedicated music module, designed to process music signals differently from speech, via different neural pathways, strongly resonates with Leontyev's insight that dated some 50 years earlier. This cognitive-neurophysiological model of music processing has recently found solid support in numerous experimental studies (Nunes-Silva and Haase 2013). Additional support for Leontyev's ideas comes from the developmental perspective on mechanisms of learning and memory in music and speech, suggesting that their modularity is emergent rather than present at the beginning of life (Mcmullen and Saffran 2004).

To test his hypothesis that pitch hearing is a functional organ that is absent at the moment of birth and forms only at a later point as a consequence of cognitive development, Chumak (1962) conducted a series of experiments aimed to exclude hearing altogether as a sensory apparatus in the discrimination of pitch, while retaining the effector apparatus of intonation (Leontyev 2009, 128–33). They built a practically noiseless vibrator machine equipped with a rod that was embedded in a soundproof shell and used it to generate vibrations in the frequency range between 100 and 160 Hz. The ratio of amplitudes in the measuring of discrimination thresholds was 1:2, and the frequency and amplitude of the stimuli were continuously controlled. At first, they established the sensitivity of those subjects who were touching the rod for the stimuli that had the same amplitude. Then, by comparing the frequency values of stimuli with different amplitudes, they established the discrimination thresholds which turned out to be two to four times as high as the differential thresholds. The subjects were then asked to match the perceived vibration with their singing in an attempt to produce the tone whose frequency they thought would correspond to the vibration perceived by their hands. Although this task initially seemed impossible to many subjects, after a considerable number of trials, Leontyev and Chumak succeeded and demonstrated the same results as found in previous experiments with singing to the auditioned tones—the initial threshold of discrimination in all the subjects throughout their singing attempts was falling within the range of 66–80%, as measured between the initial and the last attempts.

Evidently, in the process of the experiment, a completely new functional system of pitch discrimination was artificially founded and made operational. To complete the study, experimenters excluded the motoric component, while keeping the sensory component. They constructed a disk with a tensometer, the pressing of which caused a smooth change in the generated frequency that was transmitted to a frequency meter, oscillograph, and telephone. The stronger the key was pressed, the higher was the frequency that was generated by the instrument. The test was administered to three "tone-deaf" subjects (all subjects in the previous vibration experiment had good musical hearing). The objective was to establish a conditioned link between the frequency of the sound and the degree of static force of the muscles of the arm. The subject had to react to a pure tone of 100–500 Hz by the pressure of his hand without hearing the actual sound generated by the instrument. After 25–33 sessions of 40 min each, a conditioned link of "pitch degree of muscular effort" was formed in all the subjects. The thresholds of discrimination were found to reduce from 1994 cents to 700 in the first subject, and from 1615 to 248 and from 828 to 422 in the other two. In this artificially created functional system, the role of the vocal apparatus in traditional means of generating a pitch was transferred to the arm muscles.

The scope of this chapter does not allow for the description of all the important studies conducted in the Soviet Union in the investigation of "musical" versus "speech ear" beyond the most basic information that could suffice to highlight the validity of this research and the usefulness of its theoretic framework for understanding and interpretation of early children vocalizations. Leontyev's experiments go to prove that the human brain has the capacity to generate new functional sensory and motor organs based on instruction and learning. Once formed, such organs display considerable stability and can function on par with innate organs, seemingly manifesting elementary innate capacities. The principal difference is that such functional organs are not "analog" in a sense that the reflexes that comprise them do not "simply trace or copy the sequence of the external stimuli, but unite independent reflex processes with their motor effects in a single complex reflex act" (134), which is then established as a whole to become more and more automatic. In the end, this course of development installs what appears to be a fully automatic mechanism of pitch categorization that might effectively operate even when a well-trained individual is not consciously trying to identify pitches in music, making an impression that his/her ability to detect pitches is inborn.

## 7.6 The Ontogenetic Problem of Absolute Pitch and "Pitch Reductionism"

Crucial for the question of evolution of music and language is the phenomenon of the so-called "perfect pitch," or "absolute pitch," as psychologists prefer to call it (Parncutt and Levitin 2001). "Genuine" absolute pitch hearing, characterized by instantaneous automatic identification of all pitch classes across all timbres, without any ear training, is most probably a biological product of long-lasting cultivation of frequency-based music—long enough and culturally important to make it inheritable at least in some cases (Tan et al. 2014). Of course, not every type of absolute pitch has to do with inheritance—it is imperative to distinguish between different types of absolute pitch (Bachem 1937). Its most widespread form is the *passive* "*piece*-absolute pitch," when the listener hears if a familiar music piece is played in the original key or if it is transposed to some other key (Levitin 1994). Musicians, musicologists, and ear-training specialists do not consider this to constitute "perfect pitch." They give credit only to *active* "*tone*-absolute pitch." However, many possessors of the latter also remain limited in their frequency detection: succeed only in identifying the sounds of specific musical instruments (most commonly, piano), only in diatonic tones (white keys of the piano), or only within the middle range (Deutsch 2013a). Such "imperfect" cases of perfect pitch are usually handled by categorical perception and have to do with long-term memory of tuning standards (Rakowski 1993). Like "piece-absolute pitch," "tone-absolute pitch" can be acquired by those who originally did not exhibit any form of absolute pitch. This is usually achieved through consistent ear training according to the effective methodology.[10]

[10] I can attest to the latter from my own experience. Up until the age of 14, I did not possess any absolute pitch and could reliably detect degrees of a key, intervals, and chords only in a "relative" manner. After following the special methodology developed by Boris Utkin in the State Schnittke University, Moscow, for 2 years, in 1980, I developed a high-quality absolute pitch that enabled me to reproduce on the piano any simultaneous combination of up to four tones in series of up to three sonances in any register by any string, wind, or keyboard instrument tuned to A = 440 Hz (plus/minus 30–40 cents) without any mistakes. Moreover, all six of my classmates developed absolute pitch hearing of comparable reliability.

In fact, many experienced professional singers can remember pitches in absolute terms by referring them to specific neuromuscular sensations in their vocal apparatus: i.e. remembering how "middle C" feels in the throat versus a C an octave higher (Utkin 1985). Vocal training based on such neuromuscular control can reliably build an absolute pitch hearing (Teplov 1947, 138). Operationally similar absolute pitch hearing, but limited to a particular musical instrument, is found in professional instrumentalists after they acquire fine coordination between the extent of their technical action and the frequency generated by that technique (Lockhead and Byrd 1981).

The "natural" absolute pitch, when a person automatically detects all frequencies within the critical bandwidth across all timbral variations, without any training, is rather rare, found in about 0.0001% of population (Profita et al. 1988). However, such a low rate has been questioned by later studies, some of which reported substantially higher rates, up to 15% (Gregersen 1998). The estimates of ear-training specialists in Russia, where ear-training courses have been offered to both general population, as part of the kindergarten and primary school curriculum, and to musically gifted children within the State system of elementary music schools (Karasyova 2010), range within 6–9% among musically gifted and 1% in the general population (Berezhansky 2000). The rates for the acquired absolute pitch are generally much higher, substantially varying depending on the choice of the methodology of ear training and the consistency of training.[11] The discrepancy in the estimates must owe to the difficulty of defining criteria for testing the functional "tone-absolute pitch" hearing. Yet another obstacle is posed by the difficulty of finding out if absolute pitch was acquired through cultural exposure or was present at birth. Even if found in newborns, it could be a product of intrinsic learning during the last months of gestation, since the fetus can already hear sounds louder than 80 dB SPL, and even softer for lower frequencies (60 dB) (Abrams and Gerhardt 1997).

Therefore, we can make valid generalizations only by delineating between two general classes of absolute pitch: "trained" versus "non-trained."[12] Such a distinction might correspond to two different mechanisms of pitch processing (Ross et al. 2005):

1. *Simple* categorization of high/low changes in pitch is used to track the melodic contour in music and prosodic contours in speech—this mode of pitch processing is based on *tonotopic* representations of sounds and is neither limited to music nor to human perception, exceedingly common for animal communication as well (Snowdon 2003);

[11]The estimations of different methodologists and authors of ear-training courses range from about a half of their students (e.g., Dmitrii Blium, Nina Kachalina) to 80–90% (Boris Utkin, Vladimir Kiriushin)—according to personal communication. The historic review of different methodologies of absolute pitch acquisition is reviewed by Berezhansky (2000).

[12]This distinction is not exactly the same as "cultural" versus "natural"—because non-trained absolute pitch is only an indicator of the possibility of "natural" possession of it.

2. *Complex* categorization of pitch classes is used to identify fundamental frequencies and pitch intervals between the successive tones—this mode is unique to music and is based on *periodotopic* representation of auditory stimuli.

The former is much more widespread than the latter. An implicit form of absolute pitch, found in majority of population, includes remembering the "proper" frequency value for the opening of familiar tunes, dial tones, and lullabies sung by mothers to infants (Deutsch 2013a, 142–4). In fact, memorization of tunes by their absolute rather than relative pitch constitutes the preferred strategy of 8–9-month-old Western infants (Saffran and Griepentrog 2001). The same applies to the auditory signal learning by animals, especially common for birds (Hulse et al. 1984). Monkeys and rats too can learn to discriminate between a couple of tonal patterns on the basis of the absolute frequencies of 1–2 tones (D'Amato 1988).

Both of these mechanisms do not exclude each other. In fact, the current notion of musical professionalism implies the possession of both skills, usually qualified by ear developers as, respectively, "melodic" and "intervallic" hearing (Starcheus 2005). Here the specialists in ear development drastically differ from the rest of specialists and general public who tend to view "absolute pitch" as the requisite of outstanding musicality (reflected in its nickname "perfect ear"). Professional music educators consider absolute pitch hearing to present a likely obstacle in full-fledged musical development, potentially preventing the integration of one's emotional experience from hearing specific musical patterns with learning to categorize such patterns by ear (Berezhansky 2000). For Russian ear developers, possession of natural absolute pitch is first and foremost a warning that its possessor risks to have the hearing of a piano tuner rather than of a musician.

A teacher of ear training in a Russian musical school is likely to pay an extra attention to a possessor of the natural absolute pitch, helping that person overcome his/her instinct to momentarily define the pitch values of melodic, harmonic, and textural patterns of music, which all require successive "caching" and integration of changes in the flow of music. This is in polar opposition to the "natural" strategy of the possessor of natural absolute pitch, who automatically breaks the music into discrete particles. Listening to music in this way resembles listening to the spelled out verbal speech. Just like spelling of letters significantly complicates the retrieval of lexic meaning of words and sentences, "spelling out" of absolute pitch values stands on the way of the retrieval of the affective meaning of musical idioms. Teplov was one of the first psychologists to notice that "the possession of natural absolute pitch hearing can delay the development of the other forms of musical hearing (melodic, harmonic, polyphonic) because it can substitute them, thereby removing the practical need in their mastering (Teplov 1947, 153).

In essence, relying exclusively on absolute pitch constitutes yet another type of "pitch reductionism," where the quantization of pitch contours, harmonies, and parts causes the omissions or distortions in the retrieval of the melodic, harmonic, and textural grouping (respectively, motifs, chordal progressions, and counterpoint). Possessors of natural, untrained, absolute pitch "reduce" the variety of acoustic characteristics of a sounding stimulus to rather roughly defined frequency values, usually disregarding the functional relations between proximal and distant tones, which obstructs the perception of "horizontal" components, such as musical intonations, motifs, and phrases, and even some of "vertical" components, such as harmonic intervals and chords—along with their corresponding semantic qualia. Even professional musicians with natural absolute pitch after completion of ear-training courses and developing their natural gift typically have the threshold of discrimination between two different pitch classes 2–5 times higher than other professional musicians (Garbuzov 1948).

- From the perspective of an ear developer, absolute pitch is the ability to process music by the default approach of breaking the entire musical ambitus of the auditioned sounds into rather wide zones, thereby processing each pitch change as a switch from one zone to another, akin "brushstrokes" rather than "points" of pitch, as non-absolute musical professionals usually hear music (Grebelnik 1985). This extended zonal categorization relies on the *theoretical rules* of definition of pitch, according to which the variety of intonations and the variety of "sonances" (i.e., simultaneous combinations of sounds that occur in chords, double notes, and melodic counterpoint) are reduced to a few pitch classes. This, by design, introduces substantial omission of meaningful data.

In practice, the predisposition to absolute pitch needs to be capitalized on—by turning the algorithm of "pitch reduction" into a default method for music processing early in child's life. In the entire vast database of millions of people that passed through the ear-training courses in Russia in the past century, there is not a single account of the emergence of an effective absolute pitch hearing in an adult professional musician in a natural way, without vigorous training—and this is despite the ongoing enhancement of musical hearing throughout a musician's career (Berezhansky 2000). Nevertheless, an adult musician can acquire reasonably good absolute pitch hearing by means of extensive rigorous exercising based on the right methodology (Brady 1970). However, such absolute pitch evidently remains unstable, prone to deterioration and even disappearance once the regular exercises are stopped (Köhler 1915). Self-reported age when absolute pitch possessors usually discover their gift is tied to their first instruction in musical theory and constitutes the average age of 5.4 years (Gregersen et al. 1999). The earlier the age of absolute pitch activation (provided it is done through formal schooling), the more reliable the absolute pitch categorization (Deutsch et al. 2006). Critical period for absolute pitch acquisition seems to exist not only in relation to the "non-trained" gift but for its trained variety. Thus, after training, 3–6-year-old children were found to surpass adults in learning a single absolute pitch reference (Russo et al. 2003).

Absolute "pitch reductionism" is even more severe at the moment of activation of the natural absolute pitch gift than in its adult possessors. When a predisposed child is familiarized with the "names" and the target tuning of the "correct" pitch classes, the child develops a special intermediate mode of hearing that was called *black-and-white hearing* by Grigory Liubomirsky (1924). The first researcher who noticed that the majority of absolute pitch possessors have at first mastered to identify white keys and—only later—black keys of the piano was von Helmholtz (1877). He and Otto

Abraham (1901) explained this by the commonality of using piano as a first musical instrument to teach music theory to children. Miyazaki adopted the same explanation (Ken'ichi Miyazaki 1989) for his discovery that white-key pitches are usually recognized by possessors of absolute pitch with greater accuracy than black-key pitches (Ken'ichi Miyazaki 1988). The cross-sectional study of children's earliest piano instruction indeed confirmed that children mastered absolute pitch classes in the order of their presentation by the teacher during piano lessons (Miyazaki and Ogawa 2006). However, closer investigation demonstrated that non-keyboard instruction also generated the "white-key bias"—even to a stronger extent than keyboard instruction (Deutsch et al. 2011). The alternative explanation puts the "white-key" effect on the account of greater commonality of "white keys" (C major, D major, etc.) than "black keys" (F-sharp major, G-flat major), which supposedly simplified the frequency resolution of the degrees of these keys (Takeuchi and Hulse 1991). The statistical study of distribution of "black-" and "white"-pitch classes in the compositions by J.S. Bach and J. Haydn confirmed the distribution of reaction times in recognition of absolute pitch obtained by Miyazaki (Simpson and Huron 1994). This explanation appears more plausible than Miyazaki's, but it fails to explain why possessors of absolute pitch hearing testify that they experience black-key pitch classes as carrying a peculiar tonal quality that is absent in white-key pitches, and they detect the presence of this quality before they retrieve the name of the pitch class (Baird 1917). The modern practice of ear training confirms this fact and reveals its relation to mastering TO (in particular, the differences between diatonic degrees and their chromatic alterations) within Western tonality (Berezhansky 2000, 60).

Absolute hearing is not acquired "per se," as mechanical memorization of abstract pieces of information. *Absolute pitch reference is discovered by those who have predisposition to it within the context of Western tonality*—specifically, within the hierarchic tonal framework of a major key that serves as a model of functional relations between the degrees of a key. It is this modeling that is responsible for the greater accuracy of the absolute pitch recognition by the 5–6-year-olds than that of the 3–4-year-olds (Russo et al. 2003). Hearing of the tonal framework becomes much more reliable in kindergarten-age than in preschool-age children. In essence, children prototype their learning of all keys on the degrees of C major (Baird 1917). And in reverse, prolonged ear training in C major alone leads to higher rates of the acquired absolute pitch hearing (Agazhanov 1985). This is what lays the foundation for "black-and-white" hearing and its side-effect of reducing absolute pitch classes to mere seven diatonic tones (Liubomirsky 1924).

The "white-key bias" originates not in the commonality of "white keys," but in the **diatonic framework** that supports and enables the chromatic alterations. Otherwise, those youngsters who started their music education on the transposing music instruments (e.g., saxophone in B or in E flat, or clarinet in B or in A) would have not possess white-key bias. On the other hand, the errors in pitch recognition by possessors of natural absolute pitch reveal the systemic relation of absolute hearing to diatonicity. Those who were introduced into tonality by learning the circle of fifths typically mistake a pitch by the intervals of fourth and fifth rather than a semitone, while those who learned keys by breaking an octave "doh-doh" into seven "white" steps first (re, mi, fa, sol, la, ti/si) and then altering these steps by sharps or flats usually commit semitonal errors (Starcheus 2005). The latter seems to be more common, comprising about three quarters of mistakes of absolute pitch possessors (Weinert 1929).

This brings us to the very important issue. Chromatic alterations play the fundamental role in supporting the capacity of tonality to express various affective states (Nikolsky 2016, Appendix-3). Phylogenesis of chromatic music systems as a rule is accompanied by profound cultural changes; when music occupation becomes professionalized, music becomes theatric, and aesthetic appreciation of music becomes the norm of listening to music. Pitch reductionism of the absolute hearing stands on the way of the emotional functionality of music by distorting the qualia of pitch classes in chromatic pitch class sets and impoverishing the emotional communication that constitutes the backbone of music. Semitonal mistakes by absolute pitch possessors indicate that they do not fully control the emotional aspect of recognizing the chromatic alterations (discriminating various levels of tonal tension) and perceive both a degree and its chromatic alteration as basically interchangeable in their expression. However, in reality, each chromatic alteration possesses its peculiar qualia (Huron 2006a). Therefore, pitch reduction of 12 chromatic pitch classes to 7 diatonic pitch classes severely cuts down the semantic diversity of music. After all, music that is limited to seven diatonic degrees only, devoid of any chromatic alterations, according to the conventions of the common practice period, expresses either excessive simplicity, childishness, naivety, or light-mindedness (Mazel 1952).

Moreover, absolute hearing reduces the TO even more than 12:7, if to make an important correction. As a matter of fact, chromatic tonality is built not on 12 pitch classes, as it is commonly stated in English musicological literature, but on 17 pitch classes. In practice, the D-sharp and the E-flat constitute not the same pitch class in music created for string and wind musical instruments as well as for vocals: these tones are tuned differently and participate in different melodic intonations, or, how Huron puts it, constitute different "tendency tones" (Huron 2006b). Unfortunately, orthodox Western harmonic theory has left this matter obscured in shadow, so that even such experts on musicology as Christopher Longuet-Higgins erroneously believed that the ideal intonation required the D-sharp to be lower than the E-flat (Longuet-Higgins 1975). However, there is no doubt that W. A. Mozart taught his pupil, Thomas Attwood, the 17-tone model of a key, and not a 12-tone model (Chesnut 1977).[13]

[13] Although Mozart adhered to the meantone model of making a chromatic semitone smaller than a diatonic semitone, following his father, the meantone model maintains the presence of 17 pitch classes in the same way as the Pythagorean-like model of the nineteenth century. It is just that Eb is higher in pitch than D#—in accordance with Longuet-Higgins. The meantone enharmonic distinction was true for the most of the classical music created before the nineteenth century: e.g., Haydn in his enharmonic modulation in the String Quartet op. 77 No. 2 used the transition from Eb to D# with the mark "l'istesso tuono," indicating that the performer should not play Eb and D# as different pitches (Duffin 2007, 79–83). The first advocates of keeping enharmonic tones perfectly equal on

Eb has been tuned lower than D# in vocal and string solo performance practice since at least the 1800s (Barbieri and Mangsen 1991), coinciding with ultimate formulation of the theory of tonality, tonal relations, and modulation. The origin of "tonality" as we know it was proclaimed by Rameau in 1722 (Rameau 1971), and fully formulated in its functional hierarchy by Gottfried Weber in 1821 (Weber 2012). The concepts of tonal plan, recapitulation, and system of relationship between keys were laid out by Johann Albrechtsberger in the 1790s (Kholopov 1988, 237). The time frame of these three most important innovations generally agrees with the first definite indication that Ab is lower than G# coming from the fingering charts by Robert Crome (c.1740s). They probably reflected the growing preference for Pythagorean thirds in favor of the earlier meantone tuning standards advocated by Pier Francesco Tosi and Francesco Geminiani, until Bartolomeo Campagnoli (c.1797) provided the first evidence of decidedly Pythagorean tuning (Barbour 1952). The transition must have occurred by the end of the eighteenth century in England and France (Barbour 2004, 58–59). The most immediate reason for needing sharper Pythagorean distinctions between major and minor intervals was to mark melodic functionality of different intervals, especially that of a leading tone—instrumental for harmonic tension within a key (Barbieri and Mangsen 1991). Hence, Pythagorean intonation was often termed "functional" and explained by teachers as the rule of "shading pitches in the direction of their attachment," as determined by the harmonic functions of resolution (Chesnut 1977). Pythagorean intonation became normative for expressive melodic performance on non-keyboard instruments and vocals since the first half of the nineteenth century, when the music system of tonality became full-fledged. The same principle of adhering to Pythagorean intonation in the melodic context of tonal music was confirmed in acoustic measurements of expressive solo performance (Friberg 1995).[14]

- Therefore, absolute hearing at its "black-and-white" stage practically reduces 17 pitch classes to only 7, which is much more than merely 2.4 times. In reality of melody making, pitch classes create functional pairs of increasing or decreasing tonal tension. So, 7 pitch classes allow for 5040 possible combinations, whereas 17 for 355,687,428,096,000! This is indicative of massive impoverishing of the melodic intonation in the youngest possessors of the "perfect pitch" gift.

string playing, such as Spohr, became vocal about it by the mid-nineteenth century, but their impact was small, affecting only performance in small ensemble with the piano (Barbieri and Mangsen 1991). For the piano tuning, equal temperament supplanted meantone tuning only around the middle of the nineteenth century (Shepard 1999).

[14]The 17-pitch model of a key by the mid-nineteenth century has caused the discrimination between tuning standards for enharmonically spelled keys. Thus, Alexandre and Provost conducted an experiment in 1862 by asking string players to perform the identical melody in F# major and in Gb major—to find that the tones of the Gb major were modified in a manner of bringing them closer to the minor mode to project a darker sound as opposed to more "brilliant" sound of F# major (Barbieri and Mangsen 1991).

Yet another manifestation of absolute pitch reductionism is a systemic error of not recognizing the octave placement of pitch classes (Lockhead and Byrd 1981). Octave errors are more common among all absolute pitch possessors than semitonal and fourth–fifth errors and are committed even by those well-trained musicians who do not commit other errors (Berezhansky 2000, 62). Berezhansky stresses that such mistakes cannot be explained in any other way but by serious deficiency in melodic and timbral hearing: in both expressive aspects, timbre and pitch contour, the tones closest in FF, are usually more similar than the tones an octave apart. Octave errors demonstrate nearly complete overriding of *melodic* hearing by *harmonic* hearing. Octave is the greatest *harmonic consonance*, yet often constitutes a *melodic dissonance*—it disrupts the smoothness of melodic motion by a huge leap up or down (Nikolsky 2015a). Such leaps introduce melodic tension and require melodic resolution in a form of a step following a leap. This rule is well-known in practice of good voicing (Huron 2001). In a musician, frequent melodic octave errors testify about the lack of emotional reactivity to music, and in general, lack of musicality—since melody constitutes the most expressive part of music (Szabolcsi 1965).

And indeed, a number of Russian specialists in ear training view absolute pitch as fundamentally *unmusical*. Teplov underlined that absolute pitch hearing was not "musical" hearing (Teplov 1947, 117). Endovitskaya considered the initial "natural" form of infant's hearing, oriented toward the reproduction of absolute pitch values, to be essentially unmusical, since musical hearing is supposed to evaluate not stand-alone tones but semantically meaningful combinations of tones (Endovitskaya 1963). The natural tendency of possessors of absolute pitch to dissect vertical and horizontal combinations of tones comes as fundamentally unmusical, preventing music from serving one of its most important biological purposes—securing emotional contagion in order to mediate and optimize the relations between an individual and a social group. At its worst, absolute pitch can effectively "lock-in" its possessor in a restricted individual use of music: e.g., Joseph Hofmann, himself a prodigy-possessor of natural absolute pitch, testified that his father could not even recognize well-familiar pieces of music if they were played in a key, different from that of his original learning (Hofmann 1920)—which would certainly isolate him from other listeners whenever participating in social musical activities. And here it would be appropriate to point to the connection between absolute pitch and autism.

## 7.7 Absolute Pitch, Its Autistic Connection, and the Psychosocial Perspective on Its Phylogeny

Autistic individuals that exhibit extreme intellectual and communicative deficits combined with islands of specific enhanced abilities (autistic savants) are found to invariably possess absolute pitch that enables them to automatically disembed individual pitch classes from chords (Heaton 2003). Non-savant autistic individuals also have enhanced pitch discrimination. Autistic absolute pitch does not seem to interfere with pitch contour perception as long as the pitch intervals remain small (Heaton 2005). The organic connection between wide melodic leaps and increased emotional intensity, common for both music and speech (Johnson-Laird and Oatley 2010, 107), confirms the presence of restraints in emotional communication long observed in autistic individuals (Kanner 1943). The subsequent research revealed the complexity of the matter of what exactly constitutes autism and which exactly emotional tasks are affected by it (Nuske et al. 2013). Pamela Heaton demonstrated that non-savant autistic children did match auditory and visual stimuli (major/minor keys with happy/sad faces) that expressed the same basic emotion (Heaton 2009), concluding that autism did not make children unemotional but merely directed their emotional development toward a different path than in non-autistic children (Heaton et al. 2008). However, Bhatara et al. convincingly demonstrate that children and adolescents with autism spectrum disorder are impaired in their judgments of the emotional expressivity of piano performance, completely failing to differentiate the expressivity levels by different performers for either the nocturnes in major or minor keys (Bhatara et al. 2010). This raises the question if Heaton's subjects indeed felt the emotions they detected in the auditory stimuli. Perhaps, they knew that major keys were supposed to express happiness while minor keys sadness, but did not actually experience these emotional states, and were aware of only compositional but not performance cues in their auditory analysis. There is evidence that autistic children have no mirror neuron activity in the inferior frontal gyrus, which likely underlies the social deficits, preventing emotional contagion (Dapretto et al. 2006). Since much of one's understanding of emotions comes from early experience with social affective communication with a caretaker (Trevarthen 2000), a neural abnormality would prevent one from developing musical and verbal competence in auditory display of emotions. Indeed, autistic 2–4-year-old children fail to show a preference for motherese over adult speech and do not display a significant mismatch negativity (MMN) in response to hearing the motherese syllable changes (Kuhl et al. 2005).

But the abovementioned autistic musical limitations are not limited to individuals diagnosed with autistic spectrum disorder. Recent research demonstrated the presence of increased autistic traits in professional musicians with absolute pitch (Wenhart et al. 2019). The mechanism of veridical mapping was proposed to explain how functional rededication of perceptual brain regions to higher-order cognitive operations may lead to enhanced pitch perception in autism (Bouvet et al. 2014). It could be that professional musicians develop a habit of frequency categorization for any auditory stimuli similar to autistic savants, which exposes them to the same advantages and disadvantages in processing and interpreting music. Indeed, it is quite common to encounter the criticism of the performance of highly advanced classical professional musicians for "clinical" execution of what is written in score without much affection, as though just doing a good job of playing the right pitches clearly with technical perfection. Unfortunately, this performance approach is promoted by the music industry through the system of music competitions, causing formalization of advanced professional music education (Horowitz 1991). Perhaps, Oliver Sacks was right in his suggestion that the absolute hearing might differ from "normal talents" by being a "savant talent"—a precocious ability to process complex tasks in functional isolation from conceptual, verbal, and even general musical powers, independently from training and practicing music (Sacks 1995). In this respect, absolute pitch might be the backbone of the "savant syndrome" that affects many extremely talented musicians who display signs of "low-functioning autism" (Sacks 2008). This view is shared by many Russian pedagogues specializing in training of musical prodigies (Kirnarskaya et al. 2003).

The etiology of musical prodigism and its attribute, absolute pitch, strongly suggests its roots in Western schooling, rational music theory, and frequency-oriented music culture. Diana Deutsch believes that if modern infants are given the chance to associate pitches with meaningful terms during the critical period for music and speech acquisition, they as a rule might readily develop absolute pitch (Deutsch 2002). Early onset of music schooling and exposure to the ear-training methods based on "absolute doh" prior to the age of 7 are the strongest predictors of absolute pitch acquisition among musicians (Gregersen et al. 2007). The great importance placed in Western urban societies on children's academic success might have promoted the precocious onset of frequency discrimination, generating a peculiar selection for an absolute pitch gene. That such gene might be located in chromosome 8q24.21 was found by Theusch et al. in a genome-wide linkage study on 73 multiplex families of absolute pitch possessors (Theusch et al. 2009). Noteworthy is that this chromosome is related to the European ancestry and the implicated gene ADCY8 is connected to learning and memory (Tan et al. 2014). Most renowned music prodigies also come from Europe, the first accounts of their celebration dated by the late Baroque period, becoming more widespread from the 1750s on (Kenneson 1998). The popularization of music prodigism definitely had to do with the competitive pressure to secure for a child a successful musical career and financial well-being—most obvious in Jewish European population (Conway 2012). But during the past 20 years, more and more music prodigies appear in East Asia.

Absolute pitch phenomenon might be nothing but exacerbation of the general pitch reductionism trends within the frequency-oriented music cultures, fueled by the competitive advantage of earlier musical maturation for career building. Precocious children receive strong edge over "regular kids" in industrial societies, where each new generation faces an increased pressure for more and more extensive learning. And tonality, being the embodiment of hierarchic and rational organization of urban lifestyle, cultivates the preferential treatment of pitch-related aspects of musical expression (melody, harmony, texture). Here, absolute pitch stands as a form of maximizing the musician's focus on frequency at the expense of other domains (time, amplitude, timbre).

The available phylogenetic information fits this perspective. There is no mentioning of anything like absolute pitch in antiquity. Ancient Greeks did not know it despite using the fixed nominal system for all pitches afforded by their music system and long-standing practice of tuning their music to aulos (West 1992, 273). Neither the usage of tuned resonators that ensured the fixed pitches for Hellenic citharas (the preferred instrument for high art) prompted the discovery of the concert pitch standard (Hagel 2009, 69) without which no absolute pitch is possible. Ancient Romans did not know pitch standards as well—the leading third-century authority on music, Aristides Quintilianus recommended to determine the absolute pitch reference by nothing better than singing the lowest producible tone and judging the tone under question in reference to it (Stumpf 1883, 1:139). The only known ancient music culture that defined the concert pitch standard was Chinese: 60 pitches of Zeng music system were standardized to absolute frequency values represented by the battery of bells, manufactured at the court as a State standard for tuning of musical instruments in important rituals (Falkenhausen 1992). And the frequency values for each of the bells were inferred by dividing the octave in six equal whole tones and then subdividing a whole tone in halves—quite similar to the "black-and-white" method of absolute pitch acquisition. However, this Chinese system was definitely not practiced on individual level by regular musicians. Its use was limited to administration of State rituals in order to maintain "correct" harmonious proportions between the celestial, social, and individual organizations as part of the concept of "harmony of spheres," crucial for ancient Chinese as well as Greek civilizations (Daniélou 1995).

The world's first tool for standardizing concert pitch in everyday music practice, a tuning fork, was invented quite recently, in 1711, in London, albeit for medical purposes of physical examination, but Handel became the first celebrity musician to use it (Feldmann 1997). However, even with this invention, it took a while before it became really standardized. Throughout the eighteenth century, as tuning fork was earning recognition among musicians engaged in ensemble performance, its reference tone kept varying in pitch within a tone or so across Europe until the early nineteenth century (Bickerton and Barr 1987). The first reported possessor of absolute pitch was W. A. Mozart, and the earliest documented commentaries on the qualia of specific keys (in regard to temperament) that would require absolute pitch reference are found in treatises by Leopold Mozart and Johann Mattheson published toward the middle of the eighteenth century (Steblin 1987). The entirety of historic, musicological, and psychoacoustic information leads to believe that absolute pitch hearing did not exist or was considered unimportant, and therefore was not reported, prior to the nineteenth century, and its establishment was directly related to the adoption of temperament as an international standard of ensemble performance in worldwide distribution of classical music (Nazaikinsky 1993). Temperament allowed the use of all keys, equalizing them in qualia and instilling the notion of universality of 12 pitch classes (Barbour 2004). This equalization led to the invention of dodecaphonic compositional techniques by numerous composers in the beginning of the twentieth century (Kholopov 1983), marking a new era in history of Western music. Dodecaphony (and serialism that stemmed from it) became a form of TO most suitable for temperament and absolute pitch. This is hardly a coincidence that dodecaphony and serialism have been severely criticized from many angles for their formalistic approach to music and lack of affective impact on the common audiences, in general, and of diversity of expression, in particular, as compared to the conventional tonal music (Krauss 1985; Smith and Witt 1989; Lerdahl 1992; Thompson and Robitaille 1992; Born 2000; Moelants 2000; Weber 2003; Ross 2007; Taruskin 2009; Kremp 2010; Thomson 2010; Ball 2011; Daynes 2011; Tagg 2012). Such criticism was voiced out even by those composers who themselves became famous through their implementation of dodecaphonic and serialist techniques (Henze 1982; Boulez 1990; Nikolskaya and Lutoslawski 1995; Nono 1999;

Schnittke 2004; Berio 2006; Denisov 2009). The byproduct of this development was pronounced decrease of the expressive means available to the composer and the performer (Duffin 2007). This trend culminated in formulation of dodecaphonic compositional techniques by numerous composers in the beginning of the twentieth century (Kholopov 1983), which has become the form of TO that is most suitable for absolute pitch.

Absolute pitch is tied to pitch reductionism in phylogeny just as strongly as in ontogeny.

- Concert pitch standard is nothing but the reduction of the diversity of all pitch classes suitable to accept the role of tonic in a variety of possible tunings to just one pitch value within the well-tempered tuning.
- Tonality is nothing but the reduction of all possible pitch class sets to one major and one minor key.
- Frequency-oriented music is nothing but the reduction of timbre to frequency.

The very concept of absolute pitch negates the semantic contribution of timbre: the ideal possessor of absolute pitch is supposed to momentarily recognize the sameness of a pitch value disregarding by which musical instrument or vocal it is performed and even in non-musical sounds. Historically, this course of development is very pronounced in the Western civilization: many of its earliest forms of music incorporated timbre as an important expressive means.[15] However, the crystallization of tonality throughout the seventeenth century was accompanied by the reduction of the importance of timbre for TO (Scruton 1997, 77–8), so that by the eighteenth century, classical music adhered to the motto "pitch is governed by law while timbre is governed by taste" (Fales 2002). In fact, the reduction of timbre to pitch in Western tradition is captured in the official psychoacoustic definition of timbre: "attribute of sensation in terms of which a listener can judge that two sounds having the same loudness and pitch are dissimilar" (ASA 1951). This definition clearly bases the notion of timbre on frequency, therefore being unsuitable to the timbre-oriented music cultures.

Is absolute pitch possible in pretonal music systems at all? Indispensable for the activation of the absolute pitch predisposition in a young child is the presence of an explicit music theory capable of defining pitch classes in a rational manner—so that a child could prime an auditory stimulus with certain acoustic properties to a name of a certain pitch class, and thereafter automatically retrieve that name upon every encounter of the corresponding sound (Nazaikinsky 1993). For phylogeny, this means that *absolute pitch can emerge only in a music culture that supports some kind of mathematical inference of pitch classes and a purely frequency-based standard of tuning* (Nikolsky 2016). Although it was demonstrated that individuals who could not accurately name tones nevertheless possessed absolute pitch (Ross

[15]For example, French "florid organum" of the twelfth century is compositionally based on the timbral and textural contrast between the bright light "florid" upper part and the dark heavy "firm" lower part (Tischler 1956).

et al. 2004), cultivation of absolute pitch can survive only if one generation can pass the system of pitch naming down to the following generation. It is possible that a single individual would develop a skill of tuning to some kind of "concert pitch standard." Kubik reports such a case of an African xylophone manufacturer-player (Kubik 1985). However, inventions like that are doomed to stay isolated unless an inventor manages to formulate a corresponding music theory and transmit it to the critical number of users sufficient to sustain a new tradition. But chances for this are rather slim—I know of only one non-European indigenous culture that developed a proprietary concept of absolute pitch designed to reflect social status, and that is the Sundanese music of West Java (Zanten 2004). However, this system has remained culturally isolated. Cross-cultural propagation of absolute pitch would require cross-cultural adoption of technological tool for delivering the concert pitch reference. As Kubik notes, across the entire Africa, aids like tuning forks are extremely unusual despite the diversity of ensemble instrumental traditions and tuning systems (i.e., a new xylophone is tuned directly from an old one) and are used only on Western musical instruments (Kubik 1979).[16]

Normal musical development requires some kind of compensation for the pitch reductionism of absolute pitch hearing, usually achieved by at least three other forms of tonal hearing, capped in a methodological literature on ear training as "relative pitch" skills: hearing of a key, of degrees in a key, and of intervals between the degrees (Geinrikhs 1978).[17] So, in order for frequency discrimination to be "musical" (contextual) rather than "autistic" (detached), each tone demands a **binary** definition:

- As a member of the ultimate *repertory of all possible pitch-classes* (absolute framework) and as a member of a *specific key or mode* (relative tonal framework)—demanding from a musician equally reliable orientation in both coordinates (Agazhanov 1977).

Only the automatic or nearly automatic orientation in a musical key/mode can secure reliable recognition of degrees of a key or a mode, thereby opening ways for quantization of pitch intervals and their subsequent memorization (Nazaikinsky

[16]It is possible that Western folk musicians might have adopted the absolute pitch following the model of classical music, since many musicians specializing in various forms of folk and popular music these days take formal schooling and obtain graduate degrees from conservatories and universities. However, in such a situation, the absolute pitch is likely to cause detrimental effects on musical expression due to the use of temperament that has to accompany the use of absolute pitch. Other than that, folk music seems to commonly feature "song-absolute pitch," as indicated by the analysis of keys in folk song databases (Olthof et al. 2015).

[17]A number of methodological studies by ear-training specialists describe the integration of absolute and relative hearing in musicians with a well-developed musical ear (Garbuzov 1948; Seredinskaya 1962; Veis 1967; A. A. Agazhanov 1977, 1985; Geinrikhs 1978; Utkin 1985; Bytchkov 1993; Nazaikinsky 1993; Sladkov 1994; Karasyova 1999; Berezhansky 2000; Os'kina and Parnes 2001; Starcheus 2005). The evidence for alternation between absolute and relative modes of pitch processing was demonstrated even in the undeveloped autistic possessors of absolute pitch (Heaton 2003).

1977). The semantic qualia of modes, keys, intervals, and chords are all instrumental in supporting effective emotional communication through music (Cooke 1959), as confirmed experimentally (Kaminska and Woolf 2000). Absence of absolute-relative integration would make the gift of absolute pitch musically useless, causing the severe distortion in retrieving the semantic information, especially detrimental for decoding music upon listening. Therefore, in practice, *every student of music who possesses absolute pitch, even a complete beginner, is likely to possess relative pitch hearing as well*—at least the ability to detect a specific pitch class set and define sounds in reference to their membership in that set (Berezhansky 2000, 59). The integration of absolute and relative modes of hearing is to blame for the errors that absolute pitch possessors routinely make. Without the relative pitch feedback, the absolute frequency discrimination would have work as perfect as a machine. Poor integration of absolute and relative pitch data is especially prone to cause errors in frequency analysis of the multipart musical textures. Many professional ear trainers in Russia know how to throw off the possessors of immature absolute pitch to the extent that they completely lose track of pitch values in a melody in a real-time task of listening to multipart music.[18]

To sum things up, the phenomenon of absolute pitch should be regarded as a relatively late phylogenetic development, specific to Western countries and frequency-based musical cultures at the industrial and postindustrial stages. Absolute pitch might constitute a genetically selected trait designed to simplify categorization of pitch according to the well-tempered tuning. However, reliance on absolute pitch alone in auditory analysis of music is likely to severely limit one's emotional experience of music and altogether obstruct musical communication. Absolute pitch appears to work as a mere "shortcut" to the processing of the TO in frequency-based music, reducing the complexity of TO to its surrogate frequency component. In order to support fully functional musical hearing, including the perception of semantic qualia of melodic intonations, motifs, harmonic intervals, chords, and different musical modes, absolute pitch hearing must be combined with relative hearing of melodic and harmonic aspects of TO.

[18]For instance, Boris Utkin was able to consistently throw off the absolute pitch possessors with underdeveloped integration of absolute and relative hearing by improvising on the piano with long four-part harmonic progressions with the melody in the bass, engaging a chain of gradual and enharmonic modulations every 2–3 bars (e.g., the tonal plan like C major-G major-B major-*G-sharp minor-G minor*-D major-*B minor-C major* is likely to cause an untrained possessor of absolute pitch to mismatch the destination key and the initial key). Typically, if a possessor of absolute pitch names every key along the modulation path, the third or fourth enharmonic modulation (the above given chain contains two of such modulations, marked by italic) would leave him/her clueless unless they have developed a reliable relative pitch hearing. Yet another strategy for confusing a possessor of absolute pitch is to start the four-part harmonic modulation chain with strict chords and then start introducing the melodic motion in different parts by engaging non-chordal tones, with frequent changes from one part to another, effectively generating four-part polyphony while committing a chain of distant modulations (e.g., C major-E-flat major).

## 7.8 Absolute Pitch and Speech Perception

Noteworthy, the reductionist side effects of absolute pitch hearing do not impact the communication through speech. Neither its lexic meaning nor its prosodic expression depends on quantization of pitch contour into different pitch values. Lexic tones in tone languages do engage pitch relations, but the exact quantity of a pitch change does not matter for verbal semiosis. In light of this, it seems reasonable to disagree with the suggestion by Ross et al. that the current consensus is wrong, and absolute pitch might not be a strictly musical phenomenon (Ross et al. 2005).

Gregersen et al. raised the issue of a possible link between tones in tone languages and absolute pitch by demonstrating (through polling) a higher incidence of absolute pitch in East Asian music students in American universities (Gregersen et al. 1999). Gregersen et al. attributed this to the greater popularity of the "fixed doh" ear-training in Asia—combined with the cultural trait of starting music training earlier in life (Gregersen et al. 2001). Deutsch et al. provided the objective confirmation for the supposed "language effect" in the greater percentage of absolute pitch possessors at the Central Conservatory in Beijing than at Eastman School of Music by administering there frequency discrimination tests (Deutsch et al. 2006). Deutsch et al. concluded that learning early in infancy to perceive pitch flexions in lexic tones must somehow prepare East Asians to learn absolute pitch better than early speech experience of Western children. Previously, Deutsch et al. established that speakers of Vietnamese and Mandarin routinely intoned their lexic tones in the same word with nearly absolute precision—with fluctuations well within a semitone—between different sessions despite breaks up to 2 days (Deutsch et al. 1999). This prompted the authors of the study to suggest that absolute pitch must have originally evolved as a feature of speech that naturally occurred in tonal languages (Deutsch et al. 2004). Rakowski and Miyazaki put it on the account of a weaker inhibitory effect of acquisition of verbal prosody in growing infant speakers of tone languages than those of nontonal languages (Rakowski and Miyazaki 2007).

However, the totality of known data on geographic and historic distribution of absolute pitch across the world, listed above, contradicts the hypothesis of linguistic origin of absolute pitch. It is quite possible that early exposure to lexic tones provides a fertile ground for developing an absolute frame of reference for production of certain vowels, but there is a long stretch from this to the formation of the practice of breaking the octave into a set of absolute pitch classes. Of course, we know of cases where characteristic patterns of lexic tones in African languages influenced patterns of musical TO in indigenous musics of the same ethnicities (Kubik 1985). However, this influence does not include absolute pitch—speakers of African tone languages do not seem to observe the sameness of pitch levels either in the production or perception of lexic tones (Schneider 1961). Kubik, one of the leading experts on TO of African music, when he speaks of linguistic influences, brings up not the issue of absolute pitch but the "deep listening" mode designed to zoom into the spectral properties of sounding objects, following the model of human voice, which includes the speaking voice (Kubik 1985). Lexic tones might provide the model of pitch reference for "verbal ear" at the earliest stages of infant's speech development as well as the beginnings of music acquisition. But such modelling definitely does not encompass the incrementation of pitch changes and the consistency of certain pitch levels that could possibly prototype a "pitch class." Stability of pitch levels discovered in speech is sustained per individual, and only on average (substantially varying in standard deviation), which precludes the functional use of absolute pitch in verbal communication and makes its phylogenesis impossible.

What unites the acquisition of absolute pitch and the acquisition of lexic tones is their time frame—they both occur during the sensitive critical period during the first years of life. For the majority of its possessors, "natural" absolute pitch acquisition falls within what is colloquially called the "language window" that stays wide "open" throughout infancy and gradually shuts down into the adolescence, making acquisition of a second language much harder (Myers 2009, 214). Critical periods are crucial for automatic acquisition of basic conspecific skills of communication in many animal species, probably directed at securing the most accurate reproduction of conspecific auditory signals at the earliest developmental stages, when effective communication with a caretaker is most valuable. For humans, the manifestation of such accurateness is the absence of foreign accent in languages acquired during the critical period (Lennenberg 1967, 175–82).

A musical equivalent of a foreign accent (a "musical accent") was first noticed and described by Roman Jakobson in 1932 (Jakobson 1987, 455–6). Thus, attempts of Yehudi Menuhin to improvise raga on his violin are perceived by indigenous Indian musicians as inauthentic due to deviations from normative tuning and intoning of standard raga motifs, while attempts of Lakshminarayana Subramaniam to play pieces from Western classical violin repertoire appear to Western musicians as carrying a foreign "Indian" accent in violin intonations.[19]

- The presence of "musical accent" manifests the *competition between discrimination of pitch in two music systems* by a performer essentially in the same way how presence of "linguistic accent" manifests the competition between two phonemic systems in a speaker.

The evidence for the presence of "musical accent" was demonstrated by Matsunaga et al. in bimusical subjects competent in Western classical and Japanese traditional music: each form of music activated different subregions in the brain around the inferior frontal cortex and the premotor cortex (Matsunaga et al. 2012). A follow-up study showed that the more efficient processing of both music systems occurred in those listeners who were more proficient in the secondary, non-native, music system—in parallel to the similarly effective bilingual communication by those bilinguals who were more proficient in L2 (Matsunaga et al. 2014). This suggests that primary and secondary learned repertories of verbal phonemes, as

[19]This example was suggested and substantiated by Jivani Mikhailov in personal communication.

well as of musical pitch classes, interact with each other based on their spectral characteristics, supporting the inclusion of a bilingual into both speech communities as well as into both music communities.

The phonemic system of a language and a method of TO employed by a music system as "normative" both are tied by mutual reflection of grouping, metric, durational, contour, and timbral patterns of organization (Lerdahl 2001). The "parent-child" developmental relation between timbral hearing and pitch hearing transpires into a number of direct analogies in TO between native music and speech, manifested in a way how an auditory stream is parsed into meaningful patterns. Verbal intonations of English, Farsi, Mandarin, and Tamil languages share the greatest concentrations of power in the normalized spectrum of speech sounds with the most frequently used musical intervals—probably reflecting "a statistical process by which the human auditory system links ambiguous sound stimuli to their physical sources" (Schwartz et al. 2003). Amplitude—frequency pairings in human utterances tend to be distributed in a way that discloses biologically significant information about the speaker (i.e. speaker's sex, age, and emotional state) in ambiguous auditory signals—by means of TO of prosodic intonations. Over time, statistical relationship between perceptual responses and naturally occurring sounds is likely to form some kind of temperation, once they include the full sonic characteristics rather than isolated phonetic features. This is akin to the division of an octave in 12 semitones with optimal harmonic distribution of the simplest ratios of 5/4, 4/3, 3/2, 5/3, and 1/2 falling exactly on the I, III, IV, V, VI, and VIII=I degrees in a heptatonic pitch class set (Shepard 2010).

The primordial foundation for TO of frequency-based music seems to come from the voiced elements of speech (viz., the vowel), which is the most significant auditory signal to humans (Terhardt 1984). The task of parsing the speech involves extraction of virtual pitches from complex tones even where fundamental tones are not present. Musically trained listeners tend to focus on the fundamental frequency of a musical sound, whereas non-musicians are more likely to focus on its fused overtones (Seither-Preisler et al. 2007). This finding indicates that musical pitch is derived from verbal harmonicity that is projected onto music by means of learning to infer the missing fundamental from the harmonics of its harmonic series. Moreover, processing of pitch in speech shares with music yet another important feature. It is essentially hierarchic, based on listener's knowledge of contourization and generalization, with substantial abstraction in processing of the low levels of the auditory hierarchy (Terhardt 1992). Developing a sense of consonance is likely a by-product of acquisition of verbal competence, according to the virtual pitch theory by Terhardt. This must be the underlying mechanism that determines "natural" acquisition of absolute pitch and its connection to Western temperament. It is possible that each musical culture initially captures the statistically prevalent features of TO present in natural acoustic stimuli (Monson et al. 2013), and once TO becomes centered on frequency, thereafter succumbs to the process of optimization of pitch classes based on the harmonic contrasts, which results in 12-tone temperament.

- According to this scenario, Western temperament presents the optimal distribution of pitch values that satisfies horizontal (melodic) as well as vertical (harmonic) axes of TO, which might be the best trade-off for various musical cultures once they absorb Western forms of multipart music. This could explain the overwhelming adoption of Western tonality across the globe by so many countries despite their cultural and political differences, even when their governments pursued nationalistic anti-Western policies (e.g., China, India or Iran).

It seems that the discovered prevalence of absolute pitch in East Asian population is the consequence of the growing popularization of Western classical and popular music along with the Western lifestyle that promotes objectivization of auditory perception and rationalization of the schemes of its processing. The experience of perceiving and producing lexic tones plays into the musical ear development, directed by the adoption of Western music and social advantages of receiving the musical prodigy status. In the modern globalized culture, East Asian prodigies receive the same prominence that was held by Jewish prodigies during the past two centuries. The totality of the phylogenetic information in regard to absolute pitch indicates that absolute pitch was *imported* into Asia rather than generated there. The direct evidence that could clarify this issue can be attained through the comparative analysis of TO of the extensive pool of the earliest infant vocalizations across different countries and age groups.

## 7.9 The Prospects of Using the Ethnomusicological TO Analytic Methods for Infant's Vocalizations

Like adult musicians who practice timbre-oriented music, Western 0–3-year-old infants (not to speak of infants in non-Western cultures) most certainly process music not by pitch classes, which they cannot even reproduce faithfully within first 1–2 years of life. Musical "timbre classes," similar to phonemes of verbal speech, present the likely alternative, united in musical modes as suggested by the extensive research conducted on the territory of the former Soviet Union from the 1920s on. Bolsheviks dedicated an unprecedented amount of resources (specialists, institutions, programs, etc.) to the investigation of traditional folk cultures to fulfill their ideological claims of the advantages of "people's music" (Krader 1990), leading to a swift advance in the understanding of TO in folk music (Zemtsovsky 1967). Indigenous peoples on the territory of the USSR varied enormously in their lifestyles, from urban industrial to nomadic hunting/gathering. The comparisons of their lifestyles and cognitive styles common in these societies supplied researchers with invaluable comparative perspective.

In 1925, Asafyev created and headed the scientific-musicological department at the Leningrad conservatory that included a musical-ethnographic specialization (Mekhnetsov 2014, 330–1). Groups of Soviet ethnographers went on expeditions to collect samples of songs, transcribe them, publish their results, and engage in scientific discussion with other experts with the purpose of understanding and identifying the principles by which these musics were made. By the 1950s, the whole Soviet Union had dozens of institutions performing such research (Ivanova 2009). During the 1940s, dedicated centers of folkloric studies were created at all major conservatories, leading to the accumulation of substantial databases and scholarly research. The Moscow Conservatory collection alone contains over 140,000 units of folk recordings, covering almost all districts and regions of the former Soviet Union, and even many foreign territories (Giliarova 2010). The Phonogram Archive of the Institute for Russian Literature in St. Petersburg contains 150,000 records, starting with wax cylinders of the 1890s, made in expeditions to Siberia (Korguzalov and Troitskaya 1993).

All major musicologists active in the USSR territory from the 1930s onward researched folkloric music. All graduate students in musicology and composition were required to take an ethnomusicology course and participate in field studies (Mekhnetsov 2014, 309–53). The structural principles of tonal organization uncovered in the research were subsequently used in reproductions of folk music and arrangements of folk music for different vocal and instrumental settings and new original compositions that emulated folk music styles. All of these works were performed through professional concert agencies and a wide network of amateur music clubs—which by 1946 included nearly 70,000 clubs in the Russian Republic alone and in 1977 involved 15 million performers participating in the multinational annual festival of folk music (Yakovleva 1973).[20] Such a production cycle of national folk music was administered in all republics and autonomous regions of Soviet Union and involved evaluation of the authenticity of produced music by experts in native indigenous cultures.

Sound recording and distribution of folk records through LPs and radio occurred on a massive scale—the Melodiya catalog of folk albums released between 1970 and 1990 has 501 entries, and this counts only authentic performances, excluding popular, light, and classical arrangements of folk music (Zemtsovsky 1991). The access to free advanced musical education allowed indigenous population to become musicologists, supporting nearly ideal integration of emic music theory and performance practice with the etic methodology and scientific methods of research. This along with enormous breadths of sample base for ethnomusicological analysis and interactive coordination of thousands of researchers have ensured the reliability of the analytic framework of Soviet-era music theory. The 1926 Census, by the time of the start of the ethnomusicological research programs, listed 169 resident ethnicities in the USSR (Semenov 1928). It would be justified to conclude that by the time of the collapse of the Soviet Union, the analytic knowledge of TO of traditional

[20]Certainly, the research in and production of folk music were supervised and coordinated by the Communist authorities, which involved promotion of certain forms of music and suppression of some other forms (Frolova-Walker 1998). However, despite all abuses of formalistic treatment of national musics at the territories of USSR, the scale and quality of research, as well as the extent of popularization of folk music, was unparalleled by any Western country (Zemtsovsky 2002).

indigenous music was about as deep and thorough as the analytic understanding of compositions of Western classical music.[21]

The theoretical base for the analysis of early forms of pretonal music was developed in the works of Eduard Alekseyev (Alekseyev 1973, 1976, 1986, 1988, 1990, 1993, 2013), summarized in English by Nikolsky (2015a). Alekseyev's model of primordial archaic TO is based on the theories of his predecessors: musical intonation analysis by Boleslav Yavorsky (Protopopov 1930), its sociogenetic model developed by Asafiev (1952), Tull and Asafyev (2000), its psychoacoustic foundation experimentally formulated by Nikolai Garbuzov (1980) and Yevgenii Nazaikinsky (1972), the corresponding theory of music form integrated with semantic analysis by Leo Mazel (1952, 1982), the systematic theory of style and mode elaborated by Sergei Skrebkov (1967, 1973), the theory of modal evolution in comparative ethnomusicology formulated by Viktor Beliaev (1990), and the generalized systematic theory of intonation by Izaly Zemtsovsky (1980, 1987). Following the idea of Eduard Alekseyev of similarity between the earliest phylogenetic and ontogenetic forms of TO (Alekseyev 1986, 14) and similar ideas by a few Western scholars (Winner 1982; Garfias 1990), I have collected the samples of children musical vocalizations and analyzed their TO, currently preparing this material for publication as a monograph. Below are the excerpts from this research related to the development of verbal and musical specialization in earliest children vocalizations.

There are a number of studies, mostly by Western scholars (Soviet and Russian researchers have completely neglected this line of research), of patterns in spontaneous musicking and its precursors in the natural behavior of young children (Bentley 1966; Michel 1973; Ostwald 1973; Moog 1976; Moorhead and Pond 1978; Davidson et al. 1979; McKernon 1979; Papoušek and Papoušek 1981; Ramsey 1983; Dowling 1984; Flohr 1984; Hopkin 1984; Davidson 1985; Kratus 1985; Davies 1986; Hargreaves 1986; Swanwick et al. 1986; Holahan 1987; Kalmar and Balasko 1987; Kelley and Sutton-Smith 1987; Ries 1987; Merrill-Mirksy 1988; Kratus 1989; McDonald and Simons 1989; Flowers and Dunne-Sousa 1990; Fox 1990; Fujita 1990; Campbell 1991; Bjørkvold 1992; Davidson 1994; Papoušek and Papoušek 1995; Hargreaves 1996; Papoušek 1996a, b; Marsh 1997; Campbell 1998b; Campbell 1998a; Harwood 1998; Sundin 1998; Dowling 1999; Adachi and Trehub 2000; Kreutzer 2001; Révész 2001; Swanwick 2001; Burton 2002; J. Davidson 2002; Tafuri and Villa 2002; M. Barrett 2003; Young 2003; Lew and Campbell 2005; Mang 2005; de Vries 2005; M. S. Barrett 2006; Custodero 2006; Gembris 2006; Young and Marsh 2006; Miyamoto 2007; Young and Gillen 2007; Tafuri et al. 2008; Addessi 2009; Emberly 2009; Lum 2009; Marsh 2009; Forrester 2010; Gillen and Cameron 2010; Yennari 2010; M. Barrett 2011; Dean 2011; Elmer 2011; Campbell and Wiggins 2012; Elmer 2012; Koops 2012; Swanwick 2015).

[21]See the brief summary of the modal typology, elaborated within the Soviet systematic musicology, and the example of modal analysis of the indigenous music in the Appendix-1, "Taxonomy of tonal organization of modal music" (Nikolsky 2015a).

Most of them, with notable exception of Elmer[22] and perhaps Bjørkvold, suffer from the same methodological shortcoming of analyzing children musicking in terms of its similarity to typical samples of adult frequency-oriented music, especially Western tonality. Whenever the sounds of children's music violate such rules, this is put on the account of children's poor hearing or/and poor vocal control (Hutchins and Peretz 2012).

Both of these reasons are definitely valid and common in limiting children's ability to create music, but more likely than not, in practice, they *limit the reproduction of melodic contours and timbral properties of children vocalizations* rather than the *reproduction of pitch classes and pitch class sets*. The latter skills typically are formed no earlier than by the age of 4 years, when children develop intuitive competence in TO rules of their native music systems (Louhivuori 2006). Exceptions to this time frame are few and reserved for precocious children raised in families where at least one of their caretakers is a professional or an advanced amateur musician (Kelley and Sutton-Smith 1987). The rest of the 0–3-year-old children who develop within the "normal," non-expedited time frame, must follow not pitch-oriented model but pitch contour and timbral models of some kind—only gradually discovering the pitch relations at first between the adjacent tones and later between remote tones. One cannot make mistakes in the production of pitch classes, if he/she is unaware of pitch classes. What appears as infant's "poor singing" to a researcher, in fact, might be not "poor" at all to an infant, but quite satisfactory rendition of a goal set to reproduce a particular pitch contour or a combination of a few timbre classes. There is evidence that sometime between 6 and 12 months of life infants switch from the biologically driven mode of musical tuning to a culture-specific form that characterizes their native music, which involves perceptual reorganization for musical tuning (Lynch and Eilers 1992). And the earliest child songs in different musical cultures across the world might share many common features (Campbell 1991), perhaps even constitute a single universal type of TO, based on indefinite pitch/timbre.

Infants' singing is no more pitch-"defective" than their first attempts to speak are phoneme-"defective." However, philologists already by the 1920s realized that deviations from "correct" adult speech by young children are systemic and reflect the peculiarity of children thinking (Chukovsky 1963). Piaget and Vygotsky made it clear that child's speech is organized by its own principles that are different from adult speech, and this has been accepted as a status quo in academia. Russian research especially has been instrumental in uncovering the systemic principles of semantic and phonetic aspects of the earliest verbal development (Shvachkin 1948).

[22]Elmer (2012) objects the approach of most researchers who evaluate early children's songs in terms of diatonic intervallic type of Western tonality and define stages in tonal development based on correspondence of children songs to the samples of diatonic music. She proposes to use quarter-tone representation of children musicking and seeks to define stages of music acquisition without committing to diatonic quantization, relying not on compositional analysis of children vocalizations, but rather on behavioral aspects of their singing. Needless to say, the purely behavioral approach is futile for defining the features of TO and their interpretation in comparative analysis.

Unfortunately, the analogous idea that child's art does not perform the same functions as artworks of an adult (Vygotsky 1971, 240–262) has found less recognition, especially in regard to music (Bjørkvold 1992; Elmer 2011, 2012). The explanation of it certainly has to do with the incredible shortage of studies of forms of children musicking in non-Western societies. Only a handful of publications in English discuss the issue of musical ontogeny in non-Western indigenous musical cultures (Blacking 1967; Buckton 1982; Hopkin 1984; Merrill-Mirksy 1988; Garfias 1990; Campbell 1991; Bjørkvold 1992; Fernald 1992; Papoušek 1996a; Campbell 1998a; Campbell 1998b; Harwood 1998; Kreutzer 2001; Mang 2002; Minks 2002; Stige 2002; Lew and Campbell 2005; Nettl 2005; Levin and Suzukei 2006; Emberly 2009; Lum 2009; Marsh 2009; Gillen and Cameron 2010; Campbell and Wiggins 2012). The lack of cross-cultural comparisons is to blame for strong Europocentric bias, pronounced in the Western literature on music development—it is much harder for a Western developmental psychologist to catch himself/herself on circularity of their interpretation of infant's musicking than on their interpretation of infant's earliest speech, because of the first-hand knowledge of how much languages can differ in their phonemic and morphonological organization. Also, language provides a more obvious connection between the sound and meaning due to the referential nature of its semiosis, which makes it easier to see that children's earliest language uses nonconventional child-made words and ascribes unconventional meanings to conventional words.[23]

Yet another problem is that many developmental psychologists simply do not know how to approach the analysis of early children's music. What adds to confusion is the lack of clarity of what exactly constitutes "children's music." Thus, nursery ditties and rhymes do *not* constitute children's music and poetry—these are adult's products created for children within a cross-cultural folk genre (Campbell 1991). The same applies to motherese (Mampe et al. 2009). Such forms of musicking can reflect the patterns of TO characteristic to children's musical thinking albeit only indirectly, provided that caretakers who created it resorted to only those patterns that were indeed perceptible and meaningful to children. However, in practice, many of the tunes that have earned a stable place in a repertoire of "children's music" (commonly performed by caretakers and often reproduced by children), in fact,

[23]For instance, Shvachkin gives examples of a boy who used the word *dany* in reference to a bell, a clock, a telephone, and a bellflower or a girl using the word "moo" to refer to a cow and to a big bird, whereas another girl using the very same word in reference to a cow and to a big dog (Shvachkin 1948). Such polysemantism often spreads as broad as to include the opposites of the concepts: e.g., the word "boo" in reference to lighting a candle up as well as to turning it off. Shvachkin explains the origin of this polysemantism by the Vygotskian concept of *emotional experience* (*perezhivaniye*—literally, "living through"). The first words of children refer to their "emotional experience" of a particular object rather than to an object in itself. The meaning of a word is comprised by the complex set of emotional experiences from perception of objective, affective, and functional characteristics of surrounding objects in reality. Their admixture is initially syncretic and poorly differentiated, semantically diffused to the extent of resembling a vague semantic circle over a delineated center point—e.g., the word *foo* used to refer to anything that has to do with warmth (p. 102).

were composed by classical composers abiding by the rules of Western tonality and therefore are way too complex for adequate acoustic analysis by infants who are likely to misunderstand the TO of this music.[24] By no means such compositions are representative of early infant's indigenous musical thinking, and their popularity obstructs the task of identifying the proprietary children's features of TO in children's musicking.

The problem is that once an infant is exposed to an adult song, he/she is likely to try to imitate it only as much as their perceptory capacities would permit (Peery and Peery 1986). But by ethnomusicological standards, such kind of reproduction constitutes an inauthentic borrowing from a foreign musical culture, not representative of the indigenous traits of the investigated culture. It is crucial for developmental specialists willing to investigate the TO of genuine children music to distinguish between infant's reproduction of adult's music and infant's proprietary music created as a spontaneous original improvisation that elaborates a musical idea preselected by a child (McKernon 1979). This might not be a simple task, since any musical idea picked by a child is likely to originate from some sounds once overheard in the child's environment. The deciding criterion should be the quantity and the scope of the prototyped features in a vocalization under question: usually, a deliberate reproduction of a tune involves reproduction of many motifs, rhythms, and words borrowed in their totality from the source song, whereas child's original music usually elaborates just one-two musical patterns that are interesting to a child and inspire a playful experimentation. Such music, created *not* in response to the adult's singing or playing and *not* fulfilling a request to sing a particular song, but generated to own lyrics or meaningless syllables, is usually termed "invented songs," (also "self-invented"), "narrative songs," or "spontaneous improvisations" (Moog 1976).

The framework of the so-called "ekmelic mode," based on indefinite pitch and indefinite size of pitch intervals, theorized by Soviet musicologists (Nikolsky 2015a), appears to be applicable to the analysis of the earliest samples of such songs. Moog (1976), Campbell (1998a, b), and Bjørkvold (1992) left thorough descriptions of different types of spontaneous vocal improvisation by children, many of which accompany other activities, such as playing or reading, or appear to be created without any purpose, merely to amuse themselves (see the video footage supplemented in Forrester 2010). Noteworthy, in ethnomusicology, ekmelic organization is associated with "song-for-oneself" as opposed to "song-for-others" (Alekseyev 1986, 12). Songs "for-others" are consumed collectively, which promotes rhetoric principles of communication, leading to the intense development of technical skills and compliance to some conventional model. Songs "for-oneself" do

[24]For example, *I'm a Little Teapot* and *How Much Is that Doggie in the Window* contain many leaps in different directions that make these tunes hard to reproduce for very young children; *O Little Town of Bethlehem* contains many chromatic alterations and *Frosty the Snow Man*, and even some folk tunes, such as *Deck the Hall* include modulations that demand reliable recognition of multiple pitch class sets. Such songs represent more of what those adults who were brought up on the Western tonality believe sounds "childish" enough to be suitable for children musicking.

not have to convince anyone in anything. Sheikin (2002, 304) nicknames personal singing tradition, prevalent among the majority of Siberian ethnicities, as fundamentally "Cartesian": "I sing therefore I am." The manner of usage of individual song here is quite similar to "safety signals" employed by animals that live in social groups. Such songs merely accompany common occupations: riding, fishing, sawing, etc. and are produced instinctively, with little conscious control. This is quite analogous to children's "narrative songs" that accompany activities like playing, reading, drawing, etc.

Many infant's songs are essentially "songs-for-oneself." Jay Dowling describes infant's first songs with indiscrete pitch intervals that are very different from the adult music, wandering and sounding "out of tune" to an adult ear, with a high variability of interval sizes and a drift in a "tonal center" (Dowling 1984). Such melodies, usually 5–9 phrases in length, as a rule include a refrain-like repetition of the same melodic pattern. They are sung in rhythms that remain simple and steady within a phrase, making the variants of the "same" song recognizable. Multiple variants rotate in a child's repertoire for a period of no more than 6 weeks—to be replaced by a new song with its new variants. Dowling insists that these are indeed variants, and not a bunch of unrelated songs, based on the similarity of their melodic material and of their lyrics/vocables. Such songs are spontaneously composed starting from the age of 2 and differ from "cover songs" by increased concern for privacy. A child performing "song-for-oneself" is highly conscious of being observed and can easily cease singing from feeling intimidated. Burton conducted the interviews of the children who engaged into spontaneous song making: most of them could not answer the question what were they thinking about during singing—indicating that they did not consciously pursue the reproduction of any specific musical model in their choice of sounds (Burton 2002). This agrees with Moog's (1976) conclusion that spontaneous songs of 2-year-olds give an impression of a narrative directed at communication with oneself, using lyrics that make no sense to others. Barrett also describes child's song as transitional event through which a child symbolizes his/her personal feelings and articulates his/her understanding of the encounter with a world (M. Barrett 2003).

Similar descriptions of self-invented pretonal songs-for-oneselves of 2–3-year-old self-conscious singers are made by Helmut Moog (1976), Margaret Barrett (2011), and Lisa Koops (2012). The use of discrete pitches that wander around [resembling the ekmelic "stretchable" intervals] is reported by Bentley (1966), Moog (1976), Moorhead and Pond (1978), Davidson et al. (1979), and Révész (2001). In the Boston Project Zero, children were found to acquire the ability to use discrete pitches by the age of 19 months—although limited to small size intervals. According to the longitudinal study of infants' singing in home environment, 7-month-olds mostly used the major second in ascending and descending directions in legato articulation (Ries 1987). At the age of 11 months, the interval of a third became common as well, although not as much as second. The interval of a second retained its dominance in the singing of 19-month-olds, although the use of wider intervals increased. McKernon confirmed the overall dominance of a second by the age of 2, reporting that major second, minor third, and unison (in this respective order) were the most commonly used intervals (McKernon 1979). It is very likely that this reported major second-minor third was actually a single indefinite (stretchable) interval of the ekmelic complementary step, whereas unison-minor second constituted the ekmelic anchoring unison (Nikolsky 2015a).

David Hargreaves, in addition to such intervallic preferences, points out the repetitiveness of the melodic contours and of the rhythmic patterns as important features of organization in the spontaneous songs of 2-year-olds, so that the consistency of the melodic scheme compensates for the wandering pitch levels (Hargreaves 1986, 69). Alekseyev (1976) regards this *formulaic* rigidity as one of the most characteristic traits of Siberian "songs for oneselves," called to compensate for the looseness of the "stretchable" intervals. The third important point of similarity between ekmelic music and earliest attempts of music-making by infants was noted within the research project of the Boston Project Zero group in relation to the typical method of composition engaged by 2–3-year-olds. They displayed possession of some basic concept of the "frame of a song"—akin to a skeleton without flesh (Davidson 1985)—something that closely resembles the musical form of variations on the ekmelic melodic *formula* favored by many Siberian indigenous ethnoses. Formulaic organization is common in self-made songs of 2–4-year-old children—possibly, cross-culturally (Bjørkvold 1992; Mang 2005).[25] And the fourth common "ekmelic" trait of early infant's music was reported by McKernon. The most common melodic typology of the earliest formulas was *undulating* rather than simple ascending or descending shape (McKernon 1979). Papoušek and Papoušek (1981) considered undulating contours to be a general characteristic of vocalization from birth to 1.5 years of age. Based on his communication with folk performers, Alekseyev (1986) explains the strong preference of Siberian indigenous musicians for undulating formula by their strategy of "scanning" the intervallic space reserved for the ambitus of a song. Life in wide open spaces that lack landmarks promotes the adoption of such strategy of orientation in unfamiliar places (Nikolsky 2016, Appendix-7). Tracking pitch in a pitch contour can follow the model of tracking visual motion (Huddleston et al. 2008). Perhaps, young infant in a room finds himself in a situation like an indigenous Siberian travelling over an unknown terrain, therefore resorting to the same "scanning" strategy and committing to the strict application of the wavelike contour. Preceding Alekseyev, Ellen Winner drew the parallel between undulating typology of the children melodies with that identified by ethnomusicologists in the music of indigenous peoples (Winner 1982, 234).

Scalable intervals in ekmelic mode can be viewed as the compensation for increased verbal skills that put pressure on music users to increase the *pitch strength* (pitch saliency)—the relative strength of the perceived frequency component of the complex sound as opposed to the overall spectral content of that sound (Shrivastav

[25] A well-known example of the cross-cultural formula that has earned the reputation of the "universal chant" is the succession of descending third, ascending fourth, and descending second, closed by another descending third (Hargreaves 1986, 68). This formula constitutes a musical semiotic phenomenon because it is universally related to playful teasing (Bjørkvold 1992, 71).

et al. 2012). In general, pitch strength might be defined as the perceived difference between the total sound and its timbre (Yost 2009). The impulse to stress the frequency modulations in a pitch contour selected for babbling by an infant might prompt representation of that contour in terms of a succession of melodic "steps" of variable size. The pragmatics of communication directs the growing infant toward developing two communication systems, each specializing in its own type of information: referential for speech and affective for music. The necessity to distinguish verbal intonation from musical one pushes the development away from leaning on timbral TO toward reliance on pitch. Timbre still remains important—much more so than in the music of tonality: e.g., vibrato, pitch bends, contrasts in sound production, as well as in articulation styles, all contribute to significant variations in tonal quality of ekmelic degrees. However, in ekmelic mode, comparing to the earliest timbral-oriented vocalizations, more attention is attracted to pitch in transitions from one tone to another. Ekmelic songs generate *consistent musical intonations that adopt formative functions in determining the musical mode*. Even more importantly, four basic ekmelic functions of anchoring, complementing, opposing, and extreme relations between musical tones that belong to different registers each receive a distinct emotional denotation.[26] Together with the ascending and descending melodic inclinations, acquired during the first year of life, this lays the foundation for a purely musical system of emotional communication.

During the first 3–4 months of life, children employ various utterances in intuitive attempts to establish communication with their caretakers—which can be regarded as a "prelinguistic alphabet" (Papoušek 1996b). During that time, 82% of vocalizations consist of descending glissando patterns within a narrow diapason, despite the fact that infants can vocalize across a very wide range of frequencies (Fox 1990). Vocalizations become susceptible to "vocal contagion"—convergence between baby's voice and that of a caretaker (Ostwald 1973). At about 10 months of age, infants start vocalizing following their own spontaneous patterns that resemble intonations of their native language (Boysson-Bardies et al. 2008). The specifically

[26]Degrees can *anchor*, *complement*, *oppose*, or *extremize* (polarize) each other, projecting melodic attraction or repulsion. These functions do not permit conservation of pitch but do allow for rough estimation of intervallic distances between salient melodic tones. Each function becomes associated with a particular registral span in correspondence to that function's valence. The larger the span, the more dissonant the melodic relation. Thus, anchoring is the smallest in range [functionally equivalent to the melodic unison], and it always strengthens the degree, with a confirming or insisting intonation. Unlike the anchoring, complementing intonation is larger in size, and one of its tones always attracts the other, making a softer impression than the anchoring intonation. The next in line of size is the opposing intonation that always involves rivalry, either between two anchors or two complementing tones. Finally, the extreme intonation gives the largest interval, generating the relationship of maximal discontinuity between two tones, causing rivalry not only between the tones but between the margins of the register—often involving timbral contrast. Consistent usage of these functions puts in place the notions of four intervallic zones, relative in size, distinguished by melodic functionality and semantic values. For more information see "Ekmelic mode" in (Nikolsky 2015a).

"musical" features, such as phonation that produces resonant frequencies in the vocal folds, are evidenced from 2 months onward (Tafuri and Villa 2002).

David Hargreaves qualifies the first period of musical development as "sensorimotor," delimiting it by the first 2 years of life (Hargreaves 1996). The landmarks of this period are the onset of babbling and rhythmic dancing. Both of them provide fertile soil for the separation of musical sound production from speech. This distinction was first drawn by Helmut Moog in his longitudinal study of musical behavior of 50 infants along with supplementary information from their parents (Moog 1976). He distinguished musical babbling from non-musical (verbal) babbling. The latter was found to appear first, in the first year of life, and did not seem to constitute a response to the environment, serving rather as the precursor of speech in a format of playful exercises with syllables. In contrast, musical babbling occurred in response to musical stimuli heard in a child's environment. Such a reaction sets on at the age between 3 and 6 months and is triggered by awareness of some tonal unity present in the auditioned sounds (i.e., realization of a musical mode).

Musically, babbling is made of sounds of varied pitch, sung out on a single vowel. Moog characterizes their structure as "amorphous," showing no rhythm or pitch accuracy, with sporadic pauses inserted by whim. Dowling disagrees with this observation, considering rhythmicity to be a distinctive feature (albeit a fuzzy one) between speech and music babbling (Dowling 1984). Both Moog and Dowling agree that wandering pitch and stretchable intervals together structurally distinguish musical babbling from verbal. The downward melodic direction prevails, often involving glissando slides from pitch to pitch. The successive sounds are mostly proximal pitch-wise, with rare descending leaps up to fifth, unclearly intoned. Such singing is accompanied with physical movement of increasing variety, involving the feet, head, and hands—synchronizing the physical and melodic motions. This coordination grows to become consistent throughout the second year of life, after which seizes to improve.

John Holahan provides another valuable account of the development of babbling. In his longitudinal 2-year study of musical babbling in the environment of a day care center, he identifies three levels of babble (Holahan 1987). At the first level, children typically do not babble apart from musical stimulation and generally put together discrete musical elements to synchronically track the salient features of the external musical stimulus, as if trying to guess the features of their TO. At the second level, they put together the *combinations* of musical elements to sing along the external music, as well as spontaneously perform their own music apart from the external stimulus—but their music does not give rise to patterns of tonal or rhythmic organization present in adult music. At the third level, their babbling becomes more coherent and only resembles, but is not identical to the musical features of the familiar songs. At this level, the first signs of a recurring "pitch value" and a consistent "tempo value" appear, indicating that children become aware of the relationships between the successive sounds of the melody, engaging their short-term memory (this level corresponds to the emergence of "ekmelic mode").

Hargreaves (1996) reports that compositional efforts of sensorimotor children demonstrate their awareness of the melodic contours while basically presenting experimentation with sounds, durations, and their grouping. A sensorimotor child's attempts to represent the music graphically on paper produces "action equivalent" scribbling. Swanwick et al. (1986), who studied not only singing but also instrumental improvisations of children, also qualify compositions of the ages 0–2 years old (what they call "sensory level" of the "material stage") as unpredictable and explorative, based on pure "sensuous engagement with sound materials" (p. 316). Swanwick notes that the compositional choice here seems to come from taking pleasure in the sound itself, especially in timbre and opposites of loud and soft (Swanwick 2015, 113). Conversely, children pay no attention to the significance of variations in timbre—similar to their disregard for steadiness of pulse or adherence to a specific pattern.

Swanwick's account captures the essence of TO at its primordial stage. Children center either on timbre or loudness in their choice of tones for musical babbling. They might restrain the timbre from changing and vary the loudness or vice versa. Variation is processed not incrementally, but by opposition: loud sound against soft, similar timbral quality against contrasting one. It is possible that in practice one sound would come out as slightly louder or softer, thereby producing a dynamic value between the opposites of loud-soft. The same applies to timbre. Often such variations are not intentional but are by-products of experimentation. In the same vein, variations in pitch can hardly represent a compositional plan on the part of a sensorimotor child. His intelligence at this period is motoric: as is his use of his vocal folds. Just like his hands commit actions without any goals, merely exploring the resistance of the surrounding objects, so is his vocal folds constrict at certain points. The pitches most often come out as a side effect of an attempt to preserve the tonal quality of a particular timbre, and therefore fall not far from each other. The only exception would be dropping the pitch as a result of uncontrolled relaxation by the end of a phrase, at the expiration point. The importance of this stage is that it supports serendipitous forging of new musical modes, enabling production of original forms of TO.

The roots of ekmelic organization should be sought in the process of establishment of a "tonal constant"—a model of which was developed by Yevgenii Nazaikinsky (1973). According to him, the ability to perceive constants in auditory signals depends on the capacity to perceive an auditory stimulus as an object. This capacity is formed at the earliest stages of the development of musical hearing as a result of discovered correspondences between visual and auditory impressions. *A sound object* is detached from the auditory scene and is perceived and felt according to the salient features in its auditory characteristics, which quickly turn attention to timbre as the most "interesting" parameter suitable for playing with it. Timbre is usually experienced in synesthetic integration with the mechanism of sound production (action of singing or playing). Hearing a constant becomes tied to perceiving it as a sound object by the following factors:

1. Stability and permanence characterize objects more than anything else.
2. An auditory constant is easier to detect as a property of a sound source located in a certain acoustic space—where the generator of a sound becomes an object, be it physical or imaginary.
3. A constant is defined via breaking an auditory scene into foreground and background, the latter involving the knowledge of possible sound generators, of acoustics in a room, of size of a sound source, a distance from it, etc.—in a word, a context of audition that supports identification of a constant even at the absence of this or that particular piece of information.

Nazaikinsky stresses that this "sound localization" model for objectification of a musical tone separates music from non-music and establishes the foundation for commonality between perception of tonal and spatial organizations. The first advance in the development of musical perception occurs when a *relation* of two tones is realized as an auditory constant. Then a particular quality of relation of two physical objects becomes converted into a relation of two sound objects. The simplest element of such relation is a melodic contour—an "auditory line." Like a geometric line, it does not have thickness; it is not material but mental realization of unity, a grouping connection. Since music involves time, melodic contour is a *process*. The latter is very important, because melodic line is not like a line drawn by pencil. Melodic line is like an ongoing trace left by a moving body. If we were watching the motion, we see a trace. If we did not pay attention to the moving body, we see no trace. Here a sound object acts as an agent of melodic motion: its most salient characteristics charge melodic motion with certain tonal quality. The most characteristic and easily identifiable timbre is of the human voice, as well as of the musical instruments that come the closest to its imitation (e.g., oboe and bassoon). Therefore, *vocal intonation serves as a primary source for establishing the tonal constant*. In essence, the listener learns to focus on the linear connection between two musical tones, centering on its most salient perceptory feature—and takes it as a quasi-object of an auditory "observation."

It is in this sense that timbral mode at the beginning of a babbling period utilizes melodic contour without actually generating "melopoeia"—a skill of constructing a "good" melody (West 1992). No actual "composition" of tones in relation to their pitch value takes place here—what is conventionally considered "melody-making." Instead, tones are grouped in relation to their most salient features of timbre and loudness. However, these salient features are not perceived as separate entities, but rather as an organic part of the *tonal quality*, prototyped after the *voice quality*. The latter is an "auditory impression" from the "amplitudinal relationship of the partial tones of a voice," comprised of the conglomerate of pitch, loudness, register, and relative degree of adduction—in common words, a specific "sound color" (Eerola 2009, 167). A particular timbral mode establishes a tonal quality of some kind that is then maintained within a musical vocalization.

At this level, we can think of music-making at the atomistic level: defining a musical tone as an aesthetic object and manipulating it in the same way as how a child would manipulate a rattle or a ball. The transition from atomistic to molecular level constitutes the first advance in TO—construction of musical intonation that is indefinite and arbitrary in pitch. During the first year of life, infants learn to discriminate small timbral differences (Papoušek 1996a), which should enable them to combine tones that are slightly different in their tonal quality. There is experimental evidence that 6-month-old infants are capable of grouping tones into a pattern despite the presence of short pauses in-between (staccato articulation) (Thorpe and Trehub 1989). TO here is analogous to *sensorimotor play*—noteworthy, both playing behavior and musical babbling typically appear between the 12th and the 18th month of life (Gembris 2006). Papousek even considers vocal tract "the first naturalistic toy and/or musical instrument" (Papoušek and Papoušek 1995). Music comes into being from the discovery of a musical mode through playful solitary experimentation that eventually is turned into a form of emotional self-regulation.

## 7.10 Conclusion

This chapter started by reviewing the issue of the relation between ontogeny and phylogeny and showing the validity of projecting the patterns of individual acquisition of verbal and musical skills in early infancy onto the earliest stages in evolution of music and language. The lack of clear semantic boundary between music and language even in mature adult implementation makes structural psycho-acoustic distinction between their earliest ontogenetic forms the principal means of establishing their origin. And here instrumental is the analysis of their tonal organization (TO)—the manner in which music users and speakers combine tones of music and speech in order to convey their message. There is little controversy in relation to tonal organization of speech, which occurs primarily through the combination of phonemes, whose contrasts support the production of syllables and words that are assigned referential meaning via social conventions. The prosody of language, closely resembling the pitch aspect of musical TO, serves as the second important source of verbal semiosis. It is not primary because at the absence of meaningful words prosodic features alone cannot support reliable transmission of referential information, which constitutes the main reason for choosing speech rather than music as a medium of expression.

The only phonological trait that blurs the distinction between TO of speech and music is the use of lexic tones in tone languages. However, although lexic tones do employ changes in pitch levels and contours, resembling melodic pitches, the latter remain unique in their capacity to form melodic intonations that convey specific affective semantics: e.g., the ascending fanfare intonation expresses happiness, whereas the lamento intonation sadness. This turns melodic intonations into *idioms* that operate on the iconic basis. Ascending leaps of the fanfare are associated with the uplifting and energetic character of the happy state. Descending steps of the lamento are associated with the depressed and droopy character of the sad state. This is in contrast to the abstract nature of purely symbolic connection between lexic tones and the lexic meaning of words that contain lexic tones.[27]

However, TO of music presents a much greater intricacy than that of speech. The problem is that discrete incremental pitch, subdued to a single "pitch center" (tonic), which provides the clearest contrast to speech, constitutes just one of three general types of TO in music—tonality. The other one—modality—does not possess a pitch "center," featuring heterarchic rather than hierarchic TO. The third type—timbre-oriented music—does not rely on pitch classes, instead engaging continuous pitch modulations in melodic contours combined with timbre classes. This type comes the closest to speech due to its indefinite-in-pitch intervallic typology, making it hard to distinguish melodic intonations of timbral music from lexic tones in speech. The most effective way of achieving this is to use the pragmatic definition of music that postulates three criteria: (1) entrainment of psychophysiological responses to music to structural changes in music, (2) emotional contagion between the affective state assigned to a particular musical idiom and the actual affective state of a listener, and (3) transposability of intention to convey specific information by the repetitive use of the same idiomatic pattern on different occasions by different music makers. Together these criteria reliably exclude any form of speech.

The starting point in the development of vocal communication of human infants is the same as for primate cubs—the instinctive utterances that inform their caretakers of their affective state. Semantically, these cry-based vocalizations are closer to human music than to human speech. However, unlike any other species in the animal kingdom, human infants quickly pass through the profound transformation of their initial repertory of calls, based on interaction with the elderly and the experience of playing with sounds. As a result, they learn to manage modifications of vocalizations in pitch, rhythm, dynamics, and timbre within the framework of two-end communication with a caretaker. This communication gradually breaks into two general domains, determined by the pragmatics of delivering referential (speech-like) versus affective (music-like) information. The former is adjusted according to the efficacy of a receiver in retrieving the conveyed information, and therefore features a *dialogic* format, whereas the latter tends to prefer the *monologic* format. This is not a hard-bound distinction. An infant might resort to referential communication in a soliloquy format, which nevertheless usually retains dialogic traits, when the child speaks for different characters, either in imaginary play or in problem-solving situations. On the other hand, an infant typically learns patterns of affective communication via dialogic exchange with the caretaker, but then quickly

[27]The use of iconic signs in language is quite limited to phonesthemes and ideophones (Dingemanse et al. 2016). However, their contribution to semantics in modern languages is obvious perhaps only through a few onomatopoeic words. Iconicity seems to constitute the vestige of some earlier stage in the evolution of languages, supplanted by conventiolization as language develops the capacity to reflect the relations between multiple abstract concepts, which become more important than the relations between the sound of a word and its meaning (Ahlner and Zlatev 2010). The share of iconicity in languages of peoples that maintain preindustrial lifestyle and animistic ideology is rather higher than in languages of industrial countries and urban societies (Nuckolls 2004).).

switches to solitary play with the learned patterns, which seem to promote an aesthetic evaluation. In the process of ongoing playing (babbling), a child develops an expertise in production and perception of musical emotions through the repertory of dedicated idiomatic patterns of pitch, rhythm, and timbre, shaped by the dynamic enveloping (usually, a meaningful pattern is marked by a clear crescendo, diminuendo, or wave-like dynamic contour).

This leads to crystallization of two alternative ways of vocal communication—interactive dialogic propositional, characterized by rapid continuous updating of information, versus largely self-directed communication of a selected musical emotion to personal satisfaction, characterized by sustaining the same affective state for a rather long time. Dialogic communication becomes the main form of social interaction (externalized speech) and problem-solving (internalized speech) (Vygotsky 1987). Monologic self-communication, on the other hand, becomes the main form of emotional self-regulation (Saarikallio 2009). The ontogenesis of two corresponding methods of TO gives an insight into the phylogenetic origin of music and language.

However, the task of identifying the musical TO of the earliest stages in ontogeny of music finds a serious obstacle. The problem is that almost all research studies that contain the description and the analysis of the earliest forms of infant's musicking are compromised by "pitch reductionism." They evaluate acoustic features and patterns in children vocalizations in terms of the model of Western tonality as it is presented in adult's music. This retrograde application of the principles that rule the production of only one general type of music—"frequency-based" music—onto *every child's musicking* clearly constitutes a speculative assumption that is likely to severely distort our understanding of the basic principles of infant's music. It is hard to believe that infants who grow in societies that cultivate modality- or timbre-based musical cultures start their musical development with Western tonality. Even in Western countries, where tonality constitutes the main form of TO, children obtain the capacity to recognize the structural and semantic features of major and minor keys and process musical patterns in reference to those keys not earlier than by the age of 3–4 years. Earlier acquisition of "pitch class set hearing" occurs only for a few specially gifted who grow in a family of professionals or advanced amateur musicians.

A further complication is that pitch reductionism occurs not only in attempts of scholars to understand TO in early infants' musicking, but in processing of music by children themselves—in the form of the "natural" acquisition of absolute pitch. It seems that there is a genetic substrate of frequency discrimination that is common for a number of animal species, including humans, and can be activated upon exposure to Western schooling. Once children who have stronger predisposition learn to recognize pitch classes of the temperament tuning upon hearing music and to retrieve the names of these pitch classes, they rapidly, often without much training, acquire an automatic ability to reduce any musical and even non-musical sounds to the tempered pitch classes. This usually occurs more like a discovery—in contradistinction from the rigorously "learned" perfect pitch that can be attained by many, if not most, individuals after completing the course of effective ear training directed at memorizing the pitches of C major. Once internalized, they are abstracted into pitch classes to be projected on any key and any kind of music—even atonal (Krumhansl et al. 1987).

Unfortunately, the common reputation of absolute pitch as an attribute of "perfect" musical hearing that distinguishes the most talented performers and composers promotes extensive pitch reductionism. In reality, although many outstanding Western musicians, featured in history of Western music, indeed possessed "natural" absolute pitch, this gift was combined with other gifts that typically compensate for pitch reductionism. The most crucial of them are hearing of a key/mode (pitch class set), hearing of a sound as a degree in that key/mode, hearing of melodic (successive) and harmonic (simultaneous) intervals and chords, and more complex forms of hearing, such as hearing of a musical texture (polyphonic) and hearing of the progressions of chords within a key/mode as well as modulations to other keys/modes. Acquisition of fully fledged musical hearing by individuals seeking maximal proficiency in music proceeds exactly in this progression of stages: from discrimination of pitch in a single melodic line to hearing of the tonal plan for a multipart musical composition.

Absolute pitch at the absence of these other skills of processing tonal intervals, chords, counterpoints, textural modifications, and tonal plan, regrettably, amounts to unmusical understanding of music: a creation and appreciation of music is fit into the procrustean bed of tempered 12 tones like spelling out words instead of their expressive pronunciation. Needless to say, such "spelled" representation of music leaves behind meaningful melodic intonations, harmonic intervals, timbral diversity, and expressive harmonic changes, severely cutting down the emotional experience in production and perception of music. Reliance on absolute pitch alone often produces autism-like usage of music, when an individual is attracted by music yet fails to communicate affective information through it. At best, such an individual will intellectually know which musical emotion is implied by the most salient patterns of tonal organization (e.g., such basics as "happy" major versus "sad" minor), but suffer from lack of emotional reactivity, especially to fine details of TO (semantic qualia of intervals, chords, modulations).

All in all, absolute pitch appears to be a very recent cultural product of Western civilization, brought to life by cultivation of multipart music, complex music forms based on recapitulation of the original tonality in the tonal plan, and veneration of symphonic music as the pinnacle of artistic expression in music. The establishment of the tradition of orchestral performance effectively undermined the formative importance of timbre for TO, since in the orchestral music it is exceedingly common to pass the same theme from one instrument to another, which in no way induces changes in tension or relaxation in addition to melodic, harmonic, and rhythmometric aspects of TO. On the other hand, the rule of recapitulating the initial key in every composition grants formative powers to absolute pitch reference. The entire line of historic development of Western music runs toward overwhelming dominance of frequency and debasement of timbre to the secondary, mostly ornamental, role in expression. The first documented references to absolute pitch hearing are dated to no earlier than mid-eighteenth century, accompanying the institution of tonality, temperament tuning, and a tuning fork. In this light, the spread of absolute pitch in Western countries and especially its rampant dissemination in East Asia seems to reflect a cultural hype, related to social prestige and career advantages of musical prodigism in modern urban societies.

This overview of the earliest stages of music acquisition highlights the necessity for scholars of music to critically revise the traditional views on musical TO and methodology of testing musical skills. Here, about 130 years of non-interrupted research in ear development in Russia, pioneered by Rimsky-Korsakov (1963, 2: 207) and Maykapar (Maykapar 1900), come in handy in supporting the line of research of musical abilities by the Vygotskian school. Yet another valuable asset of Soviet and later Russian scholars was the enormous pool of population exposed to the standardized obligatory musical education across a huge territory that was a home to hundreds of ethnicities with the most diverse cultural and socioeconomic background. As a result, Russian researchers enjoyed much broader comparative ethnomusicological perspective and statistically grounded generalizations of most common types of TO in music than their Western colleagues. It is paramount to test children not by essentially unmusical stimuli, such as isolated pitches, especially if they are presented through pure tones, but by representing them through semantically valid musical elements, such as melodic intonations, harmonic intervals, chords, etc., preferably in meaningful musical context. Only this approach has chances to reduce the pitch reductionism and make it possible to identify the real underlying principles of early TO.

All the collected database of musicking during the first 3–4 years of life should be re-examined according to the acoustic traits and regularities that the music objectively contains rather than wish-to-be principles of Western tonality. Western notation should be altogether avoided in musicological analysis, since it introduces a clear Western bias and constitutes "pitch reductionism." A better alternative is to use prosograms for pictorial representation of music (Mertens 2004) and statistical methods for estimation of the stability of tuning of degrees in musical modes of frequency- and timbre-oriented music (Nikolsky 2017). Every study on early children musicking should be based not only on the analysis of musical behavior but on the musicological analysis of TO. This is an invaluable source of objective information that allows to accumulate extensive databases of music samples, instrumental for inferring synchronic, and diachronic variations of TO in comparative ethnomusicological research.

Ideally, every publication on music ontogeny should provide an open access to the recordings of children musicking and to the researcher's analysis of it, so that his/her findings could be re-examined by other scholars. This has to do not only with the situation where a specialist in the musicological analysis might uncover traits of TO that could have slipped away from the attention of a developmental specialist without a dedicated musicological training. Acoustic analysis very much depends on the technical resources, such as hardware recording devices and software programs, which keep substantially improving every decade. Already, modern frequency analysis is much more refined than 30 years ago. Especially rapid progress occurs in software spectral analysis. Therefore, ideally, any analysis older than about

20 years should be redone using the most recent software. But unfortunately current procedures of scientific publication stand on the way of the ideal open access cooperation. To illustrate this point, I contacted 31 scholars who referred to their analysis of early children music recordings in their publications with the request to re-examine their recordings, but received such materials only from 5 scholars. The other ones could not share their data because of the ethical issues related to the concerns for privacy of the recorded individuals. This is a regrettable state of affairs that holds back any advance in the realm of coining the general music theory of children music.

What we know about the ontogenetic development of babbling and its bifurcation into verbal and musical types is applicable to the phylogenetic perspective of the development of musilanguage and its bifurcation into protolanguage and proto-music. It seems that the primordial musilanguage was closer to music in its reliance on call as a unit of communication, its semantic specialization on displaying the affective state of the caller and its "isophonic" (non-synchronized) texture, examples of which is still possible to find in indigenous music of few ethnicities that have so far avoided Western cultural influence, such as Akia Indians of the Amazon (Nikolsky 2018). Once the climate change toward the end of the Pleistocene enabled the steady population growth, and it reached the critical effective population size sufficient to sustain the accumulation of modern behavior traits (Powell et al. 2009), verbal "dialogic" mode of communication became indispensable in such activities as hunting large prey and passing the technological knowledge to younger generations (Ambrose 2010). Employing musilanguage for such uses must have promoted the emergence of intonational prosody to distinguish between different types of verbal interaction, e.g., request, question, statement, etc. (Brown 2017). Intonational framework probably supported the morphonological process of adjusting isolated calls to conjoin into call groups ("bouts") capable of expressing more complex and more specific ideas.

Single-word holophrasic communication of infants based on the directly observable context and the "analog" correspondence between auditory signal and a specific message must be prototypical for the starting point of the phylogenesis of syntax (Johansson 2005). The extensive use of deictic reference in the infant's earliest vocabulary and gestural support of it supports the gestural model of the language origin (Corballis 2002). This model has been opposed to the vocal model (MacNeilage and Davis 2005), but their complementary coexistence seems more plausible (McNeill 2005). Verbal babbling points in the direction of onomatopoeia and phonetic symbolism as the likely origin of verbal symbolization (Svantesson 2017). The semantic analysis of the earliest attempts of infants' verbal communication leads to believe that acquisition of verbal skills proceeds from the individual self-communication through a repertory of invented words to the dialogic communication with the caretaker and thereafter the group communication (Shvachkin 1948). The efficacy of interpersonal communication directs verbal communication away from one-end expression of one's affective state and personal conventions of expression to the two-end transfer of referential information in the alternating order. There is no reason to restrict this line of development to ontogeny.

Musical babbling runs a different course, enabling an infant to maintain self-control of his/her emotional state to reduce the overall negative experience and increase the positive one. At first based on vocalization games designed to self-entertain, musical babbling sets in place the abecedary of pitch contours and timbral classes and defines the range of their semantic values. The child learns to maintain the tonal coherence between the sounds of a musical phrase by means of imposing a specific melodic contour or retaining a timbral quality. In effect, this constitutes the acquisition of musical mode—the first landmark in the development of purely musical abilities. *It is the presence of musical mode that tells melopoeia from verbal creativity*. There is no any analog to "musical mode" in speech. Music is unique in shaping important brain functions through the hedonistic effect of holistic appreciation of sounds per se (Patel 2010a).

Once humans discovered the musical mode as the principal tool of sustaining the appreciated tonal quality, their musicking probably proceeded toward building the stock of expressive means capable of representing at first the basic and then the most common complex social emotions. In the course of this development, the strong chill-like emotional experiences, instinctively triggered by the "built-in," probably genetically rooted, basic emotions received their more refined yet subtle counterpart in the set of "aesthetic emotions" (Altenmüller et al. 2013), put in place by learning to associate specific features of musical TO with specific affective states. The end result of this process is the formation of a music system designed to support the cultivation of those proprietary "musical emotions" that are of greatest importance in a given musical culture.

## References

Abraham O (1901) Das absolute Tonbewußtsein. Psychologisch- musikalische Studie Sammelbände der Internationalen Musikgesellschaft 3:1–86

Abrams RM, Gerhardt KK (1997) Some aspects of the foetal sound environment. In: Deliège I, Sloboda JA (eds) Perception and cognition of music. Psychology Press, Hove, pp 83–101

Abramson AS (1972) Tonal experiments with whispered Thai. In: Valdman A, Ling J (eds) Papers in linguistics and phonetics to the memory of Pierre Delattre. De Gruyter Mouton, Berlin, pp 31–44

Adachi M, Trehub SE (2000) Preschoolers' expression of emotion through invented songs. University of Leicester, Leicester

Adams K (2015) The musical analysis of hip-hop. In: Williams JA (ed) The Cambridge companion to hip-hop. Cambridge University Press, Cambridge, pp 118–134

Addessi AR (2009) The musical dimension of daily routines with under-four children during diaper change, bedtime and free-play. Early Child Dev Care 179:747–768. https://doi.org/10.1080/03004430902944122

Agazhanov AA (1977) On the absolute and the relative systems of the ear-training course (Об абсолютной и релятивной системах курса сольфеджио). In: Development of musical hearing (Воспитание музыкального слуха), vol 1. Muzyka, Moscow, pp 78–85

Agazhanov AA (1985) Methodology of teaching the degrees of a musical mode in the absolute system of ear-training (Методика изучения ступеней лада в абсолютной системе сольфеджио). In: Development of musical hearing (Воспитание музыкального слуха), vol 2. Muzyka, Moscow, pp 41–58

Ahlner F, Zlatev J (2010) Cross-modal iconicity: a cognitive semiotic approach to sound symbolism. Sign Syst Stud 38:298–348

Alekseyev EY (1973) Certain peculiarities of organization of pitch in traditional Yakut melody (Некоторые особенности звуковысотной организации традиционной якутской мелодики). In: Banin AA (ed) Musical folklore studies (Музыкальная фольклористика), vol 1. Soviet Composer (Советский композитор), Moscow, pp 138–173

Alekseyev EY (1976) Problems in the genesis of musical mode (on the example of Yakut folksong): analysis (Проблемы формирования лада (на материале якутской народной песни): исследование). Muzyka (Музыка), Moscow

Alekseyev EY (1986) Musical intonation in the earliest forms of folklore. The aspect of pitch (Раннефольклорное интонирование: звуковысотный аспект). Sovetskii Kompozitor (Сов. композитор), Moscow

Alekseyev EY (1988) Folklore in the context of modern culture: thoughts on the future of folk song (Фольклор в контексте современной культуры: рассуждения о судьбах народной песни). Soviet Composer (Советский композитор), Moscow

Alekseyev EY (1990) Notation of folk music: theory and practice (Нотная запись народной музыки: Теория и практика). Soviet Composer (Советский композитор), Moscow

Alekseyev EY (1993) Speaking and singing: prolegomena to anthropophonics (Пение и говорение. Основы антропофоники). Moscow

Alekseyev EY (2013) Ethnomusicological experiment: on the way of trial and error (Этномузыковедческий эксперимент: на пути проб и ошибок). In: Varlamova A, Pavlova Z (eds) Music. Performance. Education (Музыка. Исполнительство. Образование), vol 4. University of Republic of Sakha, Yakutsk, pp 162–179

Altenmüller E, Kopiez R, Grewe O (2013) A contribution to the evolutionary basis of music: lessons from the chill response. In: Altenmüller E, Schmidt S, Zimmermann E (eds) Evolution of emotional communication. Oxford University Press, Oxford, pp 313–336. https://doi.org/10.1093/acprof:oso/9780199583560.003.0019

Ambrazevičius R, Wiśniewska I (2008) Chromaticisms or performance rules? Evidence from traditional singing pitch transcriptions. J Interdiscip Music Stud 2:19–31

Ambrose SH (2010) Coevolution of composite-tool technology, constructive memory, and language. Curr Anthropol 51:S135–S147. https://doi.org/10.1086/650296

Anisimov VP (2004) The testing of musical abilities of children (Диагностика музыкальных способностей детей). Vlados, Moscow

Arthur W (2011) Evolution: a developmental approach. Wiley, Oxford

ASA (1951) Timbre. In: Leo B, Morrical KC, McNair JW (eds) American standard acoustical terminology. American Standards Association, New York

Asafyev B (1952) Selected works (Избранные труды), vol 1. Academy of Science of the USSR (Изд-во Академии наук СССР), Moscow

Ashley R (2004) Musical pitch space across modalities: spatial and other mappings through language and culture. In: Lipscomb PWS, Ashley R, Gjerdingen RO (eds) Proceedings of the 8th international conference on music perception & cognition. Casual Productions, Sidney, pp 64–71

Athanasopoulos G, Moran N (2013) Cross-cultural representations of musical shape. Empir Musicol Rev 8:185–199

Bachem A (1937) Various types of absolute pitch. J Acoust Soc Am 9:146–157. https://doi.org/10.1121/1.1915919

Baird JW (1917) Memory for absolute pitch. In: Studies in psychology contributed by colleagues and former students of Edward Bradford Titchener. Louis N. Wilson, Worcester, pp 43–78. https://doi.org/10.1037/11008-005

Ball P (2011) The music instinct: how music works and why we can't do without it. Oxford University Press, Oxford. https://doi.org/10.1007/s00146-010-0287-1

Bamberger JS, DiSessa A (2003) Music as embodied mathematics: a study of a mutually informing affinity. Int J Comput Math Learn 8:123–160. https://doi.org/10.1023/B:IJCO.0000003872.84260.96

Barbieri P, Mangsen S (1991) Violin intonation: a historical survey. Early Music 19:69–88. https://doi.org/10.1093/earlyj/XIX.1.69

Barbour JM (1952) Violin intonation in the 18th century. J Am Musicol Soc 5:224–234

Barbour JM (2004) Tuning and temperament: a historical survey. Dover Publications, New York

Barrett M (2003) Meme Engineers: children as producers of musical culture. Int J Early Years Educ 11:195–212. https://doi.org/10.1080/0966976032000147325

Barrett MS (2006) Inventing songs, inventing worlds: the 'genesis' of creative thought and activity in young children's lives. Int J Early Years Educ 14:201–220. https://doi.org/10.1080/09669760600879920

Barrett M (2011) Musical narratives: a study of a young child's identity work in and through music-making. Psychol Music 39:403–423. https://doi.org/10.1177/0305735610373054

Beliayev VM (1990) Modal systems in the traditional music of the USSR (Ладовые системы в музыке народов СССР). In: Travkina I (ed) Viktor Mikhailovich Beliayev (Виктор Михайлович Беляев). Sovetskii Kompozitor (Советский композитор), Moscow, pp 223–377

Beliayeva-Ekzempliarskaya S (1925) Musical experience in preschool age (Музыкальное переживание в дошкольном возрасте). In: Beliayeva-Ekzempliarskaya S (ed) Collection of works of the physiolo-psychological department (Сборник работ физиолог-о-психологической секции). The State Institute of Musical Science (Гос ин-та музыкальной науки.), Moscow, pp 3–29

Bentley A (1966) Musical ability in children and its measurement. Harrap, London

Berezhansky PN (2000) Absolute pitch musical hearing: its essence, nature, genesis and methods of acquisition and development (Абсолютный музыкальный слух: Сущность, природа, генезис, способ формирования и развития). Moscow State Conservatory named after Tchaikovsky, Moscow

Berio L (2006) Remembering the future. Harvard University Press, Cambridge, MA

Berry JW, Dasen PR (1974) Culture and cognition: readings in cross-cultural psychology. Methuen, London

Bhatara A, Quintin E-M, Levy B, Bellugi U, Fombonne E, Levitin DJ (2010) Perception of emotion in musical performance in adolescents with autism spectrum disorders. Autism Res 3:214–225. https://doi.org/10.1002/aur.147

Bickerton RC, Barr GS (1987) The origin of the tuning fork. J R Soc Med 80:771–773. https://doi.org/10.1177/014107688708001215

Bjørkvold JR (1992) The muse within: creativity and communication, song and play from childhood through maturity (trans: Halverson WH). Harper Collins Publishers, London

Blacking J (1967) Venda children's songs: a study in ethnomusicological analysis. University of Chicago Press, Chicago

Blackwell WH (2007) What to make of all this commentary on Haeckel? Am Biol Teach 69:135–136. https://doi.org/10.2307/4452118

Blench R (2007) From Vietnamese lithophones to Balinese gamelans: a history of tuned percussion in the Indo-Pacific region. Bull Indo-Pacific Prehist Assoc 26:48–61. https://doi.org/10.7152/bippa.v26i0.11993

Blium D (1977) The role of dictation in development of professional musical hearing (Роль диктанта в развитии профессионального музыкального слуха). In: Agazhanov A (ed) Development of musical hearing (Воспитание музыкального слуха), vol 1. Muzyka (Музыка), Moscow, pp 86–117

Bolger D, Griffith N (2005) Multidimensional timbre analysis of shakuhachi honkyoku. In: Proceedings of the Conference on Interdisciplinary Musicology (CIM05) Montréal (Québec) Canada, 10–12/03/2005, 10–12. Montréal

Born G (2000) Musical modernism, postmodernism, and others. In: Born G, Hesmondhalgh D (eds) Introduction to Western music and its others. University of California Press, Berkeley, CA, pp 12–20

Boulez P (1990) Orientations: collected writings. Translated by Nattiez J-J. Harvard University Press, Cambridge, MA

Bouvet L, Donnadieu S, Valdois S, Caron C, Dawson M, Mottron L (2014) Veridical mapping in savant abilities, absolute pitch, and synesthesia: an autism case study. Front Psychol 5:106. https://doi.org/10.3389/fpsyg.2014.00106

Boysson-Bardies BD, Sagart L, Durand C (2008) Discernible differences in the babbling of infants according to target language. J Child Lang 11:1–15. https://doi.org/10.1017/S0305000900005559

Brady PT (1970) Fixed-scale mechanism of absolute pitch. J Acoust Soc Am 48:883–887. https://doi.org/10.1121/1.1912227

Breidbach O (2002) The former synthesis – some remarks on the typological background of Haeckel's ideas about evolution. Theory Biosci 121:280–296. https://doi.org/10.1007/s12064-002-0015-6

Brown S (2000) The "Musilanguage" model of language evolution. In: Brown S, Merker B, Wallin NL (eds) The origins of music. MIT Press, Cambridge, MA, pp 271–300. https://doi.org/10.1037/e533412004-001

Brown S (2017) A joint prosodic origin of language and music. Front Psychol 8:1894. https://doi.org/10.3389/fpsyg.2017.01894

Buckton R (1982) An investigation into the development of musical concepts in young children. Psychol Music Special Issue: 17–21

Burton S (2002) An exploration of preschool children ' s spontaneous songs and chants by. Vis Res Music Educ 2:1–16

Bytchkov YN (1993) The foundations of formation of the melodic modal hearing (Основы формирования мелодического ладового слуха). Russian Academy of Music named after Gnesin, Moscow

Campbell PS (1991) The child-song genre: a comparison of songs by and for children. Int J Music Educ os-17:14–23. https://doi.org/10.1177/025576149101700103

Campbell PS (1998a) Songs in their heads: music and its meaning in children's lives. Oxford University Press, Oxford

Campbell PS (1998b) The musical cultures of children. Res Stud Music Educ 11:42–51. https://doi.org/10.1177/1321103X9801100105

Campbell PS, Wiggins T (2012) Giving voice to children. In: Campbell PS, Wiggins T (eds) The Oxford handbook of children's musical cultures. Oxford University Press, Oxford, p 636. https://doi.org/10.1093/oxfordhb/9780199737635.013.0001

Chen-Hafteck L (1997) Music and language development in early childhood: integrating past research in the two domains. Early Child Dev Care 130:85–97. https://doi.org/10.1080/0300443971300109

Chesnut JH (1977) Mozart's teaching of intonation. J Am Musicol Soc 30:254–271. https://doi.org/10.1525/jams.1977.30.2.03a00030

Chukovsky K (1963) From two to five (От двух до пяти) (trans: Merton M). University of California Press, Berkeley

Chumak AY (1962) Experiments in forming a discriminative vibration sensitivity [Опыт формирования различительной вибрационной чувствительности]. Rep Acad Pedagog Sci 3:83–89

Clarke EF (2001) Meaning and the specification of motion in music. Music Sci 5:213–234. https://doi.org/10.1177/102986490100500205

Coall DA, Callan AC, Dickins TE, Chisholm JS (2015) Evolution and prenatal development. In: Richard M (ed) Handbook of child psychology and developmental science, vol 3. Lerner, pp 1–49. https://doi.org/10.1002/9781118963418.childpsy303

Cole M, Gay J, Glick JA, Sharp DW (1971) The cultural context of learning and thinking: an exploration in experimental anthropology. Basic Books, New York

Conway D (2012) Jewry in music: entry to the profession from the enlightenment to Richard Wagner. Cambridge University Press, Cambridge

Cooke D (1959) The language of music. Oxford University Press, London

Corballis MC (2002) From hand to mouth: the origins of language. Princeton University Press, Princeton

Corrigall KA, Trainor LJ (2010) Musical enculturation in preschool children: acquisition of key and harmonic knowledge. Music Percept 28:195–200. https://doi.org/10.1525/mp.2010.28.2.195

Costa-Giomi E (2000) Young children's identification of simple harmonic accompaniments. In: Woods C (ed) Proceedings of the 6th International conference for music perception and cognition. European Society for the Cognitive Sciences of Music, Keele

Cross I (2001a) Music, mind and evolution. Psychol Music 29:95–102. https://doi.org/10.1177/0305735601291007

Cross I (2001b) Music, cognition, culture, and evolution. Ann NY Acad Sci 930:28–42

Custodero LA (2006) Singing practices in 10 families with young children. J Res Music Educ 54:37–56. https://doi.org/10.1177/002242940605400104

D'Amato MR (1988) A search for tonal pattern perception in Cebus monkeys: why monkeys Can't hum a tune. Music Percept Interdiscip J 5:453–480. https://doi.org/10.2307/40285410

Daniélou A (1995) Music and the power of sound: the influence of tuning and interval on consciousness, Rep Sub edn. Inner Traditions, Rochester

Dapretto M, Davies MS, Pfeifer JH, Scott AA, Sigman M, Bookheimer SY, Iacoboni M (2006) Understanding emotions in others: mirror neuron dysfunction in children with autism spectrum disorders. Nat Neurosci 9:28–30. https://doi.org/10.1038/nn1611

Dasen PR (1972) Cross-cultural Piagetian research: a summary. J Cross Cult Psychol 3:23–40. https://doi.org/10.1177/002202217200300102

Dasen PR (1977) Are cognitive processes universal? A contribution to cross-cultural Piagetian psychology. In: Warren N (ed) Studies in cross-cultural psychology, vol 1. Academic Press, London, pp xvii + 212

Dasen PR (2012) Emics and Etic in cross-cultural psychology towards a convergence in the study of cognitive styles. In: Tchombe TMS, Nsamenang AB, Keller H, Fülöp M (eds) Proceedings of the 4th Africa Region Conference of the IACCP, University of Buea, Cameroun, Aug. 1–8, 2009. University of Buea, Buea, Cameroun, pp 55–73

Dasen PR, de Ribaupierre A (1987) Neo-piagetian theories: cross-cultural and differential perspectives. Int J Psychol 22:793–832. https://doi.org/10.1080/00207598708246803

Dasen PR, Heron A (1981) Cross-cultural tests of Piaget's theory. In: Triandis HC, Heron A (eds) Handbook of cross-cultural psychology. Allyn and Bacon, Boston, pp 295–342

Davidson L (1985) Tonal structures of children's early songs. Music Percept Interdiscip J 2:361–373. https://doi.org/10.2307/40285304

Davidson L (1994) Songsinging by young and old: a developmental approach to music. In: Aiello LC, Sloboda J (eds) Musical perceptions. Oxford University Press, Oxford, pp 99–130

Davidson J (2002) Developing the ability to perform. In: Musical performance: a guide to understanding. Cambridge University Press, Cambridge, pp 89–97

Davidson L, McKernon PE, Gardner H (1979) The acquisition of song: a developmental approach. Report of the Ann Arbor symposium

Davies C (1986) Say it till a song comes (reflections on songs invented by children 3–13). Br J Music Educ 3:279–294. https://doi.org/10.1017/S0265051700000796

Daynes H (2011) Listeners' perceptual and emotional responses to tonal and atonal music. Psychology of Music 39. SAGE Publications, London, pp 468–502. https://doi.org/10.1177/0305735610378182

de Vries P (2005) Lessons from home: scaffolding vocal improvisation and song acquisition with a 2-year-old. Early Childhood Educ J 32:307–312. https://doi.org/10.1007/s10643-004-0962-2

Dean B (2011) Oscar' s music: a descriptive study of one three year-old's spontaneous music-making at home. In: Young S (ed) Proceedings of the 5th conference of the European. MERYC, Helsinki, pp 275–286

Denisov EV (2009) On musical language [О музыкальном языке]. In: Tsenova VS, Kiuregian TS (eds) Composers on modern composition. anthology [Композиторы о современной композиции: хрестоматия]. Moscow State Tchaikovsky Conservatory, Moscow, pp 278–288

Deutsch D (2002) The puzzle of absolute pitch. Curr Dir Psychol Sci 11:200–204. https://doi.org/10.1111/1467-8721.00200
Deutsch D (2013a) Absolute pitch. In: Deutsch D (ed) Psychology of music. Academic Press, New York, pp 142–182
Deutsch D (2013b) The processing of pitch combinations. In: Deutsch D (ed) Psychology of music, 3rd edn. Academic Press, New York, pp 249–325
Deutsch D, Henthorn T, Dolson M (1999) Absolute pitch is demonstrated in speakers of tone languages. J Acoust Soc Am 106:2267–2267. https://doi.org/10.1121/1.427738
Deutsch D, Henthorn T, Dolson M (2004) Absolute pitch, speech, and tone language: some experiments and a proposed framework. Music Percept 21:339–356. https://doi.org/10.1525/mp.2004.21.3.339
Deutsch D, Henthorn T, Marvin E, HongShuai X (2006) Absolute pitch among American and Chinese conservatory students: Prevalence differences, and evidence for a speech-related critical period. J Acoust Soc Am 119:719. https://doi.org/10.1121/1.2151799
Deutsch D, Le J, Shen J, Li X (2011) Large-scale direct-test study reveals unexpected characteristics of absolute pitch. J Acoust Soc Am 130:2398–2398. https://doi.org/10.1121/1.3654614
Dingemanse M, Schuerman W, Reinisch E, Tufvesson S, Mitterer H (2016) What sound symbolism can and cannot do: testing the iconicity of ideophones from five languages. Language 92:e117–e133. https://doi.org/10.1353/lan.2016.0034
Doğantan-Dack M (2013) Tonality: the shape of affect. Empir Musicol Rev 8:208–218
Dowling WJ (1984) Cognitive processes in the perception of art. Adv Psychol 19:145–163. https://doi.org/10.1016/S0166-4115(08)62350-X
Dowling WJ (1999) The development of music perception and cognition. In: Deutsch D (ed) The psychology of music, 2nd edn. Academic Press, San Diego, pp 603–627
Duffin RW (2007) How equal temperament ruined harmony (and why you should care). Norton, New York
Eckensberger LH, Lonner WJ, Poortinga YH (1979) Cross-cultural contributions to psychology. Selected papers from the Fourth International Congress of the International Association for cross-cultural psychology, Munich, Germany, July 28–August 5, 1978. Swets and Zeitlinger, Amsterdam
Eerola (2009) Examination of stylistic traits in sound production of the Veps lühüd pajo songs using computer-aided music analysis. In: Niemi J (ed) Perspectives on the song of the indigenous peoples of northern Eurasia: performance, genres, musical syntax, sound. Tampere University Press, Tampere, pp 160–197
Elmer SS (2011) Structural aspects of early song singing. In: Baldassare A (ed) Music – space – chord – image. Peter Lang Verlag, Bern, pp 765–782
Elmer SS (2012) Human singing: towards a developmental theory. Psychomusicol Music Mind Brain 21:13–30. https://doi.org/10.5084/pmmb2011/21/xxx
Emberly A (2009) "Mandela went to China... and India too": musical cultures of childhood in South Africa. University of Washington, Seattle
Endovitskaya TV (1959) On the pitch discriminatory sensitivity in children of the preschool age (О звуковысотной различительной чувствительности у детей дошкольного возраста). In: Reports of the Academy of the Academy of Pedagogical Sciences, vol 5. Academy of Pedagogical Sciences of Russia, Leningrad, pp 42–46
Endovitskaya TV (1963) The specificity of development of the sensitivity in frequency discrimination in children of the preschool age (Особенности развития звуковысотной различительной чувствительности в дошкольном возрасте). In: Shatskaya VN (ed) The development of child's voice (Развитие детского голоса), pp 196–203. Moscow
Fales C (2002) The paradox of Timbre. Ethnomusicology 46:56–95. https://doi.org/10.2307/852808
Fedorovich Y (2014) History of professional musical education in Russia 19–20 centuries (История профессионального музыкального образования в России (XIX–XX века)), 2nd edn. Direct Media, Moscow

Feldmann H (1997) Die Geschichte der Stimmgabel – Teil 1: Die Erfindung der Stimmgabel, ihr Weg in der Musik und den Naturwissenschaften. Laryngo-Rhino-Otologie 76:116–122. https://doi.org/10.1055/s-2007-997398

Fernald A (1992) Meaningful melodies in mothers' speech to infants. In: Papousek H, Jurgens U, Papousek M (eds) Nonverbal vocal communication comparative and developmental approaches. Cambridge University Press, Cambridge, pp 262–282

Fétis F-J (1994) Esquisse de L'histoire de L'harmonie (trans: Arlin MI). Pendragon Press, Hillsdale

Fitch WT (2006) The biology and evolution of music: a comparative perspective. Cognition 100:173–215. https://doi.org/10.1016/j.cognition.2005.11.009

Fitch WT (2010) The evolution of language. Cambridge University Press, Cambridge

Flohr JW (1984) Young children's improvisations: a longitudinal study. In: 49th National in-service conference of the music educators national conference, Chicago, IL, March, 23, 1984, 1–12. ERIC, Chicago, IL. ED25318

Flowers PJ, Dunne-Sousa D (1990) Pitch-pattern accuracy, tonality, and vocal range in preschool children's singing. J Res Music Educ 38:102. https://doi.org/10.2307/3344930

Fodor JA (2001) The mind doesn't work that way the scope and limits of computational psychology. MIT Press, Cambridge, MA

Forrester MA (2010) Emerging musicality during the pre-school years: a case study of one child. Psychol Music 38:131–158. https://doi.org/10.1177/0305735609339452

Foster MLC (1994) Symbolism: the foundation of culture. In: Ingold T (ed) Companion encyclopedia of anthropology. Routledge, New York, pp 366–395

Fox DB (1990) An analysis of the pitch characteristics of infant vocalizations. Psychomusicology 9:21–30

Friberg A (1995) A quantitative rule system for musical performance. In: Music perception. Royal Institute of Technology, Stockholm

Frolova-Walker M (1998) "National in form, socialist in content": musical nation-building in the soviet republics. J Am Musicol Soc 51:331–371. https://doi.org/10.2307/831980

Fujita F (1990) The intermediate performance between talking and singing from an observational study of Japanese children's music activities in nursery schools. In: Dobbs J (ed) Music education: facing the future. ISME, Christchurch, pp 140–146

Galperin PY (1999) In: Podolskii A (ed) Introduction to psychology (Введение в психологию). Book House University, Moscow

Garbuzov N (1948) Zonal nature of pitch hearing (Зонная природа звуковысотного слуха). Russian Academy of Science, Moscow

Garbuzov N (1980) Selected works (1925–1955) (Избранные труды). In: Rags Y (ed) Garbuzov N.A. – musician, researcher and pedagogue (Гарбузов Н.А. – Музыкант, исследователь, педагог). Muzyka (Музыка), Moscow, pp 49–263

Garfias R (1990) An ethnomusicologist's thoughts on the processes of language and music acquisition. In: Wilson F, Roehmann F (eds) Music and child development: proceedings of the 1987 Denver conference. Book Crafters, Ann Arbor, pp 100–105

Geinrikhs IP (1978) Musical hearing and its development (Музыкальный слух и его развитие). Muzyka, Moscow

Gembris H (2006) The development of musical abilities. In: Colwell R (ed) MENC handbook of musical cognition and development. Oxford University Press, Oxford/New York, pp 124–164

Giliarova N (2010) The registry of expeditional and stationary audio recordings of the main fund of the Scientific Center of Folk Music (Перечень экспедиционных и стационарных аудиозаписей фонда Кабинета народной музыки), 3rd edn. Moscow Conservatory (Московская государственная консерватория имени П.И. Чайковского), Moscow

Gillen J, Cameron CA (2010) International perspectives on early childhood research: a day in the life. Palgrave Macmillan, New York

Gould SJ (1977) Ontogeny and phylogeny. Belknap Press of Harvard University Press, Cambridge, MA

Granot RY, Eitan Z (2011) Musical tension and the interaction of dynamic auditory parameters. Music Percept 28:219–245. https://doi.org/10.1525/mp.2011.28.3.219

Graves J, Micheyl C, Oxenham AJ (2013) Preferences for melodic contours transcend pitch. Proc Meetings Acoust 19:035031–035031. Acoustical Society of America. https://doi.org/10.1121/1.4799453

Grebelnik SG (1985) Formation and development of absolute pitch hearing as a musical ability (Формирование и развитие абсолютного слуха как музыкальной способности). Institute of preschool education at the Academy of Sciences of the USSR, Moscow

Gregersen PK (1998) Instant recognition: the genetics of pitch perception. Am J Hum Genet 62:221–223. https://doi.org/10.1086/301734

Gregersen PK, Kowalsky E, Kohn N, Marvin EW (1999) Absolute pitch: prevalence, ethnic variation, and estimation of the genetic component. Am J Hum Genet 65:911–913. https://doi.org/10.1086/302541

Gregersen PK, Kowalsky E, Kohn N, Marvin EW (2001) Early childhood music education and predisposition to absolute pitch: Teasing apart genes and environment. Am J Med Genet 98:280–282. https://doi.org/10.1002/1096-8628(20010122)98:3<280::AID-AJMG1083>3.0.CO;2–6

Gregersen PK, Kowalsky E, Li W (2007) Reply to Henthorn and Deutsch: ethnicity versus early environment: comment on "early childhood music education and predisposition to absolute pitch: teasing apart genes and environment" by Peter K. Gregersen, Elena Kowaisky, Nina Kohn, and Elizabeth West. Am J Med Genet A 143:104–105. https://doi.org/10.1002/ajmg.a.31595

Gruhn W (2004) Are different types of mental representation reflected by brain activation patterns? In: Lipscomb S, Ashley R, Gjerdingen R, Webster P (eds) Music perception and cognition; ICMPC8. Causal Productions, Adelaide, pp 124–127

Gruhn W (2009) The audio-vocal system in song and speech development. In: Haas R, Brandes V (eds) Music that works: contributions of biology, neurophysiology, psychology, sociology, medicine and musicology. Springer, Vienna, pp 109–117. https://doi.org/10.1007/978-3-211-75121-3

Gussenhoven C (2002) Intonation and interpretation: phonetics and phonology. In: Bel E, Marilier I (eds) Proceedings of speech prosody. University de Provence, Aix-en-Provence, pp 45–57

Hagel S (2009) Ancient Greek music: a new technical history. Cambridge University Press, New York

Hallpike CR (2014) Constructivism and selection: two opposed theories of social evolution. In: Dux G, Rüsen J (eds) Strukturen des Denkens: Studien zur Geschichte des Geistes. Springer Fachmedien Wiesbaden, Wiesbaden, pp 183–200. https://doi.org/10.1007/978-3-658-06255-2_10

Hargreaves DJ (1986) The developmental psychology of music. Cambridge University Press, Cambridge

Hargreaves DJ (1996) The development of artistic and musical competence. In: Deliège I, Sloboda J (eds) Musical beginnings. Oxford University Press, Oxford, pp 145–170. https://doi.org/10.1093/acprof:oso/9780198523321.003.0006

Harwood E (1998) Go on, girl! Improvisation in African-American girls' singing games. In: Nettl B, Russell M (eds) In the course of performance: studies in the world of musical improvisation. University of Chicago Press, Chicago, pp 113–125

Hauser MD (1996) The evolution of communication. MIT Press, Cambridge, MA

Heaton P (2003) Pitch memory, labelling and disembedding in autism. J Child Psychol Psychiatry 44:543–551. https://doi.org/10.1111/1469-7610.00143

Heaton P (2005) Interval and contour processing in autism. J Autism Dev Disord 35:787–793. https://doi.org/10.1007/s10803-005-0024-7

Heaton P (2009) Assessing musical skills in autistic children who are not savants. Philos Trans R Soc B Biol Sci 364:1443–1447. https://doi.org/10.1098/rstb.2008.0327

Heaton P, Williams K, Cummins O, Happé F (2008) Autism and pitch processing splinter skills: a group and subgroup analysis. Autism Int J Res Pract 12:203–219. https://doi.org/10.1177/1362361307085270

Henze HW (1982) Music and politics: collected writings, 1953–1981. Translated by Labanyi P. Cornell University Press, Ithaca, NY

Hofmann J (1920) Piano playing with piano questions answered. Theodore Presser, Philadelphia

Holahan JM (1987) Toward a theory of music syntax: some observations of music babble in young children. In: Music and child development. Springer, New York, pp 96–106. https://doi.org/10.1007/978-1-4613-8698-8_5

Hood M (1971) Slendro and pelog revisited. In: McAllester DP (ed) Readings in ethnomusicology. Johnson Reprint Corporation, New York, pp 35–56. https://www.papers3://publication/uuid/557816B9-7182-4B05-B4DB-BFB4C6657056

Hopkin JB (1984) Jamaican children's songs. Ethnomusicology 28:1–36. https://doi.org/10.2307/851430

Horowitz J (1991) The ivory trade: piano competitions and the business of music. Northeastern, Boston

Huddleston WE, Lewis J, Phinney RE, DeYoe EA (2008) Auditory and visual attention-based apparent motion share functional parallels. Percept Psychophys 70:1207–1216. https://doi.org/10.3758/PP.70.7.1207

Hulse SH, Cynx J, Humpal J (1984) Absolute and relative pitch discrimination in serial pitch perception by birds. J Exp Psychol Gen 113:38–54. https://doi.org/10.1037/0096-3445.113.1.38

Huron D (2001) Tone and voice: a derivation of the rules of voice-leading from perceptual principles. Music Percept 19:1–64. https://doi.org/10.1525/mp.2001.19.1.1

Huron D (2006a) Are scale degree qualia a consequence of statistical learning? In: Costa M, Baroni M, Addessi AR, Caterina R (eds) Proceedings of the 9th international conference on music perception and cognition (ICMPC) and 6th triennial conference of the European Society for the Cognitive Sciences of Music (ESCOM). ESCOM, Bologna, pp 1675–1680

Huron D (2006b) Sweet anticipation: music and the psychology of expectation. MIT Press, Cambridge, MA

Hutchins SM, Peretz I (2012) A frog in your throat or in your ear? Searching for the causes of poor singing. J Exp Psychol Gen 141:76–97. https://doi.org/10.1037/a0025064

Ivanova TG (2009) History of Russian ethnomusicology XX century: 1900–1941 (История русской фольклористики XX века: 1900–1941 г). Dmitrii Bulanin Publishers, Moscow

Jakobson R (1987) Language in literature (eds: Rudy S, Pomorska K). Belknap Press, Cambridge, MA

Johansson S (2005) Origins of Language. Constraints on hypotheses. John. Benjamins, Amsterdam

Johansson S (2015) Language abilities in Neanderthals. Annu Rev Linguis 1:311–332. https://doi.org/10.1146/annurev-linguist-030514-124945

Johnson-Laird PN, Oatley K (2010) Emotions, music, and literature. In: Lewis M, Haviland-Jones JM, Barrett LF (eds) Handbook of emotions, 3rd edn. The Guilford Press, New York, pp 102–113

Jürgens U (1995) Neuronal control of vocal production in non-human and human primates. In: Current topics in primate vocal communication. Springer, Boston, pp 199–206. https://doi.org/10.1007/978-1-4757-9930-9_10

Kalmar M, Balasko G (1987) "Musical mother tongue" and creativity in preschool children's melody improvisations. Bull Counc Res Music Educ 91:77–86

Kaminska Z, Woolf J (2000) Melodic line and emotion: Cooke's theory revisited. Psychol Music 28:133–153. https://doi.org/10.1177/0305735600282003

Kanner L (1943) Autistic disturbances of affective contact. Nerv Child 2:217–250

Karasyova MV (1999) Solfeggio – the psycho-technique of the development of musical hearing (Сольфеджио – психотехника развития музыкального слуха). Kompozitor, Moscow

Karasyova MV (2010) The change of time at times of change: on 50 years evaluation of development of musical hearing education in Russia (Перемена времени во время перемен: к полувековым итогам развития музыкально-слухового образования в России). Sci Cour Mosc Conserv 1:27–43

Kartomi MJ, Anderson Sutton R, Suanda E, Williams S, Harnish D (2008) Indonesia. In: Miller T, Williams S (eds) The garland handbook of southeast Asian music. Routledge, London, pp 334–405

Kelley L, Sutton-Smith B (1987) A study of infant musical productivity. In: Music and child development. Springer, New York, pp 35–53. https://doi.org/10.1007/978-1-4613-8698-8_2

Kenneson C (1998) Musical prodigies: perilous journeys, remarkable lives. Amadeus Press, Oregon

Kholopov Y (1983) Who has invented the 12-tone technique? (Кто изобрел 12-тоновую технику?). In: Muginshtein ML (ed) Problems of history of the Austro-German music. The first third of the 20th century (Проблемы истории австро-немецкой музыки. Первая треть XX века). Moscow State musical pedagogical institute named after Gnesin, Moscow, pp 34–58

Kholopov Y (1988) Harmony: a theoretic course (Гармония: теоретический курс). Muzyka (Музыка), Moscow

Kholopov Y (2005) Towards the problem of mode in Russian theoretic musicology (К проблеме лада в русском теоретическом музыкознании). In: Struchalina E (ed) Harmony: problems of science and methodology (Гармония: проблемы науки и методики), vol 2. RGK (Ростовская государственная консерватория), Rostov-na-Donu, pp 135–157

Kiilu K (2011) The concept of preschool music education in Estonian education system. Procedia Soc Behav Sci 29:1257–1266. https://doi.org/10.1016/j.sbspro.2011.11.361

Kirnarskaya D, Kiyashchenko N, Tarasova K, Tzypina G (2003) In: Tzypina G (ed) Psychology of musical activities: theory and practice (Психология музыкальной деятельности: Теория и практика). Akademiya, Moscow

Koelsch S (2009) Neural substrates of processing syntax and semantics in music. In: Music that works: contributions of biology, neurophysiology, psychology, sociology, medicine and musicology. Springer, Vienna, pp 143–153. https://doi.org/10.1007/978-3-211-75121-3_9

Köhler W (1915) Akustische Untersuchungen. Zeitschrift für die Psychologie 72:1–192

Koops LH (2012) “Now can I watch my video?”: exploring musical play through video sharing and social networking in an early childhood music class. Res Stud Music Educ 34:15–28. https://doi.org/10.1177/1321103X12442994

Korguzalov VV, Troitskaya AD (1993) The phonogram archive of the Institute for Russian Literature (Pushkin House) of the Russian Academy of Sciences, St Petersburg. World Music 35:115–120

Kostina EP (2004) Kamerton: the program of musical education for children of early and preschool ages (Камертон: Программа музыкального образования детей раннего и дошкольного возраста). Prosvesheniye, Moscow

Krader BL (1990) Recent achievements in soviet ethnomusicology, with remarks on Russian terminology. Yearb Tradit Music 22:1–16. https://doi.org/10.2307/767926

Kratus J (1985) Rhythm, melody, motive and phrase characteristics of children’s original compositions. Case Western Reserve University, Cleveland

Kratus J (1989) A time analysis of the compositional processes used by children ages 7 to 11. J Res Music Educ 37:5–20. https://doi.org/10.2307/3344949

Krauss RE (1985) The originality of the avant-garde and other modernist myths. MIT Press, Cambridge, MA

Kremp P-A (2010) Innovation and selection: symphony orchestras and the construction of the musical canon in the United States (1879-1959). Soc Forces 88:1051–1082. https://doi.org/10.1353/sof.0.0314

Kreutzer NJ (2001) Song acquisition among rural Shona-speaking Zimbabwean children from birth to 7 years. J Res Music Educ 49:198–211. https://doi.org/10.2307/3345706

Krumhansl CL (1990) Cognitive foundations of musical pitch, vol 92. Oxford University Press, New York, p 1193. https://doi.org/10.1121/1.404005

Krumhansl CL, Sandell GJ, Sergeant DC (1987) The perception of tone hierarchies and mirror forms in twelve-tone serial music. Music Percept 5:31–77. https://doi.org/10.2307/40285385

Kubik G (1979) Pattern perception and recognition in African Music (eds: Blacking J, Kealiinohomoko J). Mouton Publishers, The Hague, pp 221–250

Kubik G (1985) African tone-systems: a reassessment. Yearb Tradit Music 17:31–63

Kuhl PK, Coffey-Corina S, Padden D, Dawson G (2005) Links between social and linguistic processing of speech in preschool children with autism: behavioral and electrophysiological measures. Dev Sci 8:F1–F12. https://doi.org/10.1111/j.1467-7687.2004.00384.x

Large EW (2008) Resonating to musical rhythm: theory and experiment. In: Grondin S (ed) Psychology of time. Emerald Group Publishing Limited, Bingley, pp 189–231. https://doi.org/10.1016/B978-0-08046-977-5.00006-5

Lee PX, Wee D, Toh HSY, Lim BP, Chen N, and Ma B (2014) A whispered mandarin corpus for speech technology applications. In: Meng H, Ma B(eds) Proceedings of the annual conference of the international speech communication association, INTERSPEECH, 14–18 September 2014, Singapore. International Speech Communication Association, Singapore, pp 1598–1602. 9781634394352

Lennenberg EH (1967) Biological foundations of language. Wiley, New York

Leontyev AN (2001) Lectures on general psychology (Лекции по общей психологии). Smysl, Moscow

Leontyev AN (2009) The development of mind. Selected works of Aleksei Nikolaevich Leontyev (ed: Cole M, trans: Kipylova M. Reproduction). Bookmasters, Kettering

Lerdahl F (1987) Timbral hierarchies. Contemp Music Rev 2:135–160. https://doi.org/10.1080/07494468708567056

Lerdahl F (1992) Cognitive constraints on compositional systems. Contemp Music Rev 6:97–121. https://doi.org/10.1080/07494469200640161

Lerdahl F (2001) The sounds of poetry viewed as music. Ann NY Acad Sci 930:337–354

Levin TC, Suzukei V (2006) Where rivers and mountains sing: sound, music, and nomadism in Tuva and beyond. Indiana University Press, Bloomington

Lévi-Strauss C (1969) The elementary structures of kinship (Les structures élémentaires de la parenté) (trans: Von Sturmer JR, Bell JH, Needham R). Beacon Press, Boston

Levitin DJ (1994) Absolute memory for musical pitch: evidence from the production of learned melodies. Percept Psychophys 56:414–423. https://doi.org/10.3758/BF03206733

Lew JC-T, Campbell PS (2005) Children's natural and necessary musical play: global contexts, local applications. Music Educ J 91:57–62. https://doi.org/10.2307/3400144

Li G (2006) The effect of inharmonic and harmonic spectra in Javanese Gamelan tuning (1): a theory of the Sléndro. In: Proceeding AMTA'06 proceedings of the 7th WSEAS international conference on acoustics & music: theory & applications. World Scientific and Engineering Academy and Society, Stevens Point, pp 65–71

Lieberman P (1985) The physiology of cry and speech in relation to linguistic behavior. In: Boukydis CFZ, Lester B (eds) Infant crying: theoretical and research perspectives. Springer, Boston, pp 29–57. https://doi.org/10.1007/978-1-4613-2381-5_3

Lisina MI (1966) Development of the cognitive capacity in children during their first half a year of life (Развитие познавательной деятельности детей первого полугодия жизни). In: Zaporozhets AV, Lisina MI (eds) Development of perception in early and preschool childhood (Развитие восприятия в раннем и дошкольном детстве). Prosvesheniye, Moscow, pp 16–48

List G (1987) Stability and variation. Ethnomusicology 31:18–34. https://doi.org/10.2307/852289

Liubomirsky GL (1924) Musical hearing, its development and enhancement (Музыкальный слух, его воспитание и усовершенствование). State Publishing of Ukraine (Гос. изд-во Украины), Kiev

Lockhead GR, Byrd R (1981) Practically perfect pitch. J Acoust Soc Am 70:387–389. https://doi.org/10.1121/1.386773

Longuet-Higgins C (1975) E flat and D Sharp. Music Times 116:237

Louhivuori A (2006) Tonal development of a child's song improvisations: a case study. In: Paananen P, Fredrikson M (eds) The proceedings of the first European conference on developmental psychology of music, 17–19 November 2005, University of Jyväskylä, Finland. University of Jyväskylä, Jyväskylä, pp 287–290

Lum CH (2009) Musical memories: snapshots of a Chinese family in Singapore. Early Child Dev Care 179:707–716. https://doi.org/10.1080/03004430902944296

Luria AR (1962) On changeability of psychological functions in the process of child's development (Об изменчивости психических функций в процессе развития ребенка). Voprosy Psychologii 3:15–22

Luria AR (1976) Cognitive development: its cultural and social foundations (trans: Lopez-Morillas-M, Solotaroff L). Harvard University Press, Cambridge, MA

Luria AR (2003) The foundations of neuropsychology (Основы нейропсихологии). Academiya, Moscow

Lynch MP, Eilers RE (1992) A study of perceptual development for musical tuning. Percept Psychophys 52:599–608. https://doi.org/10.3758/BF03211696

MacNeilage PF, Davis BL (2005) The frame/content theory of evolution of speech: a comparison with a gestural-origins alternative. In: Abry C, Vilain A, Schwartz J-L (eds) Vocalize to localize II. Interaction studies. Social behaviour and communication in biological and artificial systems, vol 6. John Benjamins Publishing, Amsterdam/Philadelphia, pp 173–199. https://doi.org/10.1075/is.6.2

Malloch S, Trevarthen C (2009) Communicative musicality: exploring the basis of human companionship

Mampe B, Friederici AD, Christophe A, Wermke K (2009) Newborns' cry melody is shaped by their native language. Curr Biol 19:1994–1997

Mang E (2002) An investigation of vocal pitch behaviors of Hong Kong children. In: Bulletin of the council for research in music education, vol 153/154 (ed:. Welch GF). University of Illinois Press, Champaign, pp 128–134

Mang E (2005) The referent of children's early songs. Music Educ Res 7:3–20. https://doi.org/10.1080/14613800500041796

Mang E (2006) The effects of age, gender and language on children's singing competency. Br J Music Educ 23:161. https://doi.org/10.1017/S0265051706006905

Marsh K (1997) Variation and transmission processes in children's singing games in an Australian playground. University of Sydney, Sydney

Marsh K (2009) The musical playground: global tradition and change in children's songs and games. Oxford University Press, Oxford. https://doi.org/10.1093/acprof:oso/9780195308983.001.0001

Matsunaga R, Yokosawa K, Abe J-i (2012) Magnetoencephalography evidence for different brain subregions serving two musical cultures. Neuropsychologia 50:3218–3227. https://doi.org/10.1016/j.neuropsychologia.2012.10.002

Matsunaga R, Yokosawa K, Abe J-I (2014) Functional modulations in brain activity for the first and second music_ a comparison of high- and low-proficiency bimusicals. Neuropsychologia 54:1–10. https://doi.org/10.1016/j.neuropsychologia.2013.12.014

Maykapar S (1900) Musical hearing, its significance, specialty and method of correct development (Музыкальный слух. Его значение, природа, особенности и метод правильного развития). Jurgenson (Юргенсон), Moscow

Mazel L (1952) On melody (О мелодии). Gos Muz Izdat (Гос. музыкальное изд-во), Moscow

Mazel L (1982) On certain aspects of Asafyev's concept (О некоторых сторонах концепции Б.В. Асафьева). In: Prudnikova I (ed) Essays on theory and analysis of music (Статьи по теории и анализу музыки). Sovetskii Kompozitor (Советский композитор), Moscow, pp 277–307

Mazepus VV (2009) Analysis of timbres in ethnomusicology: the articulatory tension and its acoustical correlates. In: Niemi J (ed) Perspectives on the song of the indigenous peoples of Northern Eurasia: performance, genres, musical syntax, sound. Tampere University Press, Tampere, pp 198–209

McDonald DT, Simons GM (1989) Musical growth and development: birth through six. Schirmer Books, New York

McKernon PE (1979) The development of first songs in young children. New Dir Child Adolesc Dev 1979:43–58. https://doi.org/10.1002/cd.23219790306

Mcmullen E, Saffran JR (2004) Music and language: a developmental comparison. Music Percept 21:289–311. https://doi.org/10.1525/mp.2004.21.3.289
McNeill D (2005) Gesture and thought. University of Chicago Press, Chicago. https://doi.org/10.7208/chicago/9780226514642.001.0001
Mekhnetsov AM (2014) Folk traditional culture: essays and data. 150 year anniversary of St. Petersburg Conservatory (Народная традиционная культура: Статьи и материалы. К 150-летию Санкт- Петербургской консерватории) (eds: Mekhnetsova KA, Balevskaya YA). Nestor-Istoriya
Merrill-Mirksy C (1988) Eeny meeny pepsadeeny: ethnicity and gender in children's musical play. University of California Los Angeles, Los Angeles
Mertens P (2004) The prosogram: semi-automatic transcription of prosody based on a tonal perception model. In: Proceedings of the 2nd International conference on speech prosody, pp 549–552
Metlov NA (1985) In: Cheshev SI, Nikolaicheva AN (eds) Music – to children: the handbook for the kindergarten teacher and principal (Музыка – детям: Пособие для воспитателя и музыкального руководителя деского сада). Prosvesheniye, Moscow
Metlov NA, Mikhailova LI (1935) Musical education in preschool institutions: the didactic material for the pedagogical colleges (Музыкальное воспитание в дошкольных учреждениях: Уч. пособие для пед. техникумов). Uchpedgiz, Moscow
Michel P (1973) The optimum development of musical abilities in the first years of life. Psychol Music 1:14–20. https://doi.org/10.1177/030573567312002
Minks A (2002) From children's song to expressive practices: old and new directions in the ethnomusicological study of children. Ethnomusicology 46:379–408. https://doi.org/10.2307/852716
Mishra RC (1997) Cognition and cognitive development. In: Berry JW, Poortinga YH, Pandey J (eds) Handbook of cross-cultural psychology: basic processes and human development, vol 2. Allyn and Bacon, Boston, pp 143–176
Miyamoto KA (2007) Musical characteristics of preschool-age students: a review of literature. Updat Appl Res Music Educ 26:26–40. https://doi.org/10.1177/87551233070260010104
Miyazaki K (1988) Musical pitch identification by absolute pitch possessors. Percept Psychophys 44:501–512. https://doi.org/10.3758/BF03207484
Miyazaki K (1989) Absolute pitch identification: effects of timbre and pitch region. Music Percept Interdiscip J 7:1–14. https://doi.org/10.2307/40285445
Miyazaki K, Ogawa Y (2006) Learning absolute pitch by children. Music Percept 24:63–78. https://doi.org/10.1525/mp.2006.24.1.63
Modgil S, Modgil C (1976) In: Inhelder B (ed) Piagetian research: compilation and commentary, vol 1–8. N.F.E.R. Publishing, Windsor
Moelants D (2000) Statistical analysis of written and performed music. A study of compositional principles and problems of coordination and expression in 'Punctual' serial music. J New Music Res 29:37–60. https://doi.org/10.1076/0929-8215(200003)29:01;1-p;ft037
Monelle R (1991) Linguistics and semiotics in music. Harwood Academic Publishers, Reading
Monson BB, Han S'E, Purves D (2013) Are auditory percepts determined by experience? PLoS One 8:e63728. https://doi.org/10.1371/journal.pone.0063728
Moog H (1976) The musical experience of the pre-school child (trans: Clarke C). Schott Music, London
Moorhead GE, Pond D (1978) Music of young children: Pillsbury Foundation studies. Pillsbury Foundation for Advancement of Music Education, Santa Barbara
Mukhina TK, Lisina MI (1966) The dependency of age and individual achievements in discrimination of pitch from the type of activity in preschool age children (Зависимость возрастных и индивидуальных показателей звуковысотного дифференцирования от характера деятельности детей в пред). In: Zaporozhets AV, Lisina MI (eds) Development of perception in early and preschool childhood (Развитие восприятия в раннем и дошкольном детстве). Prosvesheniye, Moscow, pp 49–73

Müller GB (2008) Evo-devo as a discipline. In: Minelli A, Fusco G (eds) Evolving pathways: key themes in evolutionary developmental biology. Cambridge University Press, Cambridge, pp 5–30

Müller GB (2013) Beyond spandrels: Stephen J. Gould, EvoDevo, and the extended synthesis. In: Stephen J. Gould: the scientific legacy. Springer, Milano, pp 85–99. https://doi.org/10.1007/978-88-470-5424-0_6

Munroe RL, Munro RH (1997) A comparative anthropological perspective. In: Berry JW, Poortinga YH, Pandey J (eds) Handbook of cross-cultural psychology, vol 1, 2nd edn. Allyn and Bacon, Boston, pp 171–214

Myers DG (2009) Psychology in everyday life. Worth Publishers, New York

Nakata T, Trehub SE (2004) Infants' responsiveness to maternal speech and singing. Infant Behav Dev 27:455–464. https://doi.org/10.1016/j.infbeh.2004.03.002

Nazaikinsky YV (1972) On psychology of human musical perception (О психологии музыкального восприятия). Muzyka, Moscow

Nazaikinsky YV (1973) On constants in perception of music (О Константности в Восприятии Музыки). In: Nazaikinsky YV (ed) Musical art and science (Музыкальное искусство и наука), vol 2. Muzyka (Музыка), Moscow, pp 59–98

Nazaikinsky YV (1977) Interconnection between the intervallic-based and degree-based representation of music in the development of a musical ear (Взаимосвязи интервальных и ступеневых представлений в развитии музыкального слуха). In: Agazhanov A (ed) Development of musical hearing (Воспитание музыкального слуха), vol 1. Muzyka (Музыка), Moscow, pp 25–77

Nazaikinsky YV (1982) Logic of musical composition (Логика музыкальной композиции). Muzyka, Moskva

Nazaikinsky YV (1988) The sonic world of music (Звуковой мир музыки). Muzyka (Музыка), Moscow

Nazaikinsky YV (1993) Asafyev's hearing (Слух Асафьева). In: Agazhanov AA, Loginova LN (eds) Development of musical hearing (Воспитание музыкального слуха), vol 3. Moscow State Conservatory named after Tchaikovsky, Moscow, pp 62–80

Nazaikinsky YV, Rags YN (1964) Perception of musical timbres and the significance of the individual harmonics in a sound (Восприятие музыкальных тембров и значение отдельных гармоник звука). In: Skrebkov SS (ed) Application of the acoustic methods in musicology (Применение акустических методов в музыкознании). Muzyka (Музыка), Moscow, pp 79–100

Nell V (1999) Luria in Uzbekistan: the vicissitudes of cross-cultural neuropsychology. Neuropsychol Rev 9:45–52. https://doi.org/10.1023/A:1025643004782

Nettl B (2005) The study of ethnomusicology: thirty-one issues and concepts. University of Illinois Press, Champaign

Nikolsky A (2015a) Evolution of tonal organization in music mirrors symbolic representation of perceptual reality. Part-1: prehistoric. Front Psychol 6. https://doi.org/10.3389/fpsyg.2015.01405

Nikolsky A (2015b) ¿Cómo funciona la emoción musical? (How emotion can be the meaning of a music work). In: Cascudo T (ed) Música y cuerpo: estudios musicológicos. Calanda Ediciones Musicales, Baleares, pp 241–262

Nikolsky A (2016) Evolution of tonal organization in music optimizes neural mechanisms in symbolic encoding of perceptual reality. Part-2: ancient to seventeenth century. Front Psychol. https://doi.org/10.3389/fpsyg.2016.00211

Nikolsky A (2017) On the methodology of the analysis of tonal organization of Jaw Harp music (К методам анализа тоновой организации варганной музыки). In: Novikova OV (ed) Systemic methods of the research on musical culture, International scientific practical conference in memory of V.V. Mazepus (Системные методы изучения музыкальной культуры, Международная научно-практическая конференция памяти В.В.Мазепуса, 31/X-1/XI 2017). Novosibirsk State Conservatory named after Glinka, Novosibirsk

Nikolsky A (2018) Commentary: the 'Musilanguage' model of language evolution. Front Psychol 9:75. https://doi.org/10.3389/fpsyg.2018.00075

Nikolskaya II, Lutoslawski W (1995) Conversations with Witold Lutoslawski. Articles. Memoirs [Беседы Ирины Никольской с Витольдом Лютославским. Статьи. Воспоминания]. Tantra, Moscow

Nikolsky A, Alekseyev EY, Alekseev IY, Dyakonova V (2019) The overlooked tradition of 'personal music' and its place in the evolution of music. Front Psychol 10:3051

Nikolsky A, Alekseyev EY, Alekseev IY, Dyakonova V (2020) How, where and when did the authentic jaw harp traditions form in Siberia and Far East (Как, где и когда складывались аутентичные варганные традиции Сибири и Дальнего Востока). Languages and folklore of indigenous peoples of Siberia 1

Nono L (1999). Historical presence in music today. In: Simms BR (ed) Composers on modern musical culture: an anthology of readings on twentieth-century music. Trans. Simms BR. Schirmer Books, New York. pp 168–174

Nuckolls JB (2004) To be or not to be ideophonically impoverished. In: Chiang WF, Chun E, Mahalingappa L, Mehus S (eds) Proceedings of the eleventh annual symposium about language and society — Austin, Texas Linguistic Forum. Texas Linguistic Forum, Austin, TX, pp 131–142

Nunes-Silva M, Haase VG (2013) Amusias and modularity of musical cognitive processing. Psychol Neurosci 61:45–5608. https://doi.org/10.3922/j.psns.2013.1.08

Nuske HJ, Vivanti G, Dissanayake C (2013) Are emotion impairments unique to, universal, or specific in autism spectrum disorder? A comprehensive review. Cognit Emot 27:1042–1061. https://doi.org/10.1080/02699931.2012.762900

Ojamaa T (2005) Throat rasping: problems of visualization. World Music 47:55–69

Ojamaa T, Ross J (2011) The perceived structure of forest Nenets songs: a cross-cultural case study. Psychomusicol Music Mind Brain 21:159–175. https://doi.org/10.1037/h0094010

Olthof M, Janssen B, Honing H (2015) The role of absolute pitch memory in the oral transmission of folksongs. Empir Musicol Rev 10:161. https://doi.org/10.18061/emr.v10i3.4435

Os'kina SE, Parnes DG (2001) Musical hearing. Theory and methodology of its development and perfection (Музыкальный слух. Теория и методика развития и совершенствования). ACT, Moscow

Ostwald PF (1973) Musical behavior in early childhood. Dev Med Child Neurol 15:367–375

Papoušek H (1996a) Musicality in infancy research: biological and cultural origins of early musicality. In: Deliège I (ed) Musical beginnings: origins and development of musical competence. Oxford University Press, New York, pp 37–55

Papoušek M (1996b) Intuitive parenting: a hidden source of musical stimulation in infancy. In: Deliège I, Sloboda J (eds) Musical beginnings: origins and development of musical competence. Oxford University Press, Oxford, pp 88–112. https://doi.org/10.1093/acprof:oso/9780198523321.003.0004

Papoušek M, Papoušek H (1981) Musical elements in the infant's vocalization: their significance for communication, cognition, and creativity. Adv Infancy Res 1:163–224

Papoušek H, Papoušek M (1995) Beginning of human musicality. In: Music and the mind machine: The psychophysiology and psychopathology of the sense of music. Springer, Berlin/Heidelberg, pp 27–34. https://doi.org/10.1007/978-3-642-79327-1_3

Paraskeva S, McAdams S (1997) Influence of timbre, presence/absence of tonal hierarchy and musical training on the perception of musical tension and relaxation schemas. In: Proceedings of the international computer music conference, Thessaloniki, Greece, September 25–30, 1997. Michigan Publishing, Ann Arbor, pp 438–441

Parncutt R (2016) Prenatal development and the phylogeny and ontogeny of musical behavior. In: Hallam S, Cross I, Thaut M (eds) Oxford handbook of music psychology. Oxford University Press, Oxford, pp 371–386. https://doi.org/10.1093/oxfordhb/9780198722946.013.11

Parncutt R, Levitin DJ (2001) Absolute pitch. In: Sadie S (ed) The New Grove dictionary of music and musicians. Macmillan, London. https://doi.org/10.1093/gmo/9781561592630.article.00070

Pashina OA, Vasilyeva YY, Danchenkova NY, Dorokhova YA, Lapina VA, Matsiyevsky IV (2005) Folk musical creativity (Народное музыкальное творчество). Kompozitor (Композитор), Sankt-Petersburg

Patel AD (2010a) Music, biological evolution, and the brain. In: Levander C, Henry C (eds) Emerging disciplines: shaping new fields of scholarly inquiry in and beyond the humanities. Rice University Press, Houston, pp 91–144

Patel AD (2010b) Music, language, and the brain. Oxford University Press, Oxford/New York

Peery JC, Peery IW (1986) Effects of exposure to classical music on the musical preferences of preschool children. J Res Music Educ 34:24–33. https://doi.org/10.2307/3344795

Perani D, Saccuman MC, Scifo P, Spada D, Andreolli G, Rovelli R, Baldoli C, Koelsch S (2010) Functional specializations for music processing in the human newborn brain. Proc Natl Acad Sci USA 107:4758–4763. https://doi.org/10.1073/pnas.0909074107

Peretz I, Coltheart M (2003) Modularity of music processing. Nat Neurosci 6:688–691. https://doi.org/10.1038/nn1083

Piaget J (1970) Genetic epistemology (trans: Mays W). Columbia University Press, New York

Piaget J (1976) Need and Significance of cross-cultural studies in genetic psychology. In: Piaget and his school. Springer, Berlin/Heidelberg, pp 259–268. https://doi.org/10.1007/978-3-642-46323-5_19

Pike KL (1948) Tone languages: a technique for determining the number and type of pitch contrasts in a language, with studies in tonemic substitution and fusion, vol 4. University of Michigan Press, Ann Arbor

Poortinga YH (1977) Basic problems in cross-cultural psychology: selected papers from the Third International Congress of the International Association for Cross-Cultural Psychology held at Tilburg University, Tilburg, the Netherlands, July 12-16, 1976. International Association for Cross-Cultural Psychology. Swets & Zeitlinger, Amsterdam

Powell A, Shennan S, Thomas MG (2009) Late pleistocene demography and the appearance of modern human behavior. Science 324:1298–1301. https://doi.org/10.1126/science.1170165

Pressnitzer D, McAdams S, Winsberg S, Fineberg J (2000) Perception of musical tension for nontonal orchestral timbres and its relation to psychoacoustic roughness. Percept Psychophys 62:66–80. https://doi.org/10.3758/BF03212061

Profita J, Bidder TG, Optiz JM, Reynolds JF (1988) Perfect pitch. Am J Med Genet 29:763–771. https://doi.org/10.1002/ajmg.1320290405

Protopopov S (1930) Elements of construction of musical speech (Элементы строения музыкальной речи) (ed: Yavorskii B), vol 1. State Edition, Musical Sector (Госуд. Изд-во Музык. Сектор), Moscow

Radynova O, Katinene A, Palavandishvili M (1994) In: Radynova O (ed) Musical upbringing of preschoolers (Музыкальное воспитание дошкольников). Prosvesheniye, Moscow

Rakowski A (1993) Categorical perception in absolute pitch. Arch Acoust

Rakowski A, Miyazaki K'i (2007) Absolute pitch: common traits in music and language. Arch Acoust 32:5–16

Rameau J-P (1971). Treatise on harmony (Traité de l'harmonie réduite à ses principes naturels) (trans: Gossett P). Dover Publications, New York

Ramsey JH (1983) The effects of age, singing ability, and instrumental experiences on preschool children's melodic perception. J Res Music Educ 31:133. https://doi.org/10.2307/3345216

Rehbock PF (1990) Transcendental anatomy. In: Cunningham A, Jardine N (eds) Romanticism and the sciences. Cambridge University Press, Cambridge, pp 144–160

Repina TA (1964) Characteristic features of matching tones in frequency to the assigned model by preschoolers (Особенности подравнивания звука по высоте к заданному эталону у дошкольников). In: Luria AR (ed) New research in pedagogical sciences (Новые исследования в педагогических науках), vol 113–II. Izvestiya of Academy of Pedagogical Sciences of Russia, Leningrad, pp 174–179

Repina TA (1966a) On the problem of the mechanisms of objectivitization of child's pitch distinctions (К вопросу о механизмах явления «опредмечивания» в звуковысотном различении ребенка). In: Zaporozhets AV, Lisina MI (eds) Development of perception in early and preschool childhood (Развитие восприятия в раннем и дошкольном детстве). Prosvesheniye, Moscow, pp 98–141

Repina TA (1966b) Perception of pitch differences in relation to organization of activity of preschool age children (Восприятие звуковысотных различий в зависимости от организации деятельности детей дошкольного возраста). In: Zaporozhets AV, Lisina MI (eds) Development of perception in early and preschool childhood (Развитие восприятия в раннем и дошкольном детстве). Prosvesheniye, Moscow, pp 74–97

Révész G (2001) Introduction to the psychology of music (trans: de Courcy G). New York: Dover

Reybrouck M, Podlipniak P (2019) Preconceptual spectral and temporal cues as a source of meaning in speech and music. Brain Sci 9:53. https://doi.org/10.3390/brainsci9030053

Richerson PJ, Boyd R, Bettinger RL (2009) Cultural innovations and demographic change. Hum Biol 81:211–235. https://doi.org/10.3378/027.081.0306

Ries NL (1987) An analysis of the characteristics of infant-child singing expressions: replication report. Can J Res Music Educ 29:5–20

Rimsky-Korsakov N (1963) Complete collection of works (Полное собрание сочинений), vol 2. Muzyka (Музыка), Moscow

Ross A (2007) The rest is noise: listening to the twentieth century. Farrar, Straus and Giroux, New York

Ross DA, Olson IR, Marks LE, Gore JC (2004) A nonmusical paradigm for identifying absolute pitch possessors. J Acoust Soc Am 116:1793–1799. https://doi.org/10.1121/1.1758973

Ross DA, Gore JC, Marks LE (2005) Absolute pitch: music and beyond. Epilepsy Behav 7:578–601. https://doi.org/10.1016/j.yebeh.2005.05.019

Russo FA, Windell DL, Cuddy LL (2003) Learning the "special note": evidence for a critical period for absolute pitch acquisition. Music Percept 21:119–127. https://doi.org/10.1525/mp.2003.21.1.119

Rutkowski J, Chen-Hafteck L (2001) The singing voice within every child: a cross-cultural comparison of first graders' use of singing voice. Early Childhood Connections 7:37–42

Saarikallio S (2009) Emotional self-regulation through music in 3-8-year-old children. In: Louhivuori J, Eerola T, Saarikallio S, Himberg T, Eerola P-S (eds) Proceedings of the 7th triennial conference of European Society for the Cognitive Sciences of Music (ESCOM). ESCOM, Jyväskylä, pp 459–462

Sacks O (1995) Musical ability. Science (New York, N.Y.) 268:621–622. https://doi.org/10.1126/science.7732360

Sacks O (2008) Musicophilia: tales of music and the brain. Vintage Books, New York

Saffran JR, Griepentrog GJ (2001) Absolute pitch in infant auditory learning: evidence for developmental reorganization. Dev Psychol 37:74–85. https://doi.org/10.1037/0012-1649.37.1.74

Sander K (2002) Ernst Haeckel's ontogenetic recapitulation: irritation and incentive from 1866 to our time. Ann Anat 184:523–533. https://doi.org/10.1016/S0940-9602(02)80092-9

Scherer KR (2013) Affect bursts as evolutionary precursors of speech and music. In: Stephen J (ed) Gould: the scientific legacy. Springer, Milano, pp 147–167. https://doi.org/10.1007/978-88-470-5424-0_10

Schneider M (1961) Tone and tune in West African music. Ethnomusicology 5:204–215. https://doi.org/10.2307/924521

Schneider A (2001) Sound, pitch, and scale: from "tone measurements" to sonological analysis in ethnomusicology. Ethnomusicology 45:489–519

Schneider A (2013) Change and continuity in sound analysis: a review of concepts in regard to musical acoustics, music perception, and transcription. In: Bader R (ed) Sound – perception – performance. Springer, Berlin, pp 71–111. https://doi.org/10.1007/978-3-319-00107-4_3

Schnittke A (2004) In: Ivashkin AV (ed) Writings on music [Статьи о музыке]. Kompozitor, Moscow

Schwartz DA, Howe CQ, Purves D (2003) The statistical structure of human speech sounds predicts musical universals. J Neurosci 23:7160–7168

Scruton R (1997) The aesthetics of music. Clarendon Press, New York

Segall MH, Dasen PR, Berry JW, Poortinga YH (1999) Human behavior in global perspective: an introduction to cross-cultural psychology, 2nd edn. Allyn and Bacon, Boston

Seither-Preisler A, Johnson L, Krumbholz K, Nobbe A, Patterson RD, Seither S, Lütkenhöner B (2007) Tone sequences with conflicting fundamental pitch and timbre changes are heard differently by musicians and nonmusicians. J Exp Psychol Hum Percept Perform 33:743–751. https://doi.org/10.1037/0096-1523.33.3.743

Semenov B (1928) Russia: territory and population: a perspective on the 1926 census. Geogr Rev 18:616–640

Seredinskaya VA (1962) The development of the inner hearing during the ear-training classes (Развитие внутреннего слуха в классах сольфеджио). Muzgiz, Moscow

Sethares WA (2005) Tuning, timbre, spectrum, scale. Springer, Berlin

Sheikin YI (1996) Musical culture of peoples of Northern Asia [Музыкальная культура народов Северной Азии]. Yakutskii Scientific Center, Yakutsk

Sheikin YI (2002) History of music culture of Siberian ethnicities: a comparative historic investigation (История музыкальной культуры народов Сибири: сравнительно-историческое исследование). Eastern Literature, Russian Academy of Science (Восточная литература РАН), Moscow

Shepard RN (1999) Tonal structure and scales. In: Cook PR (ed) Music, cognition, and computerized sound: an introduction to psychoacoustics. MIT Press, Cambridge, MA, pp 187–194

Shepard RN (2010) One cognitive psychologist's quest for the structural grounds of music cognition. Empir Musicol Rev 20:130–157. https://doi.org/10.5084/pmmb2009/20/130

Shrivastav R, Eddins DA, Anand S (2012) Pitch strength of normal and dysphonic voices. J Acoust Soc Am 131:2261–2269

Shvachkin NH (1948) Development of phonematic perception in early childhood (Развитие фонематического восприятия речи в раннем возрасте). In: Teplov B (ed) The problems of psychology of perception and cognition. Works of the institute of psychology (Вопросы Психологии Восприятия и Мышления. Труды Института Психологии.), vol 13. Izvestiya of Academy of Pedagogical Sciences of Russia, Leningrad, pp 101–133

Simpson J, Huron D (1994) Absolute pitch as a learned phenomenon: evidence consistent with the Hick-Hyman Law. Music Percept Interdiscip J 12:267–270. https://doi.org/10.2307/40285656

Skrebkov S (1967) Intonation and mode (Интонация и лад). Sovetskaya muzyka:89–94

Skrebkov S (1973) Artistic principles of musical styles (Художественные принципы музыкальных стилей). Muzyka (Музыка), Moscow

Sladkov PP (1994) Development of the intonational hearing in the course of the ear training (Развитие интонационного слуха в курсе сольфеджио), vol 1–2. Russian Ministry of Culture, Moscow

Smith JD, Witt JN (1989) Spun steel and stardust: the rejection of contemporary compositions. Music Percept 7:169–185. https://doi.org/10.2307/40285456

Snowdon CT (2003) Expression of emotion in nonhuman animals. In: Davidson RJ, Scherer KR, Hill Goldsmith H (eds) Handbook of affective sciences. Oxford University Press, Oxford, pp 457–480

Starcheus MS (2005) Hearing in musicians (Слух музыканта). Moscow State Conservatory named after Tchaikovsky, Moscow

Steblin R (1987) Towards a history of absolute pitch recognition. Coll Music Symp 27:141–153

Stefanics G, Háden GP, Sziller I, Balázs L, Beke A, Winkler I (2009) Newborn infants process pitch intervals. Clin Neurophysiol 120:304–308. https://doi.org/10.1016/j.clinph.2008.11.020

Stige B (2002) Culture-centered music therapy. Barcelona Publishers, Barcelona

Stumpf C (1883) Tonpsychologie, vol 1. S. Hirzel-Verlag, Leipzig

Stumpf C (1890) Tonpsychologie, vol 2. S. Hirzel-Verlag, Leipzig
Sundin B (1998) Musical creativity in the first six years: a research project in retrospect. In: Sundin B, McPherson GE, Folkestad G (eds) Children composing. Research in music education. Malmo Academy of Music, Lunds University, Malmo, pp 35–56
Svantesson JO (2017) Sound symbolism: The role of word sound in meaning. Wiley Interdiscip Rev Cogn Sci 8:e01441. https://doi.org/10.1002/wcs.1441
Swanwick K (2001) Musical development theories revisited. Music Educ Res 3:227–242. https://doi.org/10.1080/14613800120089278
Swanwick K (2015) A developing discourse in music education: the selected works of Keith Swanwick. Routledge, London
Swanwick K, Tillman J, Maccoby EE (1986) The sequence of musical development: a study of children's composition. Br J Music Educ 3:305. https://doi.org/10.1017/S0265051700000814
Szabolcsi B (1965) History of melody. Barrie & Rockliff, Budapest
Tafuri J, Villa D (2002) Musical elements in the vocalisations of infants aged 2–8 months. Br J Music Educ 19:73–88. https://doi.org/10.1017/S0265051702000153
Tafuri J, Welch GF, Hawkins E (2008) Infant musicality: new research for educators and parents. Ashgate, Aldershot/Burlington
Tagg P (2012) Music's meaning: a modern musicology for non-musos. Mass Media's Scholar's Press, Larchmont, NY
Takeuchi AH, Hulse SH (1991) Absolute-pitch judgments of black and white-key pitches. Music Percept Interdiscip J 9:27–46. https://doi.org/10.2307/40286157
Tan YT, McPherson GE, Peretz I, Berkovic SF, Wilson SJ (2014) The genetic basis of music ability. Front Psychol 5. https://doi.org/10.3389/fpsyg.2014.00658
Tarr B, Launay J, Dunbar RIM (2014) Music and social bonding: "self-other" merging and neurohormonal mechanisms. Front Psychol 5. https://doi.org/10.3389/fpsyg.2014.01096
Taruskin R (2009) The danger of music and other anti-utopian essays. University of California Press, Los Angeles, CA
Teplov B (1947) The psychology of musical abilities (Психология музыкальных способностей). Academy of Pedagogical Sciences of Russia, Moscow
Terhardt E (1984) The concept of musical consonance: a link between music and psychoacoustics. Music Percept Interdiscip J 1:276–295. https://doi.org/10.2307/40285261
Terhardt E (1992) From speech to language: on auditory information processing. In: Schouten ME (ed) The auditory processing of speech: from sounds to words. Walter de Gruyter, Berlin, pp 363–380
Theusch E, Basu A, Gitschier J (2009) Genome-wide study of families with absolute pitch reveals linkage to 8q24.21 and locus heterogeneity. Am J Hum Genet 85:112–119. https://doi.org/10.1016/J.AJHG.2009.06.010
Thompson WF, Robitaille B (1992) Can composers express emotions through music? Empirical studies of the arts 10. SAGE Publications, Los Angeles, CA, pp 79–89. https://doi.org/10.2190/NBNY-AKDK-GW58-MTEL
Thomson WE (2010) Empirical musicology review: serialist claims versus sonic reality. Empir Musicol Rev 5:36–50. https://doi.org/10.18061/1811/46748
Thorpe LA, Trehub SE (1989) Duration illusion and auditory grouping in infancy. Dev Psychol 25:122–127. https://doi.org/10.1037/0012-1649.25.1.122
Tischler H (1956) The evolution of the harmonic style in the Notre-Dame Motet. Acta Musicol 28:87. https://doi.org/10.2307/931976
Tiulin YN (1937) The doctrine of harmony (Учение о гармонии). Muzyka (Музыка), Moscow
Trainor LJ, Hannon EE (2013) Musical development. In: Deutsch D (ed) Psychology of music, 3rd edn. Academic Press, New York, pp 423–498
Trehub SE (2013) Erratum: music processing similarities between sleeping newborns and alert adults: cause for celebration or concern? Front Psychol 4:644. https://doi.org/10.3389/fpsyg.2013.00644

Trehub SE, Unyk AM, Trainor LJ (1993) Maternal singing in cross-cultural perspective. Infant Behav Dev 16:285–295. https://doi.org/10.1016/0163-6383(93)80036-8

Trevarthen C (2000) Musicality and the intrinsic motive pulse: evidence from human psychobiology and infant communication. Music Sci 3:155–215. https://doi.org/10.1177/10298649000030S109

Tsougras C (2010) The application of GTTM on 20 th century modal music: research based on the analysis of Yannis Constantinidis's "44 Greek Miniatures for Piano". Music Sci 14:157–194. https://doi.org/10.1177/10298649100140S108

Tull JR, Asafyev B (2000) B.V. Asaf'ev's musical form as a process: translation and commentary (trans: Tull JR). Photocopy, 3 vols. University Microfilms International, Ann Arbor. 610363518

Ukhtomsky AA (1978) Selected works [Избранные труды]. Nauka, Leningrad

Utkin BI (1985) Development of professional hearing for a musician in musical college (Воспитание профессионального слуха музыканта в училище). Muzyka (Музыка), Moscow

van der Veer R, Valsiner J (1991) Understanding Vygotsky: a quest for synthesis. Blackwell Publishing, Hoboken

van Zanten W (2004) Perception of Sundanese music: an experimental approach. In: Niles D (ed) 37th world conference of the International Council for traditional music. International Council for Traditional Music, Fuzhou, pp 278–279

Veis PF (1967) Absolute and relative solfa (Абсолютная и относительная сольмизация). In: Ostrovsky AL (ed) Problems of the ear-training methodology (Вопросы методики воспитания слуха). Muzyka, Leningrad, pp 67–107

Vetter R (1989) A retrospect on a century of gamelan tone measurements. Ethnomusicology 33:217–227. https://doi.org/10.2307/924396

Volodin AA (1972) Psychological aspects of perception of music (Психологические аспекты восприятия музыки). The Institute of Evolutionary Physiology and Biochemistry named after Sechenov, Moscow

von Falkenhausen L (1992) On the early development of Chinese musical theory: the rise of pitch-standards. J Am Orient Soc 112:433–439. https://doi.org/10.2307/603079

von Helmholtz H (1877) On the sensations of tone as a physiological basis for the theory of music (trans: Ellis AJ). London: Longmans, Green.

von Hornbostel EM (1913) Melody and scale. C.F. Peters, Leipzig

von Hornbostel EM (1919) Ch'ao-t'ien-tze, eine chinesische Notation und ihre Ausführungen. Arch Musikwiss 1:477–498

Vos J (1988) The perception of pure and tempered musical intervals. J Acoust Soc Am 84:2290–2291. https://doi.org/10.1121/1.397031

Vygotsky LS (1971) The psychology of art (trans: Scripta Technica Inc). MIT Press, Cambridge, MA

Vygotsky LS (1987) The collected works of L.S. Vygotsky. Child psychology (eds: Rieber RW, Carton AS), vol 5. Plenum Press, New York

Vygotsky LS (1994) The problem of the environment. In: van der Veer R, Valsiner J (eds) The Vygotsky reader. Wiley-Blackwell, Hoboken, pp 338–354

Vygotsky LS (2001) In: Korotayeva GS (ed) Lectures on pedology (Лекции по педологии). Udmurt State University Publishing Press, Izhevsk

Walker R (1987) Some differences between pitch perception and basic auditory discrimination in children of different cultural and musical backgrounds. Bull Counc Res Music Educ:166–168

Walker R (1997) Visual metaphors as music notation for sung vowel spectra in different cultures. J New Music Res 26:315–345

Weber W (2003) Consequences of canon: the institutionalization of enmity between contemporary and classical music. Common Knowl 9:78–99. Duke University Press. https://doi.org/10.1215/0961754x-9-1-78

Weber G (2012) Theory of musical composition: treated with a view to a naturally consecutive arrangement of topics (trans: Warner JF). Nabu Press, Charleston

Weinert L (1929) Untersuchungen über das absolute Gehör (Studies in absolute pitch). Archiv für die Gesamte Psychologie 73:1–128

Wen-Chung C (1971) Asian concepts and twentieth-century Western composers. Music Q 57:211–229

Wenhart T, Bethlehem RAI, Baron-Cohen S, Altenmüller E (2019) Autistic traits, resting-state connectivity, and absolute pitch in professional musicians: shared and distinct neural features. Mol Autism 10:20. https://doi.org/10.1186/s13229-019-0272-6

Wermke K, Mende W (2009) Musical elements in human infants' cries: in the beginning is the melody. Music Sci 13:151–175. https://doi.org/10.1177/1029864909013002081

Wermke K, Leising D, Stellzig-Eisenhauer A (2007) Relation of melody complexity in infants' cries to language outcome in the second year of life: a longitudinal study. Clin Linguist Phon 21:961–973. https://doi.org/10.1080/02699200701659243

West ML (1992) Ancient Greek music. Oxford University Press, New York/London

Winner E (1982) Invented worlds: the psychology of the arts. Harvard University Press, Cambridge, MA. https://doi.org/10.1080/07421656.1984.10758756

Yakovleva LA (1973) From the worker clubs to people's collectives (От рабочих кружков к народным коллективам), Moscow, Iskusstvo (Искусство)

Yennari M (2010) Beginnings of song in young deaf children using cochlear implants: the song they move, the song they feel, the song they share. Music Educ Res 12:281–297

Yost W (2009) Pitch perception. Atten Percept Psychophys 71:1701–1715

Young S (2003) Music with the under-fours. In: London. Routledge, New York

Young S, Gillen J (2007) Toward a revised understanding of young children's musical activities: reflections from the "day in the life" project. Curr Musicol 84:79–99. https://doi.org/10.7916/D81N7ZR0

Young S, Marsh K (2006) Musical play. In: The child as musician: a handbook of musical development. Oxford University Press, Oxford, pp 193–212

Zagretdinov RA (1997) In: Zinovyeva T, Alkin M (eds) The school of playing Kubyz: a practical methodological aid (Школа игры на кубызе: Учебно-Методическое Пособие). Belaya Reka, Ufa

Zaporozhets AV (2003) The development of sensations and perceptions in early and preschool childhood. J Russ East Eur Psychol 40:22–34. https://doi.org/10.2753/RPO1061-0405400322

Zeleny M (1981) Autogenesis: on the self-organization of life. In: Zeleny M (ed) Autopoiesis: a theory of living organization. Elsevier, New York, pp 89–115

Zemtsovsky I (1967) Russian Soviet musical folk studies (Русская советская музыкальная фольклористика). In: Raaben LN (ed) The issues of theory and aesthetics of music (Вопросы теории и эстетики музыки), vol 6–7. Muzyka (Музыка), Leningrad, pp 215–263

Zemtsovsky I (1980) Asafyev and methodological foundations of intonational analysis of the folk music (Б.В.Асафьев и методологические основы интонационного анализа народной музыки). In: Kolovskii OP (ed) Criticism and musicology (Критика и музыкознание), vol 2. Muzyka (Музыка), Leningrad, pp 184–198

Zemtsovsky I (1987) Tracing Vesnyanka from Tchaikovsky's Piano Concerto: historic morphology of a folk song (По следам веснянки из фортепианного концерта П. Чайковского). Muzyka (Музыка), Moscow

Zemtsovsky I (1991) Musical folklore of the USSR peoples in LP records (Музыкальный фольклор народов СССР на грампластинках). USSR Miinistery of Culture, Moscow

Zemtsovsky I (2002) Musicological memoirs on Marxism. In: Qureshi RB (ed) Music and Marx: ideas, practice, politics. Routledge, New York/London, pp 167–189

# Chapter 8
# "Talking Jew's Harp" and Its Relation to Vowel Harmony as a Paradigm of Formative Influence of Music on Language

**Aleksey Nikolsky**

**Abstract** A very popular, yet barely researched, musical instrument is Jew's harp (a.k.a. Jaw harp, JH). Its earliest archeological occurrences date back to the early Bronze Age, but its simplest constructions, made of tree twigs and bark, along with its cross-cultural connection to shamanic beliefs, suggest its prehistoric use. Archeological evidence points to Northeast China and the Amur basin as the source of JH's early dissemination (BCE) over a vast area, from the Volga steppes to Japan. JH has been an important musical instrument in most local traditional musical cultures within this area. For some indigenous ethnicities such as the Yakuts, this is the only non-percussive musical instrument. Its uniqueness is manifested in the peculiar tradition of articulating speech-like sounds. The importance of this tradition is evident by its enormous geographic span from Western Europe to Melanesia. Its use to camouflage romantic communication between a young male and a young female is especially common. Little known is the similarity between the vocal system of articulations in the "talking JH" tradition and vowel harmony found in most languages of the Turkic, Mongolic, and Tungusic families of the Transeurasian (a.k.a. "Altaic") family, as well as in many Uralic languages – all of which are spoken by peoples that regularly use the JH. This paper outlines a possible scenario in which the spread of the cult of ancestral plants across this vast region, from Altai to Sakhalin, and the cult of the "singing mask" of the Tuva-Amur area may have given special importance to musicking on the JH, initiating its spread along pastoralism to the neighboring regions. Once established, the JH tradition may have bifurcated into two types: the framed idioglot, usually made of organic materials, which was sustained in north and northeast, and the bow-shaped heteroglot metallic type that spread to west and southwest. The ethnicities that preferred the framed JH construction retained the JH as their sole (or one of the very few) musical instruments within their timbre-oriented music culture. The ethnicities that adopted forms of frequency-oriented music developed a rich assortment of musical instruments which transformed their JH traditions and reduced the importance of the JH in their

A. Nikolsky (✉)
Independent Researcher, Austin, TX, USA


N. Masataka (ed.), *The Origins of Language Revisited*,
https://doi.org/10.1007/978-981-15-4250-3_8

music cultures. It is JH's unique status as a primary traditional instrument that may have granted it formative influence over the existing languages of the Transeurasian family and the neighboring Uralic family.

**Keywords** Jew's harp · Protomusic and protolanguage · Musilanguage · Tonal organization · "Timbre-oriented" music · "Timbre-class," vowel harmony · Agglutination · Morphonology · Transeurasian family · "Altaic" languages

Language and music have been seen by generations of scholars as semiotic systems unique to human species. Recent research, however, has questioned this view. Not only the delineation between animal communication and human language/music became blurred out, but the traditional opposition between "referentially oriented" language and "affect-oriented" music as well. Music's capacity to convey referential information is employed in military, hunting, and herding calls (Monelle 2006). Language's capacity to convey emotional information is high enough to justify it as language's "second-order" function (Jensen 2014), especially important in poetic speech (Kraxenberger and Menninghaus 2016). Cross-disciplinary investigation of speech, music, and animal communication has led to conceptualization of a single communicative system used to convey referential and affective information, parental to both music and language and derivative of animal communication – **musilanguage** (Brown 2000a). Since both language and music possess shared and domain-specific features, it would be logical to accept that the shared features were the first to evolve, followed by domain-specific, turning language and music into homologues (Brown 2001). This paper examines the shared features between the languages and the music systems practiced in northeastern Eurasia and argues that the indigenous tradition of Jew's harp musicking that remains vital, important, and surprisingly homogeneous across many local ethnoses constitutes a common ground for both tonal organization of music and vocal systems of languages of this region.

## 8.1 Current Views on Emergence of Language and Music

The musilanguage model allows to distinguish six possible ways for music and language to be formed, as determined by the differences in vocal learning and known ethological typology of auditory signals. These six ways are as follows (Cross et al. 2013):

(a) Autonomy of music and language components within musilanguage and after its extinction
(b) Their mutual origin in musilanguage, followed by their bifurcation and autonomous development
(c) Their inseparable origin and development

(d) Their autonomy within musilanguage, followed by their convergence
(e) Music emerging as language's offshoot (Spencerian view)
(f) Language emerging as music's offshoot (Darwinian view)

Of these possibilities, (b) (Brown 2000b), (f) (Fitch 2010), and to a lesser extent (e) (Bowling 2013) find support among modern scholars. The consensus leans toward the (b) scenario (Cross and Morley 2009). The recent update of the musilanguage model further elaborated the (b) framework, distinguishing two evolutionary stages (Brown 2017):

1. Formation of the **affective prosody** from the innate conspecific animal calls, produced collectively in an uncoordinated manner by a group of animals based on emotional contagion
2. Formation of the **intonational prosody** from common heterospecific traits of affective prosody, produced individually and branching into "musical" and "linguistic" domains, as determined by different approaches to temporal organization of frequency modulations:
   (a) **Musical intonation** is faceted by tightening the synchronization of sustained pitch levels in collective vocalization, aimed primarily at mediating the affective state of each of the participants within a group.
   (b) **Linguistic intonation** is faceted by the alternation of vocalizations in a duetic setting in a freer timing, without strict timekeeping, aimed primarily at exchanging propositional information and only secondary at manifesting speaker's affective state and attitude toward the conveyed data.

Temporal organization here plays the formative role, directing the entire **tonal organization** (**TO**) of protomusic and protolanguage, including frequency, amplitude, and timbre-related aspects of vocal expression (e.g., pitch contour, dynamic contour, registers) toward different targets. Inherently uncoordinated collective production of "affective prosody" engages a peculiar musilinguistic texture – "isophony": brief calls, continuously reproduced by multiple performers with irrational deviations in timing and pitch, where each participant retains idiosyncrasy of rhythmic, timbral, and directional attributes of the pitch contour (Nikolsky 2018). "Intonational prosody" imposes TO on "jumbled" isophonic texture, transforming it into synchronized and definite-in-pitch progressions of musical pitch classes or asynchronous and essentially atonal indefinite-in-pitch progressions of linguistic phonemes:

1. The paradigmatic model of "musical texture" becomes **monodic choral** – simultaneous production of the same melody by multiple singers,[1] expressing the same

[1]Not all forms of traditional non-Western music adhere to the collective format of musicking. The vast part of Eurasia, between the Volga Plateau and Japan/China, has nourished the cross-cultural tradition of personal music that is based on cultivation of timbre classes and pitch contours as opposed to pitch classes of the frequency-based music, conventional for the Western European music cultures as well as many non-European cultures (Nikolsky et al. 2020). Nevertheless, this fact

musical emotion in a certain timing. Most known forms of music adhere to this paradigm even in complex heterophonic, polyphonic, and homophonic textures. Unlike language, any music usually contains some kind of a recognizable thematic material (Drabkin 2001) confined to a single melodic line (stream) and reproducible by numerous performers: synchronously or solo – at some other point in some other suitable circumstances (Val'kova 1992). *"Musical" TO specializes in* ***integrating*** *the expression of multiple performers throughout multiple performances, called to repeatedly reproduce the "same" affective expression in redundant and long-lasting way.*

2. The paradigmatic model of "verbal texture" becomes **individualized dialog** – successive production of different sentences by two interlocutors in loose timing, in response to each other, where prosodic features similar to musical rhythm and pitch are used without incremental quantization and where interlocutors focus on informing each other of the constantly changing whereabouts. This paradigm retains its formative power over the individual use of speech, e.g., in internal speech (Winsler 2009). The dialogic setting promotes dialogic thinking – "ongoing interplay between different internalized perspectives on reality" (Fernyhough 2009). *"Verbal" TO specializes in* ***segregating*** *the expression of few performers: "actual" in conversation or "imaginary" in internal speech. In both cases, opposing viewpoints stay non-redundant, constantly updated, and relatively compact.*

**Distance** plays a key role in transforming musilanguage into protomusic and protolanguage. Close proximity favors audio signals that are brief, diverse, and rapidly changing, relying on quick processing, innate reflexes, short-term memory, and integration of multi-sensory data – since auditory expression is often accompanied with non-auditory cues (Wallin 1991). Long-distance communication relies exclusively on the auditory data, requiring long-term memory, retrieval, and interpretative skills bound to a particular environmental frame of reference (ibid.). Such communication forges sender/receiver conventions that operate via ritualization of behavior (Wiley 1983). Conventionalization opens doors to dishonest signaling that works to the advantage of the sender, thereby devaluating the acoustic markers characterizing a particular call type and causing the signaler to eventually modify the signal. The chain of ritualization and devaluation goes on until a state of stationary equilibrium is reached within a sizeable population, at which point the convention becomes permanently fixed (Maynard-Smith 1976). Close-distance communication

should not take away from the overall importance of the criterion of synchronization in defining music in contradistinction to speech. Although traditions of personal music make collective performances exceedingly rare, makers of timbre-based music are capable of synchronizing their performance with multiple participants (which normally occurs during few of the most important annual festivities). Moreover, the historico-morphological analysis of the indigenous genres of personal music indicates that timbre-based forms have been gradually replaced by frequency-based forms. Combined with the current overall prevalence of collectively-made frequency-based music all over the world, this suggests that personal music might have constituted an intermediary form of musilanguage cultivated prior to the divergence of music and language.

allows receivers to verify the auditory information by non-auditory cues and detect the dishonest use of a signal.

Both protomusic and protolanguage seem to represent two different methods of identifying dishonesty by means of cross-relating the acoustic markers of a signal type to the prosodic traits that convey the affective state of signaler and to the framework of a "speech act" or "music act" within which a signal was produced. If *animal* communication tends to be relatively *long-range*, stereotypical, with coarse gradations of very few expressive parameters – depending on the acoustic properties of the environment – *human* communication is mostly *close-range*, diverse, using incremental gradations of multiple parameters, and quite independent from the environmental acoustic properties (Morton 1977). Semiotically, protolanguage was shaped by complexity of production and reproduction, intention to communicate, sound symbolism, parity of encoding and decoding, hierarchical organization, accounting for past and pre-planning future speech acts (Arbib 2012). All of these fit into the "face-to-face" dialogic exchange that refers to the shared environment, supported by deictic gesticulation and mimics – such communication is found in human infants as well as among primates (Kita 2003).

Protomusic differs from protolanguage in engaging gestures of another kind – not deictic but locomotive: histrionic bodily expression and technical motions necessary for sound production (Arbib and Iriki 2012). Hence, "music act" constitutes gestures that are inherently expressive on their own, not needing real objects to which to point (Godøy 2004). Such gestures become abstracted early in infancy throughout the baby's interaction with the caretakers and form the foundation for perceiving melody, harmony, and rhythm (Trevarthen et al. 2011). These gestures are appreciated holistically on par with "sound objects" (Nazaikinsky 1973), which requires extra time for their semantic processing, thereby setting an upper limit for how fast and for how long the music can go before its intelligibility becomes compromised (Nazaikinsky 1972). Gestures that accompany speech do not require such mediating interpretation. They are not autonomous "objects," but rather aids in support of faster transmission of the propositional information (Lieberman 2008):

- The goal of verbal communication is to maximize information output, whereas musical communication limits it in sake of achieving emotional contagion.

Protolanguage opposes language across "the divide between biology and culture" (Arbib 2012), where "closed" auditory qualities of phonemes are more limited than "open" qualities of natural auditory sources in establishing parity between the sender and receiver (Liberman and Mattingly 1989). Likewise, music differs from protomusic by abstracting the auditory qualities of pitch and rhythm and using them to construct histrionically expressive musical movement in a virtual space of vertical (harmonic) and horizontal (melodic) axes (Nikolsky 2015). Abstraction of phonemes and using them to construct syllables and words trigger the syntax creation process that marks the transformation of protolanguage into language, opening a great number of roads that all lead to the genesis of a great variety of languages. Similarly, abstraction of pitch and rhythm classes generates a great wealth of music systems – in contrast to homogeneity of protomusic. The gist of this diversification can be grasped from observing how music skills advance throughout childhood (Hargreaves 1986). All newborns sound more or less the same in their narrow assortment of cries and coos (Loewy 1995) which becomes progressively diversified (Wermke and Mende 2009) until the child's vocal musicking obtains clear traits of its native musical culture (Louhivuori 2006). There is no reason to believe that this course of development is limited to ontogeny. Anatomic organs necessary for production and perception of language and music are present in *Homo heidelbergensis* (Wurz 2010). Its newborns must have shared music/language capacities with the modern newborns, except that their parents did not expose them to musicking as we know it today. However, the latter circumstance would simply delay rather than revert the development of vocal skills from its direction toward greater diversity and cultural specificity of vocalizations.

## 8.2 Jew's Harp and Its Sound Production

Traditional forms of Jew's harp (**JH**) music fit the second, "intonational," stage of the musilanguage model, containing traits that are likely to have a formative influence on the genesis of both TO of music and language.

JH is a unique musical instrument (Fig. 8.1) that somehow evaded the attention of scholars until the mid-twentieth century despite the fact that for at least 21 indigenous ethnicities of northeastern Eurasia (the largest of which is Yakut) JH still remains a principal musical instrument in the scarce instrumentarium that features primarily the instruments of "timbre-oriented music" (Nikolsky et al. 2020). Such music engages timbre as the primary expressive means and differs from "frequency-oriented music" that prevails in the West by relying on "timbre classes" rather than "pitch classes" in its TO. A JH music piece is built from the limited set of sounds made from onomatopoeic imitations of environmental sounds, speech-like articulations, isolated harmonics, and special effects. Changes in pitch in timbre-based music are usually indefinite: performers take into consideration only the direction and the rough estimate of the interval of the pitch change (e.g., step vs. leap). Evidently, people habituated to such music have no need in "frequency-oriented" musical instruments, so that JH completely satisfies their musical needs.

Amazingly, JH remains one of the very few instruments whose sound generation is not yet fully understood, leaving problems in its organological classification. According to Fox (1988), already the earliest scholarly attempt to classify JH remained inconclusive: in 1636, Mersenne pronounced JH "percussive" (Mersenne 1957), whereas circa 1640 he called it "pneumatic." This controversy has not yet been resolved (Kolltveit 2016). The classic work by von Hornbostel and Sachs (1914) qualified JH as a plucked idiophone, based on the plucking action of the hand on the lamella (Fig. 8.1) (Sachs 1917). This classification lasted until Crane (1968) pointed out that the lamella alone is incapable of generating melodies made of the sounds of the harmonic series, therefore reclassifying JH as a free aerophone. Not

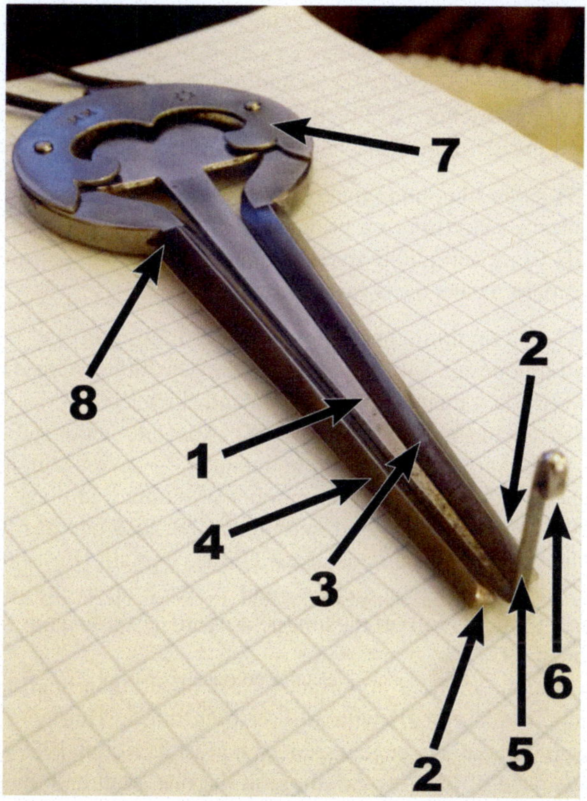


**Fig. 8.1** *The typical construction of JH* on the example of the favorite instrument by Ivan Alekseyev, made by Ivan Kolodeznikov (Khampa, Vilyi ulus). All parts are numbered according to their contribution to typical sound production of the JH. Their Yakut names are provided by Shishigin (2015): (1) Lamella or "JH's tongue" (*tyl*) is usually 1–2 mm thick, 20–25 mm wide, and 9–11 cm long. (2) Left and right arms (*syngaakhtar*) are on average 5 mm thick and are separated from the lamella by the mean gap of 0.1 mm (0.04–1.6 mm). (3) The inner cheek (*iss iedes*) is usually the widest of all four sides of an arm and comprises a wide angle in relation to the upper front side of the lamella. (4) The outer cheek (*tas iedes*) usually forms an acute angle in relation to the inner cheek. (5) The knee of the lamella (*khakhuora*) is typical for metallic instruments, where it supports stronger striking, and usually makes about a quarter of the lamella's length. (6) The tip of the lamella (*chyychaakh*) enables more energetic plucking as compared to its knee. (7) The ring (*tierbes*) can be held with various extents of firmness/looseness. (8) The frame (*iedes*) can be held in a variety of ways. Together, these eight parts determine the timbral characteristics of the JH sound

all organologists accepted Crane's correction (Montagu and Burton 1971). Ledang (1972) confirmed Crane's classification by experimentally demonstrating that the distance between the JH's arms and the lamella is no less critical for sound production than plucking of the lamella, which per se produces a bleak monotone when plucked away from the mouth, and that the range of JH's tones is determined by the configuration of the player's mouth. Adkins created an electric analog of the

JH and demonstrated that JH's sound is generated by the resonances of the mouth cavity which can be varied continuously at a high-frequency range (Adkins 1974). Nevertheless, Kolltveit argues that JH should not be regarded as a reed instrument because it requires hand action (Kolltveit 2016). Such arguments zoom into the mere mechanics of playing JH, ignoring what kind of music is produced on JH:

- Qualifying JH as an idiophone makes it a **percussive** instrument, which subsequently emphasizes **rhythm** as a primary means of its expression
- Qualifying JH as an aerophone makes it a **wind** instrument, which emphasizes **pitch** as a primary means of its expression

The use of JH as a rhythmic instrument is exceedingly rare and is limited to the South Indian and Indonesian ensemble music (Morgan 2008). Across the vast region of northeastern Eurasia – its most likely homeland (Fox 1988, 45) – JH is used as a melodic solo instrument (Yesipova et al. 2008). Absolute majority of ethnicities of former Soviet Union consider JH purely melodic – obvious in its use to play traditional songs (Alekseyeva 1986). A number of ethnic traditions, even those far away from north, consider JH a wind instrument, because one has to blow into it, e.g., Khmer (Sam 2008). Huli of Papua New Guinea explain JH's sound production as "thoughts formed in a person's chest" and "carried up to the mouth" through breathing, where "thoughts roll off the tongue as words," spoken out with the help of JH (Pugh-Kitingan 1977).

Breathing technique is more important for JH music than hand technique. It is simply impossible to play on JH without breathing into it. But there are many JHs that can be played without any engagement of hands (Mazepus 1989). Interaction of the strength and speed of inhaling/exhaling on the one hand and the frequency and amplitude of the lamella's vibrations on the other hand can stop the sound or excite it without plucking (Shishigin 2015, 47). There are entire traditions of handless performance on JH. Thus, Altaic women used to play JHs while doing housework or handicraft, when their hands were busy, holding the instrument by their jaw and controlling the lamella with their tongue (Konchev 2004, 13). The JH designed for such performance was called *t'ak komys* (Tiukhteneva et al. 2006). Its name, translated from Altaic as "jaw harp," reflects its handless design (Sheikin 2002, 469).[2]

Breathing rhythm is not any less important for JH articulations than plucking rhythm (I. Alekseyev 1988). In Indonesian tradition, JH players even employ a special solfege to teach JH, where pitch levels are named according to whether they are produced by inhaling or exhaling (Morgan 2008). Similar approach is used in Bashkiria, where performers master the repertory of syllables in combination with various patterns of breathing (Zagretdinov 1997). The hand rhythm constitutes merely one of four and the most rudimentary means of rhythmic expression – after the rhythms of breathing, mouth articulations, and pitch changes (Alekseyev 1991a).

[2]One of the leading JH players of today, Aigul Abysheva from Kirghizia, often plays without using her hands. Many of her performances can be found on YouTube.

JH's sound is generated in **three stages**. Sound production starts with the excitation of the lamella generating transverse waves (Ledang 1972). Unlike longitudinal vibrations, transverse cantilever vibrations do not form harmonically related modes, though they produce standing waves (Fletcher and Rossing 1998, 58–67). Their propagation depends on the vector of the shear force applied to the lamella, causing different states of polarization (Bettini 2017, 235–236). Each of the modes of transverse, longitudinal, and torsional vibrations features its own frequency, altogether comprising a complex tone that consists of numerous partials of different intensities, some of which form periodic relations, while others do not (Fletcher and Rossing 1998, 624). This can be thought of as a "spectral chord" where some component tones are more salient than others. The performer can activate a partial vibration by varying the angle of excitement of the lamella, the tightness of fixing the clamped end, and the jaw/lip pressure on the JH's ring.

The slit between JH's arms and the lamella acts as a filter, dampening most of the vibration frequencies that lay above the lamella's FF (Ledang 1972), comprising the second stage of sound production. Gaps >0.2 mm cause the reduction of the periodic content in the spectrum, impoverishing it beyond any musical use at gaps exceeding 1 mm. Absolute majority of commercially available JHs have the clearance size between the lamella and the arms about 0.1 mm.[3] Ledang specifically states that it is this geometry that explains why *the enormous variety of JHs made of different materials retains the same characteristic JH sound, easily recognizable by the ear.*

The mechanism of the second stage was identified by Adkins (1974). The turbulence of air, excited by the lamella's vibrations, interacts with the tightly spaced arms as in an accordion where the valve drives the air through the reeds. The lamella pumps air through the slot of the frame, increasing the air pressure in the front and decreasing it in the rear, and then goes through that slot – at which point the air begins to stream into the low-pressure area. The thinner the arms, the faster the lamella travels through the frame and the fewer partial vibrations are dampened by the arms, enriching the sound. As the amplitude of vibrations recedes with the decay of sound, the lamella's oscillations slow down, and the radiated sound loses its harmonic richness – from the smaller vibration modes to the larger.

The third stage in sound production is initiated when the air turbulence along the frame reaches the player's mouth and finds the Helmholtz resonator in the oral cavity. The coupling of sound generator to the resonator occurs through the air rather than the solid material of the JH body (Crane 1968). The player continuously reconfigures his/her oral cavity, tuning it to different resonant frequencies: the

[3]There are 236 JH retail models sold at the Oberton store, in Saint Petersburg, Russia, one of the world's biggest retailers of JHs (as of 12/9/18), offering a variety of JH constructions developed by various ethnic traditions all over the world (https://www.oberton.pro/en/). The lab at this store publishes the metrics for all instruments that have been featured in its catalog (https://www.oberton.pro/en/lab.htm), as well as the information on how these metrics relate to the instrument's sound. All JHs have a mean gap of 0.1 mm (SD 0.06 mm), 4x ratio of the lamella's overall length to the distance from its elbow to the tip, and 5 mm arm thickness – which corresponds to the wavelength of 68.6 Hz. The average FF of these 236 models is 78 Hz (SD 24 Hz).

more reduced the cavity's volume, the higher the frequency (Dournon-Taurelle and Wright 1978). JH is peculiar in its design that delegates the frequency and dynamic controls to different stages: the lamella generates partials but cannot selectively alter any of the partials' amplitude (only all of them together), whereas the player's vocal apparatus amplifies or attenuates specific partials while unable to generate any additional frequency components (Trias 2010):

- The **musical instrument** itself is responsible for producing changes in frequency content within the JH's spectrum, many of which are not audible even to a close observer.
- The **player's vocal apparatus** is responsible for making a particular frequency audible to observers and the player.

Integration of instrumental and vocal production makes JH a hybrid musical instrument – what Eduard Alekseyev, one of the main specialists on JH music, calls "centauric" (1991b) – akin a centaur that inseparably combines anthropomorphic and non-anthropomorphic features. What this merging introduces is *a paradoxical combination of the monotonous complex tone, generated by JH's lamella, and extremely diverse articulations, produced by the vocal apparatus.* In essence, the performer uses his/her mouth in a way identical to speaking a language or singing a song – but silently, without engaging the vocal cords – instead of relying on the JH's monotone. This has two very important consequences (Nikolsky et al. 2017):

1. JH's monotone replaces the player's voice, which does not allow the listener to retrieve personal information about the player (gender, age, state of health, etc.). Effectively, JH makes every person sound "the same," thereby making a player anonymous. JH also conceals the affective state of a player by cutting off expressive modulations of the player's voice. *JH replaces individualized human conspecific "affective prosody" with generic "intonational" prosody of JH.*
2. JH player's articulations compensate for JH's monotony by engaging a much wider range of motions of vocal anatomical organs than that of singing and speaking. JH *maximally enriches vocal articulation to support all sorts of naturally prototyped as well as "fantastic" sounds.*[4]

[4]JH players routinely imitate an enormous repertoire of natural sounds, e.g., animal calls, sounds of animal gait, rain, wind, and even streaming water (E. Alekseyev and Levin 1990). This trait is the most cross-cultural (Maslov 1911; Beliayev 1933; Roux and Charles Constant 1950, 2:507–8; Picken 1957, 186; Koizumi et al. 1977; Zagretdinov 1997; Alekseyeva 1986; Alekseyenko 1988; Bulgakova 2001; Sermier 2002, 103; Alexeyev and Shishigin 2004; Mamcheva 2005; Canave-Dioquino et al. 2008; Yesipova et al. 2008; Suzukei 2010; Kartomi 2012, 159) – in fact, it is hardly possible to name a single indigenous JH tradition that would not include onomatopoeic imitations. JH also uses "experimental" articulatory devices that find no analogs in the player's native language or sound environment, but are the result of exploring the JH's sonic capacities, such as a special device of traditional Yakut technique, *khos yrya* ("2-part singing"), characterized by the ongoing opposition of different registers in JH musicking (I. Y. Alekseyev 1988).

**Audio Example 1 Onomatopoeic imitations** of bird songs, horse trot, etc., in the khomus improvisation "A summer morning" by Peter Ogotoyev. Used with his permission. http://chirb.it/AszLqB.

## 8.3 Jew's Harp as the World's Earliest Vocoder and Its Cultural Ramifications

The combination of monotone and diverse articulations, in effect, makes JH into a vocoder (Leipp 1963). The lamella generates the "carrier" signal, whereas the vocal apparatus passes it through the "modulator filter." Such vocoding is related to secrecy. Electric vocoders were invented to conceal the speaker's identity while delivering those speaker's words in military communications (Tompkins 2010). Adoption of vocoders in Western popular music also answered the need to conceal singer's personal traits (gender, age, ethnos, class, temper), perceived as potential vulnerability (Dickinson 2001). The same applies to two cross-cultural uses of JH, typical for many indigenous traditional music cultures. From North-Western Europe to Papua New Guinea, across the entire Eurasia, JH is employed as a ciphering device in private romantic serenading of young couples (Hauser-Schäublin 1995; McLean 1999, 265; Hsu 2001; Levin and Suzukei 2006, 116–17; Sam 2008; Uchida and Catlin 2008; Kartomi et al. 2008; Matusky 2008; Canave-Dioquino et al. 2008).[5] Among the speakers of tone languages, JH is effective in imitating not only linguistic phonemes but tone intonations as well (Poss 2012).

Another cross-cultural tradition, probably of greater antiquity, is using JH to camouflage one's voice in rituals dealing with supernatural forces. JH excels in the production of "grammelot" effect (Jaffe-Berg 2001). Such enigmatic "foreign" speech, interspersed with onomatopoeic imitations, is used by Siberian shamans to represent messages from spirits (Alekseyenko 1988). Common people, too, use JH as a "vocal mask" for protection from evil. The animistic belief system, surviving in Altai, holds that masters of landmarks (mountain, river, forest) can steal a person's voice, which equates with that person's death (L'vova et al. 1989, 90). In many indigenous cultures of Eurasia, voice is regarded as an immanent live object, tied to the person through breathing and vulnerable to invasion by a spirit (Pashina et al. 2005, 50). Hence, JH is seen as a shield protecting its possessor.[6]

---

[5]Typical applications include concealing the young man's identity in his confessions of love, articulated on JH just outside of the maiden's house, while she is not allowed to look out (Canave-Dioquino et al. 2008, 437), or the reversal of the serenading roles between both genders (Hsu 2001). A variation of this use is in engaging JH to "encode" romantic confessions in order to prevent possible eavesdropping (T. Levin and Suzukei 2006, 117).

[6]Thus, Yakuts believed that *abaasy* (evil spirits) liked the sounds of JH and tried to imitate them but became confused, upset, and therefore left the playing human alone by himself/herself (Popov 1949, 265). Similar beliefs would explain the tradition of using JH as a talisman to repel bad fortune, still surviving in Siberia and Altai. In Tuva, a JH is sewn into a person's attire to prevent

"JH magic" most likely comes from the totemic ideology that is still very much alive in Siberia and Far East, especially in Altai (Nikolsky et al. 2019b). Many ethnicities draw their ancestry from a specific plant: Nivkhi, from larch; Oroks, from birch; Ulchi, from cedar; and Ainu, from fir (Duvan 2003). Plants were believed to breathe and to possess soul (L'vova et al. 1989, 89), therefore capable of nourishing or even giving birth to human kin-founders (Tadysheva 2018). As of today Khakassian and Altaic kins still hold different tree species for their ancestors (Sagalayev and Oktiabr'skaya 1990, 50–59). A dying tree near a human settlement is seen as an ill omen for the death of someone from the corresponding kin: Kuzen, for pine; Tubalar, for aspen; Jyc, for fir; and Todosh, for honeysuckle (Kypchakova 2006). Therefore, cutting of ancestral trees is tabooed across Altai.

In Siberia, Ugric and Turkic ethnicities use trees to identify genealogical "trees": they believe that bones of people from the same kin are made of the same "wood" type in accordance with the paradigm of the tree species and the forest, so that every person corresponds to some tree, while their kin to the same tree species within the forest that represents a particular ethnos (Sagalayev and Oktiabr'skaya 1990, 43–57). Traditional identification system involves a totemic animal, an ancestral tree, and a sacred mountain (Tadysheva 2016). This system is still in use: in 2010, there were 82 kins (*seoks*) registered by the census in Altai Republic (Tyukhteneva 2015).[7] Turkic cosmology envisages the Seok Tree as the global origin of life (Turan 2006). In such belief system, JH made of the ancestral tree would secure its possessor against misfortune with its ancestral protective power. Therefore, unauthorized use of someone else's JH was seen as potential danger of triggering revenge from an ancestral spirit. Thus, in Buryatia, after a shaman died, no one played his personal JH – it was hung on the sacred tree nearest to the shaman's place of burial (Agapitov and Khangalov 1885).

The association of JH with supernatural powers might have originated in the "Aeolian harp" naturally made by the lightning striking an ancestral tree. Turkic ethnicities of Siberia consider such trees purified of evil spirits and have a custom of keeping their chip as a talisman, using such chips to make JH (Duvan 2003). Modern Yakut master-makers of musical instruments explain that the lightning strike dries and fragments the tree so that its wood can "sing" unlike the regular wood (Sheikin

---

evil spirits from approaching that person in his/her sleep (V. Dyakonova 1981). Some informants explain JH's power by the evil-repelling properties of metal of which the JH is made and the blacksmith who has forged that metal – akin to beliefs in the horseshoe protection, common in Europe (Tadagawa 2017b). However, this cannot explain why JHs made of organic materials are believed to possess supernatural power. Thus, Ulch shamans use not only metallic but also wooden JHs (Duvan 2000).

[7]The Altaic word "seok" has been traditionally used to refer to "kin," "bone," "remnants," and "generation" in the context of a specific form of societal exogamic patrilineal organization (Verbitsky 1865). These meanings are all united under the umbrella of the general symbolic meaning of "cemetery," where bones are metaphysically understood as "the quintessence of an alive matter, capable of supporting future births" – native Siberians think that children of the same kin "grow" out of their parental bone ingredients (Sagalayev and Oktiabr'skaya 1990, 39–40).

2002, 68). Yakuts call such JHs *mas-khomus* (Yakut, "chip harp") and today use it as a children's toy (Dyakonova 2017).[8]

Siberian myths attribute the invention of JH to the bear, a sacred animal, related or ancestral to humans, who accidentally discovered that a splinter of a tree makes a peculiar sound (Startsev 2017, 104). Ugrian ethnicities worship a bear-looking deity Kaigus who made the first JH from the birch's chip and used it to imitate animal calls in order to gain power over these animals – later teaching human hunters to play JH before hunting to secure prey (Sheikin 2002, 126). It is quite possible that this myth indeed faithfully reflects the invention of JH – albeit by humans – Udege hunters of Primorye still use wooden JHs right after setting hunting traps despite its reputation of a children's toy, unlike the "artistic" metallic JHs used to make "serious" music (131).

The vocoding capacity of JH is perhaps of greatest value to shamans whose job in traditional societies consists of dealing with evil spirits on behalf of their clients (Balzer 2016). In fact, the idea of putting on an "auditory mask" complements putting on a facial mask that Siberian and Altaic shamans wear for self-protection (S. Ivanov 1975). Not surprisingly, JH has earned the reputation of shamanic accessory all over the Eastern part of Eurasia: from Bashkiria (Shchurov 1995) to Malaysia (Canave-Dioquino et al. 2008).[9] A distinguished turkologist, Saul Abramzon, held that ancient Turkic shamans used two musical instruments: JH and tambourine (Suleimanova 1994). Across Altai shamans receive their "standard" musical instrument, a tambourine, only after reaching maturity in their supernatural skills, prior to which they were allowed to use only JH (Rouget 1985, 126). This applies to Buryat and Mongol shamans (Pegg 2001), as well as Ulchi in Amur and Sakhalin (Duvan 2000). Noteworthy, Chukchi, Kereks, and Koryaks call JH "a mouth tambourine" – probably because of their shared shamanic affiliation (Podmaskin 2003).

---

[8]In the mid-twentieth century, mothers still taught little children to make mas-khomus by splintering a chip from a young tree and holding the thicker end by the hand while bringing the thinner end to the mouth and pinching it by the finger, treating it like the JH's lamella (Tchakhov 2012). Beliayev (1933) considers the JH's adoption as children's toy by Turkic peoples in Uzbekistan and Turkmenistan as one of the typical cases of "infantilization" of an archaic musical instrument after the ideological aspects of its original music culture have become forgotten.

[9]The only large Siberian ethnos that does not unequivocally regard JH as a traditional shamanic instrument is Sakha. Thus, Maria Czaplicka categorically stated that, in contrast to the surrounding peoples, for Sakha JH was a purely secular instrument (Czaplicka 1914). However, her predecessor, Ivan Khudiakov, reported that Yakut shamaness Dyereliyer always carried a few JHs with her (Khudiakov 1969, 362). Few surviving rites and superstitions of Yakuts in regard to JH indicate that JH might have been a part of the archaic shamanic cult of Mother-Beast (Vasilyev 2016). It seems that Czaplicka's informants misunderstood her questions. "White" shamanesses, "udagan," who unlike "black shamans" did not commit sacrifices and specialized in contacting benevolent spirits, have been using JH to cure diseases and foretell fortune (V. Dyakonova and Grigoryeva 2017). It could be that Czaplicka's informants were speaking about "black shamans" ("oyuun"). Khudiakov, most likely, had a better rapport with local indigenous population since, unlike Czaplicka, who just briefly visited Yakutia (1914–1915), he spoke fluent Yakut and resided there for 10 years in exile (1867–1876) while serving his sentence.

Some shamans never qualified to receive a consecrated tambourine and were left with a JH until their death (Vainshtein 1991, 248). This circumstance led Vainshtein to believe that JH was the primordial musical instrument of shamanism and tambourine was introduced at a later point (273). Vasilevich (1969, 209) considered JH to have proto-Tungusic origin and belong to female shamanism. Among Ugric ethnicities, JH retains the reputation of female shamanic accessory (Alekseyenko 1988). The reputation of JH, especially its more ancient, non-metallic varieties, as belonging to the female and children domain, is exceedingly strong among nearly all Middle Asian, Ugric, Turkic Altaic (Emsheimer 1986), and Far Eastern ethnicities (Mazepus and Galitskaya 1997) – to the extent that male players would abstain from playing such instruments (Sheikin 2002, 131). This could be the remnant of an ancient shamaness cult, rooted in a special need for vocoding by wives within the circle of the husband's relatives. The traditional practice among Uralo-Altaic ethnicities, ancient enough to have left a genetic impact, is for men to marry foreign women (Pakendorf 2007). According to a widespread behavioral code, the wife remained a stranger to her husband's kin and therefore was tabooed from calling his kin's totemic entities by name under the threat of harming the kin (Tadysheva 2018). Using JH to articulate their names or hint at them by sound imitations would circumnavigate this taboo. This may explain why most archeological finds of JHs in Volga-Kama and Ural region of Turkic and Finno-Ugric cultures occurred in female burials (Aleksandrova 2017).

If JH indeed served as the oldest auditory shamanic tool, then the entire area occupied by Turkic diaspora can be regarded as the basis of cultivation of the JH's vocoding capacities within the framework of ancestor cult that was served by shamanism, thereby culturally uniting a vast territory by establishing JH as a model for all forms of instrumental and vocal musicking (Suzukei 2010). The available archeological data seems to support this perspective: across most part of Eurasia, JH remains one of the few musical artefacts consistently found in burials, which must be taken as a proof of its connection to supernatural and ancestral beliefs (Oleszczak et al. 2018). The similarity of forms of JH usage and its religious functions over Central Asia, Siberia, and Mongolia indicate the existence of a vast archaic culture based on common shamanic rites (Emsheimer 1986).

One of the leading specialists in religious beliefs of Siberian peoples, Nikolai Alekseyev, points out that shamanism constituted a pan-Turkic religion that since antiquity has been providing a "state" ideology for the confederation of Turkic tribes all over Siberia (Alekseyev 1992). Shamanism absorbed the principal traits of the earlier pan-Siberian ideology of Paleo-Siberians: cult of spirits of local landmarks, hunting rituals, beliefs in reincarnation of hunted animals, and magic. It seems that the transition from this ideology to shamanism corresponds to the transition from using JH as an instrument of magic to its use as a musical instrument, appreciated for the sounds it makes rather than its talismanic power. This transition was probably concurrent with the tambourine becoming the principal shamanic instrument in Siberia and Far East. At this point the vocoding capacities of JH must have found different, non-magic uses: of private romantic serenading, of representing one's environment in its aesthetic appreciation, of entertaining oneself during everyday chores, and of providing children with a versatile sound-toy.[10] Such uses forged the specific "musical" patterns of playing the JH, causing *trifurcation of JH music from singing and speaking and a generation of proprietary JH vocal system.*

## 8.4 The Repertory of JH's Sounds

All styles of playing JH share the same repertory of sounds determined by JH construction. The frequency range of JH is limited by the player's ability to control his/her vocal apparatus. The upper range remains quite universal for different performers, since the minimal volume of the resonance camera is about the same for every human. Even unskilled players can produce 11 distinct harmonics to support pitch changes within 3.5 octaves above the FF (Nikolsky 2017). Skilled performers reach even the 13th harmonic for melodic use (Emsheimer 1986). Non-melodic use of harmonics for timbral recolorations extends much higher. Metallic instruments bring out up to the 145th harmonic (over seven octaves above the FF) above the noise floor – 7.6 times higher than the highest salient 19th harmonic of the same articulation pronounced in a spoken manner and 6.3 times higher compared to the sung version of the same articulation (Nikolsky 2017).

The lower limit of JH's range is much more variable, depending on the player's selective control of four cavities: oral, nasal, laryngeal, and thoracic (Exs. 2–6) – maximizing their volume and coupling their resonances can generate frequencies well below the normal speaking voice of a player (Mazepus 1989).

**Audio Examples 2–6**

Ex. 2. **Sürün dorğoono** (the main tone) – the device of traditional Yakut khomus performance technique that is characterized by the isolation of the player's oral cavity in sound production and fixing the position of a tongue (Alekseyeva 1986). Demonstration by Ivan Alekseyev, used with his permission (http://chirb.it/Fd1P7J).

Ex. 3. **Uos dorğoono** (the lip tone) – the device of Yakut JH traditional technique that is characterized by extending the oral cavity by means of stretching the lips out. Demonstration by Ivan Alekseyev, used with his permission (http://chirb.it/BnIcFa).

Ex. 4. **Beles dorğoono** (the throat tone) – the device of Yakut JH traditional technique that is characterized by engaging the laryngeal cavity as a resonance chamber by means of quick closing and opening of the performer's larynx together with contracting and relaxing the tongue's root. Demonstration by Ivan Alekseyev, used with his permission (http://chirb.it/tqC87w).

[10] Historically more recent use of vocoding's secrecy was found in those Turkic cultures, such as Bashkir, which adopted Islam – since the latter restricted musical entertainment, especially for Turkic women who constituted the base for using JH made of non-organic materials (Zagretdinov 1991).

Ex. 5. **Kõñköleỹ** (chest voice) – the device of Yakut JH traditional technique that is characterized by engaging the additional resonance of thoracic cavity while excluding the nasal resonance. A performer relaxes muscles of the mouth, throat, and trachea in an attempt to maximize the resonance chamber while avoiding breathing with the nose. Demonstration by Ivan Alekseyev, used with his permission (http://chirb.it/9e0E8H).

Ex. 6. **Murun dorğoono** (the nasal tone) – the device of Yakut JH traditional technique that is characterized by adding the nasal resonance by closing the oral air passage while opening the nasopharynx. Demonstration by Ivan Alekseyev, used with his permission (http://chirb.it/P6eMEB).

In Tuvan and Bashkir traditions, players can transpose JH's FF an octave lower (Ikhtisamov 1988). Here, the difference between JH traditions becomes important. Limiting the resonance to the oral cavity, common for Western European performers, restricts the low end of JH's range to about 500 Hz (Dournon-Taurelle and Wright 1978, 21). However, Norwegian *munnharpe* tradition apparently supports the range down to 330 Hz by engaging laryngeal technique (Trias 2010). Yakut tradition utilizes eupnea and expands the low end down to 218 Hz. So, all in all, the JH's range for Siberian lay-players who possess chest technique includes harmonics nos. 3–11 (or 240–880 Hz for the JH tuned to 80 Hz).

Eduard Alekseyev (1991b) lists three principal types of vocalizations on JH (Table 8.1).

Each type calls for a different analytical approach to its TO: (1) soundstage analysis, (2) musicological analysis of aspects of musical expression (pitch, rhythm, dynamics, etc.), and (3) phonemic and morphonological analysis. The first two kinds of analysis are well known. A glimpse of the latter is provided by new method books of traditional indigenous JH playing. Thus, Robert Zagretdinov (1997) offers the biggest list of the following "phonational exercises" for mastering kubyz (the Bashkir JH) in the order of increasing complexity:

I. The principal vowel positions [a], [o], [i], [ɘ̆], [ĕ], [u], [ja], and [ju]
II. The principal diphthongs [ai], [aj], [au], [ie], [jʊ̯], [juja], [ui], [ou], and [ɯja]
III. The principal consonant-vowel combinations [va], [vɯ], [la], [li], [vi], [vu], [vo], [za], [ta], [tɕa], [gi], [gu], [lja], [lĕ], [na], [no], [ni], [χa], [ku], [ko], and [kĕ]
IV. The principal vowel-consonant combinations [ar], [in], [ol], [ip], [ĕs], [uk], and [jud]
V. The principal syllable combinations [laa], [vaj], [ana], [ali], [ara], [aki], [asi], [ati], [api], [azi], [aka], [atja], [gaja], [guju], [ila], [inĕ], [ina], [eʒi], [ora], and [χat]
VI. The principal vocable "words" [aj-aj-aj], [ala-la-la], [au-vau-vau], [ana-na-na], [ali-li-li], [ara-ra-ra], [aki-ki-ki], [asi-si-si], [api-pi-pi], [azi-zi-zi], [ga-lja-lja-lja], [gu-lja-lja-lja], [gu-sja-sja-sja], [jala-la-la], [χrr-raj-raj], [χrr-ta-tan], [jʊ̯la-la-la], [eʒi-ʒi-ʒi], and [ĕgĕt-ɘ̆gɘ̆t-gĕt]

**Table 8.1** Classification of vocalizations produced on JH within indigenous musical traditions of Eurasia, according to Eduard Alekseyev (Alekseyev 1991a, b)

| Environment-like | | Music-like | | Speech-like | |
|---|---|---|---|---|---|
| (1) Timbro-registral non-melodic playing, based on contrasts and similarities between holistic spectral complexes in regard to their coloristic (sonoristic) aspect of sounding – focusing on realism of the imitation, a desired animistic effect, or fun of play/game | | (2) Melodic playing, based on dynamic stressing of a single partial and its integration with the consecutive partial to form a chain of melodic intervals – focusing on aesthetic or ritualistic expression of a certain affective state by means of changes in pitch levels | | (3) Phonemic playing, based on emulation of sounds similar to syllables of speech – focusing on aesthetic or ritualistic expression of a certain affective state by means of articulatory changes | |
| Depictive program-based compositions (like "tone poems" of Western classical music) built on imitation of naturally occurring sound sources (often comprising "music stories") (Ex. 7) | Free exploration of the sounds producible on the JH in a form of extemporaneous improvisation, usually to entertain a performer or a listener (similar to a "fantasy" and/or an "etude" of classical music) (Ex. 8) | Reproduction or emulation of pre-existing tunes from songs or instrumental music, which might reuse the original TO or adapt melodies to the JH's harmonic series (Ex. 9) | Original melodic improvisations that find no prototypes in the repertoire of songs or instrumental tunes, usually based on JH's harmonic series, often with the addition of proprietary JH technical devices of playing (e.g., "droplets" or "slides) (Ex. 10) | Encoding of actual words, phrases, and/or sentences of the player's native language, ascribed a new (in relation to a verbal prototype) semantic value (Ex. 11) | Production of speech-like sounds that do not conform with the vocabulary of any language but express based on phonetic symbolism of vowel and consonant articulations (resembling grammelot) (Ex. 12) |

The typology breaks into three domains, music-like, speech-like, and environment-like, each of which consists of "fixed" imitative versus "free" creative uses. Most likely, imitations preceded creative improvisations, supplying them with the patterns of TO. Environmental imitations likely preceded musical ones, which in turn preceded speech-like ones. Audio examples are provided for each of the sub-classes

Ex. 7. **JH "program music."** This is an example of the composition based on the idea of "sound painting" of a particular situation typical for the indigenous lifestyle. It is called "On my way" and contains the representation of how one is preparing to travel, takes a horse, and finally arrives at his destination. Improvisation by Ivan Alekseyev, used with his permission (http://chirb.it/9P0kN8)

Ex. 8. **Timbral fantasy.** This improvisation demonstrates how players typically entertain themselves by exploring a specific technical device and/or special effect (rather than reproducing a pre-existing melody or speech-like patterns). Improvisation by Ivan Alekseyev, used with his permission (http://chirb.it/yvOgBJ)

Ex. 9. **Adaptation of a folk song "Mary had a little lamb."** Demonstration by Erkin Alekseyev, used with his permission (http://chirb.it/HNcBsq)

Ex. 10. **Melodic fantasy on few Yakut folk tunes** – "Ysyakh medley." This composition by Ivan Alekseyev develops the selected melodic ideas in a free style. Used with permission of the author (http://chirb.it/4Ok8bM)

(continued)

Ex. 11. “**Talking khomus**” – “Tangalay.” This is a humoristic description of a woman who is too fond of men. Performed by Ivan Alekseyev, used with his permission (http://chirb.it/31Je43)

Ex. 12. **Romantic JH duet**. This is an example of the pan-Eurasian genre of employing JH by a young man and young woman for confessions in love in an attempt to conceal the exact content of communication from eavesdroppers (Morgan 2008). Such duets rely on the “vocoded”-like distortion effect that supports the interpersonal conventions of both “speakers” while making them incomprehensible to witnesses. Characteristically, one performer usually reproduces the playing of another performer (Haid 1999). Erkin Alekseyev and Nastia Petrova, used with their permission (http://chirb.it/rH6bFD)

Zagretdinov's system clearly differs from the phonological system of Bashkir language (Kiyokbayev 1958), which indicates that JH's articulations do not merely reproduce Bashkir speech on JH, but present some autonomous form of expression. Similar assortments of phonemes and vocables are used in other influential neighboring JH traditions: Tartar, Kyrgyz, Yakut, and Tuvan. They all contain the tradition of "talking JH" (Grigoryan 1957) that is highly regarded by indigenous population (Ivanov 1999). For Yakuts this tradition is called *syyia tardyy* (Yakut: "syyia," slow and graceful; "tardyy," to pull) – the tradition of "moderated playing" – ancient solitary style of musicking, characterized by restricting the diversity of hand strokes, imitations, and special effects in order to focus on the clarity of phonemic articulations (Shishigin 2015). This style has influenced other styles. Today it is implemented wherever the performer feels the need to emphasize the vocal articulations, such as when expressing an elevated affective state (Zhirkova 1991).

Noteworthy, the Yakut tradition does not oppose the "singing" and "speaking" styles of JH playing – most probably, because they share the same articulation technique (Alekseyev 1991b). Nevertheless, each is distinguished by TO. One of the leading masters of JH playing, Peter Ogotoyev, testifies that TO of "singing *khomus*" (Yakut JH) is governed by the **harmonic series** of its FF – in contrast to "speaking khomus" with its "articulatory series" (or "phonematic series") that tend to distort the just tuning (Ogotoyev 1988). According to Ogotoyev:

- "Singing JH" requires the player to select degrees of the "**just mode**" = a set of lower harmonics from the harmonic series (e.g., f4-f10, or f6-f11), built from the FF to which the JH lamella is tuned;
- "Speaking JH" engages degrees of the "**articulatory mode**" = a set of tongue positions used to produce specific vowels throughout the continuum of changes in vowel height and frontness/backness (e.g., [a]-[u]-[e]-[i]).

Something similar to JH's articulatory mode has been noted by phonologists. Edward Sapir (1929) discovered what he called "gradated size scales" – the consistent cross-linguistic grading of vowels in regard to their perceived size, representable in the manner of an ascending musical scale in the order of the decrease in size. For English, such scale (from the lowest to the highest sound) is [a] > [æ] > [ɛ] > [i]. Stanley Newman (1933) extended such scale for English to eight vowels, where two pairs equaled in size: [ɔ] = [o] > [u] = [a] > [æ] > [ɛ] > [e] > [i]. More modern studies unveiled even more extensive scales in vowel-rich languages, e.g., 13 vowels and diphthongs of Kammu: [u] = [ua] > [ɨ] > [ɨa] > [o] > [ə] > [ɔ] > [a] > [ia] > [ʌ] = [i] > [e] > [ɛ] (Svantesson and Tayanin 2003). Although the "degrees" of such scales are relative in their frequency values, nevertheless each of the degrees was found to fluctuate within a certain range of musical intervals per speaker, e.g., major third–major seventh, placed consistently higher in the speaking frequency range for [i] and lower for [a] (Lehiste and Peterson 1961). This phenomenon was called "intrinsic vowel pitch" and was found to occur as a result of tongue action, increased for high vowels, pulling on the larynx and raising the tension of the vocal cords, thereby raising the pitch of the voice (Ohala and Eukel 1987).

The overall musicality of the resultant distinctions between vowels of a given language prompted Peter Ladefoged to compare a set of vowels in a language with an ensemble of different musical instruments – vowels retain their tonal qualities, including their typical pitch range, essentially in a way similar to the way in which a musical instrument retains its characteristic timbral quality enough to be recognized upon hearing it despite all variations in pitches (Ladefoged and Disner 2012, 32). Each language can be thought of as an "orchestra" of vowels supporting its capacity to convey the speaker's affective state by means of expressive aspects common for human music (pitch, rhythm, dynamics, etc.). Each vowel here contributes its own specific semantic value, determined by its phonetic symbolism, most evident in ideophones and phonoaesthemes (Svantesson 2017). The gist of this was captured by Diedrich Westermann (1937) in his description of the systemic opposition of high ([i], [e], [ɛ]) and low ([u], [o]) vowels of three West African languages as small/large, narrow/broad, light/heavy, quick/slow, and bright/dark.

JH players observe similar symbolic associations while producing and perceiving vowel articulations (Nikolsky et al. 2020). Indigenous traditions often combine the articulatory and just modes, enriching them with onomatopoeic imitations and special effects in a single piece of musicking. **Musicality** remains *the key feature of any JH style* (Table 8.2). "Speaking JH" and "onomatopoeic JH" also engage aspects of expression that are traditionally related to music – and carry them out in a systemic and interactive manner. So, it would be justifiable to conceptualize each of the aspects of expression into a dedicated "musical mode" and gradations within it – as "degrees" of that mode. That would result in six types of modes: melodic, harmonic, rhythmic, metric, tempo, and dynamic – the first three of which are already recognized by many music theories.

## 8.5 Modal Nature of Tonal Organization of JH Vocalizations

At this point, it is necessary to stress the difference between the concept of "mode," suggested by Ogotoyev, and the concept of "scale" already in use by linguists. Musical mode is much more than a mere "scale" – the very same pitch classes can be organized differently, e.g., G-A-B-C-D-E can function as G major or A minor (von Hornbostel 1913). As Sachs warned, scale is nothing but attempts of scholars to logically reconcile what musicians do in practice according to some theory (Sachs 1960). Unfortunately, the voluminous entry on "mode" in the Grove dictionary does not provide a clear distinction between mode and scale (Powers and Wiering 2001). Russian Musical Encyclopedia is more instrumental in its definition: mode is a euphonic set of tones, with one or few anchor tones, which systemically relates certain pitch classes by keeping in place specific melodic intonations (Kholopov 1976). Evidently, "mode" is greater than "scale": it incorporates not only pitch classes per se, but also their functions within the set and certain melodic rules – what Huron called "the tendency tones" (2006, 160) – those pitch classes that have tendency to proceed to certain other pitch classes. In Russian musicology, tones that

**Table 8.2** The correspondence between nine aspects of musical expression recognized by Western classical music theory and the expressive aspects of traditional indigenous Eurasian JH performance as manifested in three principal styles of JH playing – after the classification by Eduard Alekseyev

| Acoustic attributes | Expressive aspects | Range of an aspect | (1) Timbral imitations and elaborations | (2) Song-like melopoeia | (3) Speech-like articulations |
|---|---|---|---|---|---|
| Frequency | (1) Melody | High/low relation of successive tones ("linear" tonal components), incremented into "pitch classes" of a melodic mode | Sounds are distinguished by their relative lightness or darkness depending on the salience of high- or low-frequency components in their spectrum, which contributes to their semantic value in addition to that of the imitated source | Sounds are distinguished by their position in pitch in relation to each other, wherever they form melodic intonations (the connected tones that change in pitch); the resultant intervals carry various semantic values | Sounds are distinguished by relative differences in high/low and front/back articulations of vowels in JH syllables, forming pitch-related "degrees" of phonemic scales – each degree assigned with its own semantic value based on its phonetic symbolism |
| | (2) Harmony | Concordant/discordant relation of simultaneous tones ("concurrent" tonal components), incremented into "interval classes" of a harmonic mode | Player chooses the specific range in JH's ambitus where the adjacent partials of the JH's harmonic series contain the sustained intervallic relations (e.g., arpeggio for f1–f6, anhemitonic f6–f11, diatonic f8–f14, chromatic f14–f21, microchromatic >f21) against the drone tone – suitable for the needed expression | Player chooses and brings out a specific harmonic interval (in just tuning) between the tone of a principal melody and a pedal tone in the bass of the JH texture based on the desired harmonic tension and the melodic function of that upper tone within a melodic phrase (e.g., initiation, climax, cadence) | Player chooses such phonemes for syllables that match the configuration of salient harmonics of the adjacent phonemes within the same syllable, word, or vocable, thereby supporting the vowel harmony by means of tweaking the intensity of specific harmonics in connected sounds |
| Time | (3) Tempo | Fast/slow metric pulse, sustained over a sizeable portion of a music piece, incremented into | Player increases or decreases the rate of changes or repetitions of specific timbral devices | Player chooses a discrete tempo appropriate for the expression of a specific affective state according to the density of pitch changes and uses | Player speeds up or slows down the rate of JH articulations similarly to the rate of speech – |

(continued)

**Table 8.2** (continued)

| Acoustic attributes | Expressive aspects | Range of an aspect | (1) Timbral imitations and elaborations | (2) Song-like melopoeia | (3) Speech-like articulations |
|---|---|---|---|---|---|
| | | "tempo classes" of a musical "movement" | (e.g., imitation of the sounds of droplets or of a gait) to show changes in the intensity of the expressed affective states | tempo inflections or tempo changes to convey changes in the intensity of that expression | usually in a flexible and gradual rather than discrete manner – to show a gradual increase or decrease in relative excitement or relaxation |
| | (4) Rhythm | Short/long relation of tones, incremented into rhythmic values according to some division ratios ("rhythm classes") of a rhythmic mode | Player groups consecutive tones into patterns based on the similarity of their timbre, where longer tones anchor shorter tones; then the group's size reflects the extent of excitement or relaxation – the longer the group the greater the relaxation | Player uses rhythmic values to mark subdivision of the thematic material into musical motifs, phrases, and sentences – their relative size reflects fluctuations in excitement or relaxation as well as simplicity or complexity of the expression | Player groups consecutive tones into patterns based on similarities in articulation, where longer vowels generally mark distribution of stresses within a pattern to reflect the relative stability or instability of an affective state |
| | (5) Meter | Short/long periodicity of rhythmic patterns, marked by stressed patterns of beats and incremented into "metric classes" of a musical "movement" (i.e., a "metric mode") | Metric organization here remains the most elementary, reduced to mere regularity or irregularity of the average size of a rhythmic group, reflecting the relative stability (long) or instability (short group) of the expressed affective state | Metric organization is elaborate in marking discrete changes of different metric pulses, whether symmetrical or asymmetrical, as well as their gradual inflections – and can include even the use of essentially ametric (irregular) musical movement, as long as it remains patterned by the same metric subgroups | Metric organization usually stays very basic in supporting a certain pulse, similar to a metric foot in poetry, or clearly defies the regularity of the pulse as in free verse, the choice of which reflects the dominance of excitement |
| | (6) Articulation | The styles of attaching/ detaching of successive tones can be incremented | A great variety of articulations: legato, non-legato, staccato, | The overall economy of articulation styles is determined primarily by the syntactic function; general emphasis | General emphasis on separating the rhythmic groups of vocables, |

| | | | | | |
|---|---|---|---|---|---|
| | | into groups of contrasting articulations | mezzo staccato, tenuto, marcato – selected based on suitability of expression of a particular onomatopoeic imitation or a special device of tone production | on joining the tones together in a musical motif, phrase, and sentence, which gives prevalence to legato in the expression of relaxation versus staccato – of tension, the rest of articulation fitting in between | giving greater importance to staccato than legato; the gradations in the extent of this separation are indicated by the relative duration of a pause (caesura) and the use of staccato on the last tone in a group; the stronger the separation, the less excitement |
| Amplitude | (7) Dynamic | Loud/soft relation of tones, sustained throughout a musical piece and incremented into a set of "dynamic classes" of a dynamic mode | A great variety of dynamic gradations, from pianissimo to fortissimo, including accents of various strengths, sudden drops in intensity (subito piano), crescendo, and diminuendo – to directly reflect the subtle changes in the intensity of the expressed affect | Prevalence of soft and medium soft dynamic gradations, with the use of dynamic envelopes to mark the boundaries of a melodic phrase by imposing a wave-like shape – to maintain the coherence of the affective expression within a phrase | Prevalence of loud and very loud dynamics, with little dynamic gradations, called to enhance the intelligibility of vocal articulations and make the phonemic expressions clearer – in order to secure the communication of a "secretive" or confidential message |
| Timbre | (8) Register | Light/dark and/or thin/thick relation of tonally homogenous groups of tones within the available ambitus, no increments | Prevalence of registral diversity, with many timbral recolorations of sounds produced within the same part of JH's ambitus – called to diversify the expressions and to refer to the imitated sound sources in the player's cultural environment | Prevalence of registral homogeneity in retaining the same tonal quality within a musical phrase – called to euphonize the overall sound and to facilitate the parsing of melodic phrases by confining timbral changes to the margins between the phrases | Prevalence of deliberate monotony of the drone tone – called to direct attention to the contrasts in articulation of phonemes, syllables, and vocables/words and to reduce the emotionality in perception of each of the elements |

(continued)

**Table 8.2** (continued)

| Acoustic attributes | Expressive aspects | Range of an aspect | (1) Timbral imitations and elaborations | (2) Song-like melopoeia | (3) Speech-like articulations |
|---|---|---|---|---|---|
| | (9) Harmonicity | Periodic/non-periodic spectral content, no increments | Overall diversity of harmonious, harsh, rich, transparent, stable, modulating, and noisy sounds, with many gradations – depending on the harmonicity of the imitated natural sounds as well as the desired vocal expression | Overall avoidance of harshness, complete dominance of harmonious sounds, varying in the extent of their harmonic richness – depending on tension or relaxation associated with their position within a melodic phrase | Prevalence of harmonious sounds, varying in configuration of their most salient harmonics in vowel articulations – called to stress vowel contrasts – combined with noisy consonants whose articulation complements the vowel's articulation |

Each style is numbered according to the overall commonality of its use among the indigenous musical traditions of northeastern Eurasia. The table demonstrates that playing JH, in general, qualifies to be a form of music, whether it follows the model of a song (2), of a speech (3), or of an elaboration of onomatopoeic imitations and fantastic or experimental sounds (1). Six out of nine aspects of expression usually possess incremental organization that justifies qualifying the TO of that aspect as a "musical mode," where increments generate the "degrees" of that mode. In addition to the recognition of pitch (melodic and harmonic) and rhythm modes, which is traditional for many music theories, it is possible to conceptualize tempo modes, metric modes, and dynamic modes

tell one mode from another are called "modal" (Grigoryev 1981).[11] It is exactly the characteristic "modal tones" and "modal intonations" that allow listeners to recognize a familiar mode upon hearing the music (Nazaikinsky 1977). Melodic modes often coincide with harmonic modes, but not always.[12] Melodic mode is the primary form of musical mode in overwhelming majority of Western and non-Western music cultures (Powers and Wiering 2001; Porter et al. 2001; Kholopov 1976). Harmonic modes ("harmonic," "melodic" and "natural" varieties of keys) are peculiar to Western tonality (Mazel 1972).

Rhythm too is often conceptualized into the "mode" – a patterned succession of long and short values within a certain ratio where the value and relative duration of each tone are determined by its position within a group (Roesner 2001).[13] Rhythmic modes are well known in Western music (Roesner 2001) but are also important for some non-European music systems, such as raga (Clayton 2000) and maqam (Touma 1996).

In a similar vein, metric organization can be conceptualized into a "metric mode" as grouping of beats based on the perceived periodicity of stresses generated by longer and louder tones as well as by changes in melodic direction and harmonic pulse (London 2004). In this case, a "metric mode" would be a specific progression of symmetric or asymmetric binary and/or ternary groups regularly used within a music work – each of such groups constitutes a "metric degree". In non-Western

[11]For example, the characteristic "modal" tone in the harmonic minor mode is the sharpened seventh degree that distinguishes the harmonic mode (e.g., A-B-C-D-E-F-G#-A) from the natural (A-B-C-D-E-F-G-A). This degree is nicknamed the "leading tone" because of its tendency to be resolved by the ascending melodic motion to the upper "tonic" degree – which prompted Huron (2006, 160) to introduce the term "tendency tone." The characteristic modal intonation of the harmonic minor mode is the interval of augmented second that descends down to the V degree of a minor key (G#-F-E). More rare is the ascending intonation (F-G#-A), directed at the upper tonic (Nikolsky 2016, Appendix-4).

[12]A well-known example of this discrepancy is the subdivision of the minor key of Western classical music into 3 modes: harmonic, melodic and natural. A "harmonic mode" is indeed "harmonic" in a sense that it is defined by "vertical" relations of tones in chords and double notes. In absolute majority of cases in Western classical music, "harmonic mode" is not displayed melodically – there is not a single occurrence of an augmented second (between the sixth and seventh degrees) in the melody, and this second is confined to the accompaniment for the melody (usually, in chords, but possibly engaged in melodic figuration). Melodic use of the harmonic minor in the Western tradition usually has to do with referencing non-Western music of Jewish, Turkish, Arabo-Persian or Gypsy origin (Nikolsky 2016, Appendix-4). In contrary, a "melodic mode" is genuinely "melodic," i.e., defined by the succession of two "modal tones" (sharpened sixth and seventh degrees) in melody rather than harmony. Harmonically, the use of major subdominant triad in a melodic minor is exceedingly rare. Finally, a "natural mode" in the Western tradition is harmonic as well as melodic. its modal tone, "natural" seventh degree, occurs with equal frequency in harmonic progressions and melodic motion.

[13]In Western classical tradition, modal rhythm constituted the backbone of temporal organization in early Medieval music (Ars Antiqua) (Hughes 1954). Modal rhythm had formative influence on the compositional practice, most obvious in "isorhythmic motets" that often featured different rhythmic modes reserved for different parts in multi-part settings. However, in general, modality of rhythm by no means was limited to Western music systems, e.g., classical Indian music theory also observed rhythmic modes despite employing completely different metric principles (Clayton 2000).

cultures, "movement" – a musical term for one's impression of a specific pattern of motion (akin to a gait) adopted for an entire musical work or its substantial portion (Sadie 2001) – is not often limited to a single clear-cut pattern of regular stresses, expressible by time signature, because the latter emerged as means of keeping time in a multi-part music (Houle 1987). Therefore, "fixed" simple meter is poorly compatible with many non-European musical traditions, posing difficulties for those Western ethnomusicologists who are trying to notate metrically flexible music (Clayton 1996). Usually, the procedure for rhythmo-metric analysis of folk music closely follows the lexic and syllabic structure of a song's lyrics, taking into consideration the performer's breathing cycle, the overall expression of tempo, and the repetition of structural patterns – all of which are cross-related in order to establish the boundaries of motifs and phrases (E. Alekseyev 1990). The resultant notation usually comes out as a continuous change of alternating meters rather than a single time signature that holds from the beginning to the end of a musical work, standard for Western popular and classical musical compositions of the Common Practice Period.

The reason for this is the deep methodological divide between the stress-based **divisive** metric standard of Western civilization and the duration-based **additive** standard of many non-Western civilizations such as Indian (Clayton 2000, 37). In comparative ethnomusicology these have been termed, respectively, qualitative and quantitative and are related to the formative influence of accent-based and quantitative languages (Sachs 1953, 24–26).[14] Western tactus, and "tonic versification," corresponding to it, evolved from the additive metric theories of Ancient Greeks and Romans, forged by the strategy of ordering the rhythmic versatility within the poetic prosody by means of imposition of rhyme and propelled by the necessity to coordinate parts in multi-part music, as polyphonic textures gained more and more importance from the eleventh century onward (Kharlap 1978). Hence, the adequate way of thinking of metric organization of non-Western music would be to represent it as changes in the number of "pulses" (beats/syllables) that are united into a metric group [="metric degree"] – while taking into consideration that such pulses can be of unequal sizes.[15] Incrementation by virtual "dance steps" of different sizes does to metric organization in instrumental music what incrementation by syllabic quantity does to lyrics in vocal music. In both cases, the metric mode constitutes the formula for constructing rhythmic pulses within the movement.

[14]Additive meter should not be confused with compound or composite meters of Western music theory (Read 1969): the latter two remain divisive in their concept by marking the downbeat, thereby dividing musical time into a succession of equal-size tactus, although this tactus could consist of symmetric as well as asymmetric parts which seem to be added together (London 2004).

[15]Thus, in many Eastern European, especially Balkan, traditions, the metric group is created by adding contrasting (viz., even versus odd) "long" and "short" pulse-groups, e.g., the meter in a Greek Levendikos is defined as a 17-beat period made of long-short-short-long-short pulses (i.e., 4 + 3 + 3 + 4 + 3) estimated by performers in terms of the relative sizes of dance steps (Joukowsky 1965, 68–75).

By the same token, the tempo aspect of a musical work can be conceptualized as a mode made up of discrete changes between a number of characteristic gaits (e.g., running, walking, mincing, skipping, etc. – each constituting a "degree" in the "tempo mode"). The interaction of the beat rate and the rhythmic patterns establishes the relative density of pitch changes per metric unit of time (Madison and Paulin 2010), which sets the limit for how fast or slow the music can proceed in order to optimize the perception of the music flow – in other words, defines an ideal "perfect tempo" (Gabrielsson 1999). Evidence for its existence comes from empiric studies of "absolute tempo" consistently preferred for a particular music piece within a rather narrow range of beat rates by musicians (Garbuzov 1950) and non-musicians (Levitin 1994). The assortment of the most common gaits usually forms a music system, e.g., 12-tempi system of the Common Practice Period (Nazaikinsky 1972), where each tempo constitutes a discrete class, recognizable by the ear (Garbuzov 1950). Like chromatic alterations of pitch classes, tempo class can also have inflections: incremental (e.g., piu mosso or meno mosso) and gradual (ritenuto or accelerando).[16]

Similarly, the dynamic aspect includes transitions between discrete dynamic levels that refer to specific intensity levels of excitation and relaxation – as well as their gradual changes. Thus, a piece of music can be limited to only two "dynamic degrees" (e.g., forte and piano) and the inflection of diminuendo. Although performers do not conserve the exact dynamic values for such gradations interpersonally, each performer remains consistent intrapersonally, in his own performances (Garbuzov 1955). Music practice generates standard dynamic distinctions that support the affective intensity of music by an assortment of dynamic increments (e.g., fortissimo, mezzo forte, forte, mezzo piano, piano, pianissimo) spread over the entire dynamic range from the softest to the loudest acceptable levels (Berndt and Hähnel 2010). The use of such sets of dynamic levels is found not only in Western classical music but in many music cultures that came after the Ancient Greek music (Thiemel 2001).

All in all, the TO of a piece of JH musicking can be adequately characterized by defining six concurrently applied modes: melodic, harmonic, rhythmic, metric, tempo, and dynamic. Such analysis applied to JH vocalizations and to singing or speaking vocalizations is going to reveal their differences. Thus, speaking and singing of vowels of Yakut languages by native speakers contain pronounced glissando inflections, whose frequency span and timing phases characterize specific vowels, while rendition of the same vowels on JH by the same performers has no such inflections (Nikolsky 2017). This suggests that clarity of pitch levels' distinctions is more important for JH articulations than for speech or singing and that stressing of specific harmonics for specific vowels, responsible for pitch distinctions,

[16]The exact pace of each tempo can be adjusted without changing its main character. For example, in Western classical tradition, applying "meno mosso" in Presto does not turn Presto into Allegro, since meno mosso (Italian "less motion") only refers to the pace but not the character of the music – it does not cancel the sense of hastiness and nervousness that characterize Presto. Similarly, a gradual inflection of "ritenuto" (Latin, "retain") does not modify the character of the original movement. Ritenuto would only hold Presto back, generating tension – by no means easing the tempo (which would be required by Allegro that is characterized by a natural "lively" and jolly feel).

plays a formative role in the vocal system of JH's articulations – unlike those of speaking and singing.

Aleksey Aksyonov (1964), an expert on Tuvan culture, believed that JH provided a model of TO for the famous deep throat "solo-polyphony" of Altai and Mongolia by means of spectral "dispersion" of a complex sound into its constituent partials. Khamza Ikhtisamov (1988) nicknamed this "spectral refraction" and compared JH with a prism. Furthermore, he noticed that JH articulations tied pitch levels that marked specific speaking vowels with specific harmonics of JH's FF: "by hearing JH's tone, we can figure out which vowel it renders and by which mouth configuration it is achieved" (p. 208). Ikhtisamov concluded that this registral correspondence between the just mode of harmonics and the articulatory mode of tongue positions was intuitively discovered by JH players and turned into a principle of vocal and instrumental music-making by Turko-Mongol population of Middle Asia, Siberia, and Mongolia.

This correspondence seems to also exist in other music cultures where JH occupies a central role. Thus, in Indonesia, children are taught to manufacture, tune, and play palm-made JH, using the solfege system where the JH range is broken into pitch classes, each of which bearing a syllabic name with a unique vowel (McPhee 1955, 79).[17] Ikhtisamov's mode ([a] > [o] > [u] > [e] > [i]) comes close to Ogotoyev's mode, possibly disclosing similarity between the constructions of metallic bow-shaped Bashkir *kubyz* and Yakut *khomus*, as well as between the Bashkir and Yakut vowels, belonging to the same Turkic family – in contrast to the construction of a wooden frame-shaped *genggong* and vowels of the Balinese (Exs. 13–14).

**Audio Examples 13–14**

Ex. 13. **Ascending articulatory scale of Yakut vowels on metallic khomus**: o, a, ö, u, y, ä, ü, i. The vowels are ordered according to the convention adopted among Yakut khomus players (Ogotoyev 1988), with the correction by Shishigin (2015) that places "o" rather than "a" at the bottom of the scale. The only degree that clearly violates this order is the sixth – "ä" (perhaps, due to its greater variability on different khomuses). Demonstration by Erkin Alekseyev, used with his permission (http://chirb.it/6kyEEL).

Ex. 14. **Ascending articulatory scale of Yakut vowels on bamboo angkuokh**: o, a, ö, u, y, ä, ü, i. This recording of the same player in the same room shows more differences rather than similarities with Ogotoyev's scale in the example above: the first four degrees appear descending rather than ascending, and the sixth degree ("ä") seems lower than the fifth degree ("y"). In general, it is harder to

[17]These syllables correspond to the following pitch values, expressible in Western diatonic tuning as "ning" = E, "nong" = G, "neng" = Bb, "nung" = C, and "nang" = D, if FF=C (Morgan 2008). Although this mode ([i] > [o] > [e] > [u] > [a]) differs from Ogotoyev's ([a] > [o] > [ö] > [u] > [y] > [å] > [ü]-[i]), the very correspondence between pitch and articulatory degrees indicates that the JH tradition had at least some influence on singing and speaking practices within the vast geographic area from Bashkiria to Indonesia.

judge whether one degree is higher or lower than the other one on the bamboo instrument compared to the metallic instrument. Demonstration by Erkin Alekseyev, used with his permission (http://chirb.it/I62kr1).

Comparison of the available recordings of Eurasian indigenous JH traditions leads to the belief that articulatory modes vary between different constructions, different materials, and, to a less extent, native language of a player. Thus, two steel khomuses played by two native Yakut speakers, nevertheless, feature different lowest degrees: [a] for Ogotoyev (1988) and [o] for Shishigin (2015). The difference most likely owes to the contribution of roundness of [o] that lowers the resonance bandwidth of the frontal portion of the mouth cavity, influencing the pitch level of the second formant (Zsiga 2013). Shishigin's khomus must have been more responsive to lip rounding than Ogotoyev's, causing the rounded [o] to appear lower than unrounded [a], despite the latter engaging the maximal volume of the mouth cavity. It is plausible to expect a considerable variability of articulatory modes between different instruments and performers – akin to variability of pitch modes for songs of the same genre and tradition.

## 8.6 Where and When Did Jew's Harp Emerged as a Principal Instrument

The current consensus among the specialists in JH archeology holds that the JH tradition must have originated in Northeastern Asia, become cross-cultural, and spread westward to Western Europe (Fox 1988; M. Wright 2004; Kolltveit 2006; M. Wright 2011; Honeychurch 2015; Kolltveit 2016; Aleksandrova 2017; Turbat 2017; Oleszczak et al. 2018), as well as eastward toward Japan (Tadagawa 2007; Tadagawa 2016; Tadagawa 2017a) and Austronesia (Blench 2004) (Fig. 8.2).

Although the archeological evidence remains fragmentary in the global spread of JH, it is, nevertheless, consistent in showing that in Americas, Africa, and Australia there are no traces of the usage of JH prior to European colonization (de Ramón and Rivera 1982; Beck et al. 1983; Kungurov 1994; Barr 1994; Yakovlev 2001; Pignocchi 2002; Pignocchi 2004; Wright and Impey 2007; Crane 2007; M. Wright 2011; Honeychurch 2015, 20; Whitridge 2015). The oldest archeological find belongs to the Inner Mongolia cluster – two bone idioglot JHs from Shuiquan, Jianping, in Liaoning, dated 2146–1029 BC (Kolltveit 2016). The next in line is a bone idioglot JH from Xiajiadian, Chifeng, dated 1200–600 BC (Honeychurch 2015). To the same cluster belong four bamboo idioglot JHs from Yanqing, Beijing, 770–403 BC (Tadagawa 2016) (Table 8.3).

According to the comparative ethnographic data by Sheikin, this Inner Mongolia cluster comes exceedingly close to the epicenter of the genesis of JH pan-cultural tradition in Siberia and Far East. Sheikin regards the Amur region as the transition zone where the primordial "bamboo JH technology," elaborated within the warmer oceanic climate (e.g., Ainu Mukkuri) and transferred westward, became transformed

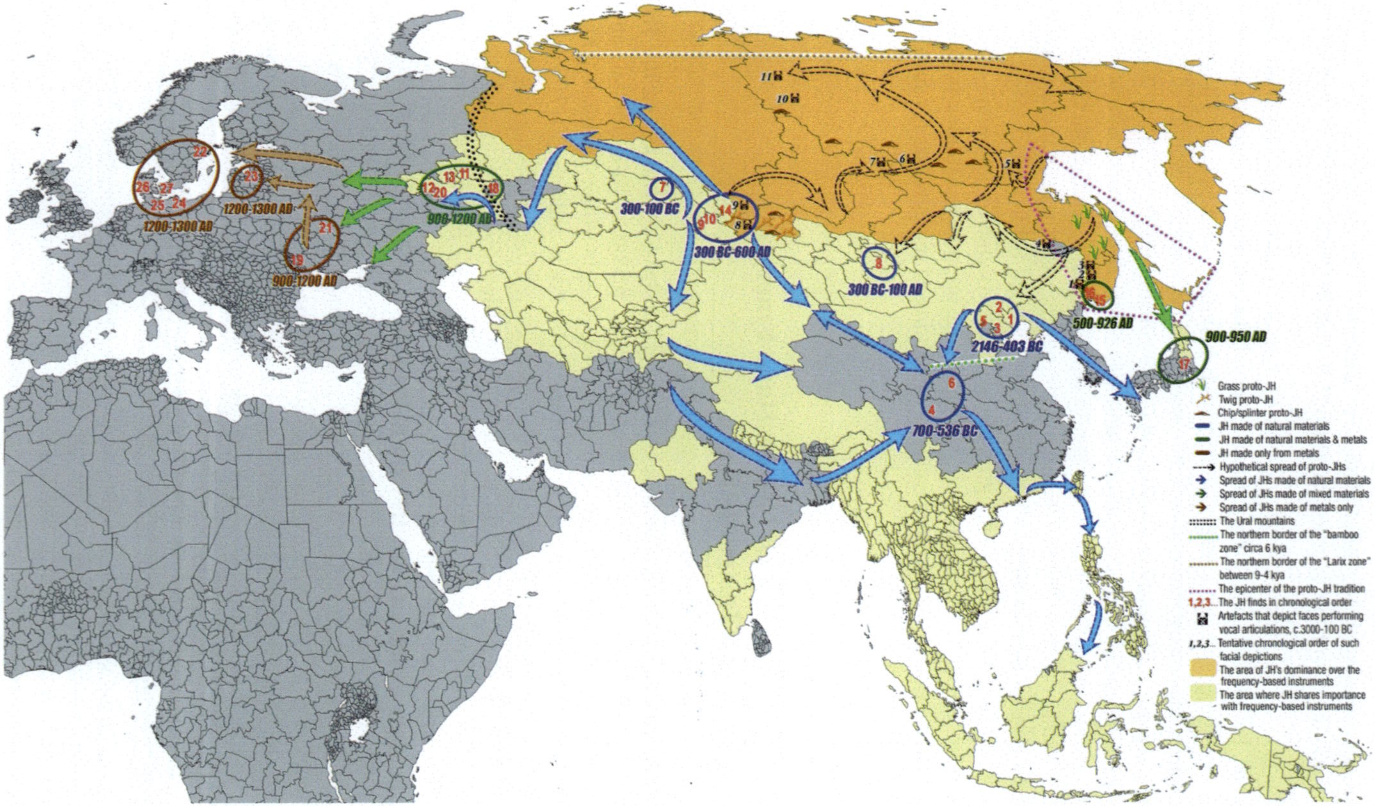


**Fig. 8.2** *The location of archeological finds of JH, dated prior to 1400 AD, plotted against the geographic areas where JH playing still constitutes an important part of indigenous tradition and the tradition of making "proto-JH" from grass, twigs, and chips still survives* (the sources are listed in Table 8.3). The orange

color marks the territory where JH continues to remain the principal musical instrument unchallenged by any of the frequency-based instruments (e.g., flute- or violin-like). The yellow color marks the areas where JH coexists with other musical instruments, therefore occupying less prominent position in the indigenous instrumentarium but supporting a distinct indigenous tradition of JH music. The green icons of grass indicate the area in Sakhalin and Primorye where grass JH is used by Nivkhi and Ulchi (Mamcheva 2012, 50); the light brown icons of twig, the area in Tuva and Altai where twig JH is used by Tuvans and Altaians; and the dark brown icons, the area in Tuva and Yakutia where chip JH is used. The black-dashed outline arrows indicate the possible routes of distribution of "proto-JHs": westward from the Far East coastline for grass JH, eastward from Tuva to Yakutia, and then northwest for chip JH (Sheikin 2002, 132). The arrowed points indicate the locations where non-frame "proto-JHs" were likely to be upgraded into the framed constructions of bamboo, wood, or bone JHs. The colored ovals show the clustering of pre-industrial archeological sites of JH. Red bold numbers reflect the chronological order of their dating

Blue oval surrounds the sites where JHs made of organic materials were found; green ovals, of organic as well as metallic materials; and brown, of exclusively metallic JHs. The colored arrows indicate possible ways of spreading the JHs from the archeological clusters, as suggested by the known ancient routes of trading and/or migration of the population associated with the archeological sites – clustered mostly by rivers: (1) Amur, Songhua, Argun (Zagoskin 1910) and Shilka/Onon in Primorye/Mongolia (Kaner and Taniguchi 2017); (2) Maya and Aldan across the northern coast of the Okhotsk Sea (Zagoskin 1910); (3) Angara, Lena, and along the coast of East Siberian Sea (Zagoskin 1910); (4) northeast Inner Mongolia, then bifurcating to the Hexi corridor (Christian 2000) and to Guangzhou, towards Taiwan (Hung and Chao 2016); (5) crossroads of Altai along Yenisei, along Ob' (Zagoskin 1910), across southern Tarim and toward Central Asia (Kuz'mina 2008); (6) from Ob' to Tobol, then Irtysh, and then crossing Ural mountains, from the source of Belaya River going to Kama (Zagoskin 1910); and (7) from Kama region trifurcating to northwest by the upper Volga to the Baltic Sea, by Seim to Desna toward Ukraine and by Don to the Azov Sea (A. Leontyev 1986). The Middle Asian branch [from (5)], in turn, bifurcates into the Southern Tian Shan route to China and the loop-around route via Taxila and Pataliputra (Hansen 2012)

The brown-dashed line indicates the northern border of the *Larix* forestation during 9–4 kya (Kremenetski et al. 1998). The green-dashed line shows the northern border of the "bamboo zone" c. 6 kya (Winkler and Wang 1993). The magenta-dotted square encloses the most likely area of origin of the proto-JH tradition. The double black-dotted line indicates the position of the Ural Mountains separating the JH cradle area from Europe – with the trading route crossing Ural in the south – to Bashkiria. The black icon of a skull indicates the location of petroglyphs depicting vocal articulations (see Fig. 8.5), dated from the early third millennium BC for Amur area (nos. 1–3) (Okladnikov 1971, 83–89) to the late first millennium for Olenek River area (no. 11) (Arkhipov 1989; Kochmar 1994), probably related to JH articulations of the time of transition from proto-JH to JH and important for ritual communication with ancestral spirits (Sheikin 2002, 259). The black italic numbers indicate tentative chronological order of these images, based on archeological dating and the hypothetical routes of cultural exchange. The locations of "singing mask" images coincide with the JH finds in two regions: Tuva (Devlet 1976, 27–43) and Primorye (Okladnikov 1971, 83–89), the latter being by at least a millennium older of the two

**Table 8.3** The earliest pre-industrial JH fsound by archeologists

| No. | JH type | Location | Country | Dating | Source |
|---|---|---|---|---|---|
| 1 | 2 bone idioglot | Shuiquan, Jianping, Liaoning | China | 2146–1029 BC | Kolltveit (2016) |
| 2 | 1 bone idioglot | Xiajiadian, Chifeng | China | 1200–600 BC | Honeychurch (2015) |
| 3 | 4 bamboo idioglot | Yanqing, Beijing | China | 770–403 BC | Tadagawa (2016) |
| 4 | 5 bamboo idioglot | Yuhuangmiao, Sichuan | China | 700–600 BC | Shulga (2015) |
| 5 | 4 bone idioglot | Jundushan, Hebei | China | 700–500 BC | Honeychurch (2015) |
| 6 | 1 bamboo container | Baojiang, Shaanxi | China | 576–536 BC | Tadagawa (2017a, b) |
| 7 | 1 bone idioglot | Dubrovinskii Borok, Kolyvanskii District, Ob’ | Russia | 300–100 BC | Borodovskii 2007 |
| 8 | 1 bone idioglot | Morin Tolgoi, Altanbulag soum, Tuv aimag | Mongolia | 300 BC–100 AD | Honeychurch 2015 |
| 9 | 4 bone idioglot | Tcheremshanka and Tchultukov, Altai | Russia | 300 BC–600 AD | Borodovskii (2017) |
| 10 | 1 bone idioglot | Aimyrlyg Burial Ground XXXI, central Tuva | Russia | 200 BC | Tadagawa (2016) |
| 11 | 1 bone idioglot | Makhoninskoye, Chastinskii District, Perm’ | Russia | 200 BC–300 AD | Aleksandrova (2017) |
| 12 | 1 copper idioglot | Ust’-Brykinskii, Laishevsky District, Tatarstan | Russia | 300–500 AD | Aleksandrova (2017) |
| 13 | 7 bone, 22 bronze idioglot | Prikamye burials and settlements, Udmurtia, Bashkortostan, Permskii and Kirovskii regions | Russia | 300–1300 AD | Golubkova and Ivanov (1997) |
| 14 | 2 bone idioglot | Sakhsar, Khakassia | Russia | 400–500 AD | Tadagawa (2016) |
| 15 | 1 iron idioglot | Andrianov Kliuch, Partizansky District, Primorye | Russia | 500–600 AD | Leshchenko and Prokopets (2015) |
| 16 | 3 iron idioglot | Nikolayevskoye, Shaiginskoye, and Smol’ninskoye, Mikhailovskii district, Primorye | Russia | 698–926 AD | Leshchenko and Prokopets (2015) |
| 17 | 2 iron heteroglot | Hikawa Shrine, Omiya, Saitama | Japan | 900–950 AD | Tadagawa (2007) |
| 18 | 1 silver idioglot | Idelbayevskiye burial mounds, Salavatskii district, Bashkortostan | Russia | 900–1000 AD | Mazhitov (1981) |
| 19 | 1 iron heteroglot | Echimauti, Rezina district | Moldova | 900–1000 AD | Feodorov (1954) |

(continued)

**Table 8.3** (continued)

| No. | JH type | Location | Country | Dating | Source |
|---|---|---|---|---|---|
| 20 | 2 bone, 2 bronze idioglot | Tankeyevskii, Bulgur nekropol, Spassky District, Tatarstan | Russia | 900–1100 AD | Aleksandrova (2017) |
| 21 | 1 iron heteroglot | Hlukhiv, Sumy district | Ukraine | Thirteenth century | Pashkovskii (2012) |
| 22 | 1 iron heteroglot | Uppsala | Sweden | Thirteenth century | Kolltveit (2006) |
| 23–27 | 5 copper heteroglot | Riga, Skanor, Ribe, Greifswald, Hamburg | Latvia, Sweden, Germany | Thirteenth–Fourteenth centuries | Kolltveit (2009) |

The numbering reflects the chronological order of the JH dating. The "JH type" column shows the number of finds per site, the material of making, and the construction type

into the "wooden JH technology" by the continental ethnicities living in taiga environment (Sheikin 2002, 132).[18] Following this model, the Lower Xiajiadian bone JH from Liaoning represents a relatively late development in JH-making that succeeded the mastering of wooden and bamboo frame-based constructions. The latter, in turn, succeed the frameless constructions of what Sheikin calls "proto-JH" – an easy-to-make sound-producing device that is activated by breathing and manipulated by hands to form a peculiar geometric configuration with the player's mouth that modulates sounds by silent articulations (Sheikin 2002, 118). Such devices are picked from natural sources almost in a ready-made condition, akin to gathering of berries or nuts, and are used only once – disposed after playing. Sheikin draws the evolution of JH from the frameless chip to the frameless twig and then to the chip in frame, to the framed idioglot, and eventually to the bow-shaped heteroglot (Sheikin 2002, 124).[19]

This line of development (Fig. 8.3) corresponds to the general shift from the "holistic" sonoristic concept of sound to the "analytic" harmonic, also affecting the

[18]Sheikin (2002, 125–132) concluded that the non-metallic framed JH constructions were cultivated in two zones: by the Ugric ethnicities of northwestern Siberia (which Sheikin names "continental mountainy") and by the Tunguso-Manchurian ethnicities of Far East ("oceanic coastal"). Sheikin holds that wooden instruments preceded bone instruments, and bone manufacturing was introduced as JH spread from Amur region and Primorye toward North and Northwest – to substitute for the shortage of forest in tundra environment. This transition must have occurred at the territory of modern Yakutia and had direct relation to the institute of *seok*. Yakut ethnos belongs to the Turkic family and descends from the seok system of Khakass (Sagai seok) – as witnessed by Russian explorers of the eighteenth century (Ushnitskiy 2016). The totality of the available linguistic, anthropological, and genetic data confirms the Altaic origin of Yakuts (Pakendorf 2007). And there is evidence that in the nineteenth century Verkhoyansk Yakuts were still using bone and wooden JHs (Khudiakov 1969, 153).

[19]Suzukei (2010, 10–11) elaborated Sheikin's model by adding two transitionary substages between the frame- and bow-shaped constructions: 1) idioglot wooden or bamboo fork-like frame JH, similar in shape to bow, and 2) heteroglot bamboo or copper JH without a bow and with a kneeless lamella.

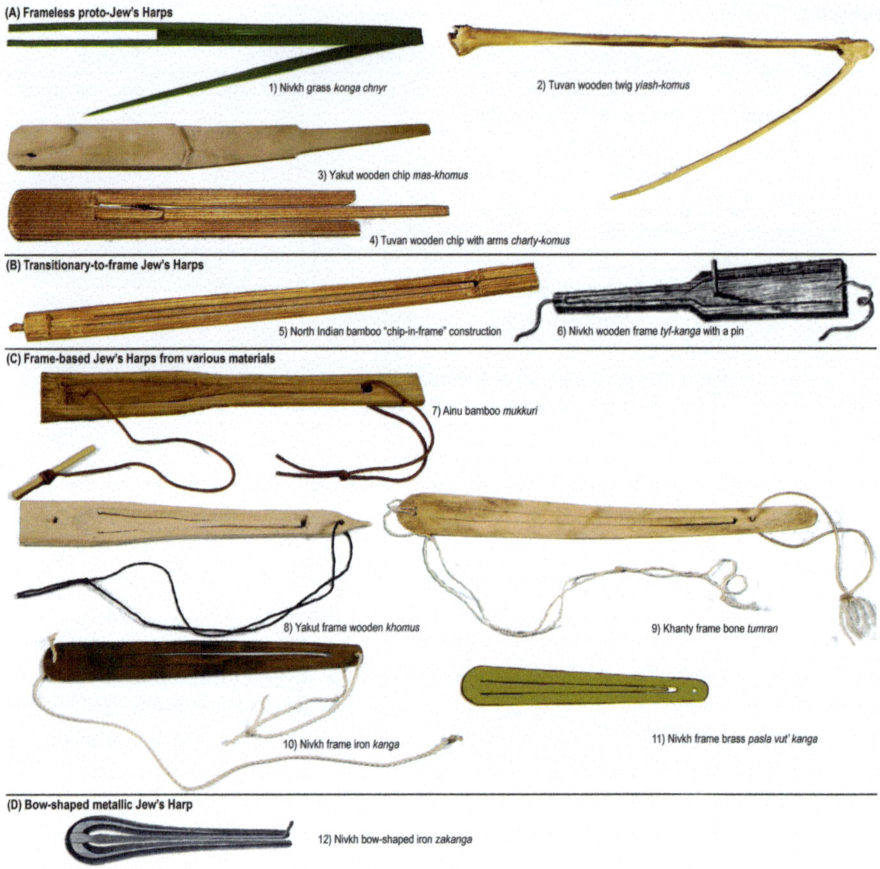


**Fig. 8.3** *Twelve types of JH most common in northeastern Eurasian indigenous musical traditions, arranged in the hypothetical order of their evolution*. This figure summarizes the reviewed data of organological development of JH constructions, prevailing method of sound production and related to it sound quality, technological complexity of their manufacturing, availability of the materials for making JHs and time/effort required for this, and paleo-ecological changes in the region of Siberia and Far East. (**a**) *Frameless proto-JHs.* This group is characterized by the one-time use of dispensable musical instruments, almost ready-made from natural resources, frameless design, playing technique based on gently tapping with a finger and very soft hollow clicking sound, rather poorly controlled, and lacking the fixed-in-pitch drone. (**b**) *JH transitionary from frameless to framed design*. This group features the frame that encloses the lamella, which makes the instrument more sturdy, suitable for multiple use, while changing the playing technique to plucking the lamella. This, in turn, affects the acoustic characteristics: JH obtains its signature trait – a drone in the bass – plus louder sound with richer and more selectively controlled spectrum. The construction seems to have originated in the idea of extending the arms of a chip proto-JH (e.g., *charty-komus*) into the framed enclosure. The added pin was probably designed for better plucking of the lamella. (**c**) *"Classic" frame-based JHs from various materials.* This group adapts the frame to the hand action of jerkily pulling the lamella by means of a rope, which requires more robust construction. Such sound production makes JH even louder, highlighting greater spectral details and promoting greater diversity of technical devices. Music becomes more complex, making spectral components distinct (up to two to three simultaneous melodic patterns), but with lack of homogeneity. Bamboo JH rattle:

playing technique that becomes more selective in controlling the acoustic properties of the overtones. This lineage builds on the foundation of Sachs' model (1917) of evolution of JH from simple to complex construction (idioglot to heteroglot), advancing from East Asia to Western Europe. Sheikin's model also resembles the five-stage model (bamboo idioglot, metallic idioglot, inward heteroglot lamella, outward heteroglot lamella, and heteroglot hairpin shape) formulated by Dournon-Taurelle (1975) based on the construction complexity as well as the playing technique. Sheikin's approach involves additional criteria: the playing technique, the use of JH music and the comparative etymology of names of different JH constructions. Sheikin tracks three principal stages in evolution of JH music: from mostly pneumatic mouth manipulation of proto-JH to hand-jerking of the rope tied to the lamella of the idioglot frame JH and then to hand-striking of the lamella of the metallic bow-shaped constructions. Similar view is shared by Emsheimer (1986).

The images are taken from the Museum of Musical Instruments of the Peoples of Northern Asia at the Arctic State Institute of Culture and Art, Yakutsk (personal collection of Yurii Sheikin); from the Museum and Center of The Khomus of The People of the World, Yakutsk; and from the personal archive by Natalia Mamcheva, used with their permission.

Further elaboration in classifying JHs came from the field research of Sakhalin and Amur JH music by Natalia Mamcheva (Mamcheva 2005). The first proto-JH most probably was *koka chnyr* – the "grass JH" made from *Leymus* (Poaceae family), common for nearly the entire Eurasia – still in use by Nivkh, Ulch, and Orok populations (Podmaskin 2003). The thicker bottom part of the leaf (about 20 cm long and 3 cm wide) is torn by the nail to make a lamella which is then bent to the reverse side of the leaf and gently touched by the finger while breathing into the leaf. It takes about a minute to make this primitive seasonal instrument that is discarded after a single use, although it retains its distinct "voice" in hands of different performers. *Leymus* was important for Nivkh rituals (Mamcheva 2012, 50) and was mythologically anthropomorphized into a woman (Kreinovich 1927, 1928, 192) – most likely ancestral to some clan, like ancestral trees in Altai.

What confirms the grass origin of JH is that the construction of *koka chnyr* sets a prototype for the local frame-based JHs (Kolosovsky 1986), reflected not only in their form but also in their names – all of them are derivatives of the word "kongon" which in Nivkh language exclusively refers to the young *Leymus* plant and the onomatopoeic imitation of ringing (Mamcheva 2012). The old *Leymus*, barely

**Fig. 8.3** (continued) bone ones, rasp; wooden, clang; and metallic, buzz. (**d**) *Bow-shaped metallic JHs*. This group is represented by a single model, since northeastern traditional Eurasian bow-shaped JHs are always metallic (usually iron or steel) and vary little in their design. It is distinguished by hitting or strumming the lamella with a great variety of hand (or finger) actions, holding positions, and lip (teeth) contact with JH. The instrument is much more responsive to their variations than framed JHs. Its tone is significantly more harmonious and homogeneous, with prolonged ringing, which not only provides superb separation of spectral components but also highlights contrasts in their thematic material. This JH type supports the greatest variety of textures but favors a single melody that can be placed anywhere in the texture except the bass

usable for JH, is called *nunnun* (Kreinovich 1927, 1928, 52). The names of Nanai, Evenk, Negidal, Ulch, Udege, Orok, and Oroch JHs are also derivative of "konga" (Mamcheva 2005), indicating their descent from the Nivkh grass JH.

In fact, Nivkh musical culture is unique in featuring the greatest variety of materials for JH-making, which allows to arrange them to a timeline and to consider the direction of their evolution (Mamcheva 2012, 53–5). Mamcheva points out that each of Sheikin's principal stages in the evolution of JH correspond to a different method of sound production that results in a different sound quality:

1. Grass JH requires gentle touching and pressing by the finger, while breathing plays the main role, which is why the informants consider *koka chnyr* a pneumatic instrument (Mamcheva 2005) – and it sounds like a soft homogeneous hollow clicking that is free from percussiveness.
2. Framed idioglot bamboo/bone/wood/metallic JHs require pulling of the rope, which generates a loud rich rattling or rasping sound that can be quite diverse in its timbral qualities, once the player has mastered the use of the rope (which is quite tricky).
3. Bow-shaped heteroglot metallic JHs require hitting or strumming the lamella with the hand, generating very loud ringing or buzzing sound with pronounced harmonics, well-separated parts in a musical texture, and nearly unlimited variety of timbral recoloring.[20]

Mamcheva's observation can be further substantiated by the fact that all frameless constructions of JH and proto-JH are characterized by a hollow uniformed sound, free of a drone monotone in the bass (Nikolsky et al. 2020). This is an important acoustic trait that distinguishes frameless constructions from framed and bow-shaped ones. Their evolution proceeds toward a growing timbral diversity and complexity of sound, increasing clarity of subdivision of JH tone into its salient spectral components, which culminates in the capacity of the metallic bow-shaped JHs to selectively control each of the lower 11th–13th harmonics of the harmonic series. This line of development in sound production justifies the periodization of JH TO into five stages (Nikolsky et al. 2017) (Table 8.4).

The oldest archeological find – the Lower Xiajiadian JH – falls past the middle of the timeline of JH's evolution (type 6 of 9 total types). The invention of the earliest proto-JH can be dated only roughly by the available paleoecological data. Xerophyte *Leymus*, used for making grass JH, constitutes a part of the Northern Hemisphere-steppe plant community, which is also present in cryophyte tundra-steppe landscapes – comparatively uniform throughout a vast area spanning from Central Asia to Wrangel Island (Yurtsev 2001). Tundra steppe has been a mesic ecotone

[20]To these three stages, it would be plausible to add the transitionary stage between (1) and (2) – exemplified in a frame-shaped idioglot JH with a little nail-like notch cut at the lamella's tip – such notch reduces vibrations while facilitating pinching and plucking (Sheikin 2002, 125). This construction indicates the rising interest in increasing the loudness and richness of JH sounds by enabling the use of few fingers in lamella's excitation. Sonorically, such nail-like JH surpasses *konga chnyr* but yields to the frame-based JH with a rope.

**Table 8.4** The correspondence of stages in evolution of the JH's construction, sound type, and TO of music in reference to the timeline of paleoclimate change and cultural evolution

| JH material and making time | Material's availability in the JH's area of use | Traditional JH of this type | Instrument class | Sound production technique | Salient timbral quality | The principal method of TO |
|---|---|---|---|---|---|---|
| (1) Grass leaf (*Leymus*) requires about a minute to make, no expertise | *Leymus* grows in Northern Hemisphere in sand and salt marshes, used in deserts and semideserts for fodder and livestock grazing, edible by humans (Tsvelyov 1976, 177); it is a part of the tundra-steppe plant community of Beringia (Yurtsev 2001) that was in place c. 30–20 kya (Brigham-Grette et al. 2004) | Nivkh and Ulch *koka chnyr* – the lamella is bent against one's arms; a disposable JH, single-time use (Mamcheva 2005) | Proto-JH, idioglot plucking idiophone, frameless lamella (frame-shaped but open on one side) | Holding the leaf between the second and third fingers of the left hand while gently touching the leaf with the right hand's finger(s), sliding and pressing down at the needed place, holding the JH close to the mouth, and breathing into it | Non-percussive very soft hollow clicking sound, homogenous, and simple (the spectral components are poorly isolated from each other, no way of playing melodic modes based on harmonic series, no sustained drone component) | Absent TO – JH is used as a totemic fetish object that possesses its own "natural voice," valuable for ideological reasons rather than specific acoustic features and therefore retained intact, without pursuing deliberate tonal modulations (Nikolsky et al. 2019) |
| (2) Bark chip or splinter of a tree (*Larix* or *Betula*) requires about a | *Larix* is one of the most cold-tolerant trees, today growing in the Arctic Circle | Yakut *mas-khomus* (Dyakonova 2017) and | Proto-JH, idioglot plucking idiophone, frameless lamella (frame-shaped but open on one side); this model was | Gentle pinching or tapping on the lamella | Clacking hollow sound with quite bright resonance, more sonorous than | Absent TO – as above |

(continued)

**Table 8.4** (continued)

| JH material and making time | Material's availability in the JH's area of use | Traditional JH of this type | Instrument class | Sound production technique | Salient timbral quality | The principal method of TO |
|---|---|---|---|---|---|---|
| few minutes to make, minimal knowledge | and c. 11,800–4500 BP, together with *Betula*, reaching as far as 70°N in northwest Siberia (Kremenetski et al. 1998) but absent in the northeast until 7300 BP (Anderson et al. 2002); in Beringia *Betula* became widespread 14–12 kya (Hopkins et al. 1982) | Tuvan *charty-komus* (Suzukei 1989, 65–66) – the splinter-like lamella sticks out between the forked arms; a disposable JH, possibly supporting multiple uses | advanced into the unframed JH with equal size arms and lamella in Altai c. the third century | with a finger while squeezing the thickest arm with the teeth, touching the lamella with the lips, and articulating the mouth cavity | grass, homogenous, and simple (limitations as above) | |
| (3) Twig of a shrub or a bush (stems of *Lonicera*, *Filipendula*, or *Caragana* and twigs of *Salix*), wishbone shape – requires about a few minutes to make, minimal knowledge | *Lonicera* and *Salix* are part of the taiga and cold forest biota that replaced tundra c. 11 kya in Primorye and Amur basin (Mokhova et al. 2009); *Filipendula* and *Salix* spread northwest to China even earlier, by 14 kya (Stebich et al. 2009) and reached north to Kamchatka | Tuvan *yiash-komus* and Telengit *taia-komus* – one of the bifurcated arms is used as a lamella; a disposable JH, possibly supporting multiple uses (Suzukei 2006, 159–161) | Proto-JH, idioglot plucking idiophone, totally frameless lamella (both ends of a wishbone shape are completely free) | Tapping on the thinner arm with an index finger while holding the thickest end with the other hand and placing the second arm in the mouth, pressing it | Quiet singular clattering sound, brighter and denser than the grass and chip, rather homogenous and simple (limitations as above) | Absent TO – as above |

| | | | | | | |
|---|---|---|---|---|---|---|
| | by 6 kya (Klimaschewski et al. 2015); *Caragana* became prominent in westward area of Mongolia later, c. 6 kya (Schwanghart et al. 2009) | | | with the teeth, and articulating the mouth cavity | | |
| (4) Bamboo stem (*Bambusa*, *Dendrocalamus*, *Schizostachyum*, *Phyllostachys*) would require something like 15 min to make with stone tools (Bar-Yosef et al. 2012), little expertise | Bamboo grows in warm and humid climates, unlike the materials used for proto-JHs, reaching only up to 36° in Paleolithic China (Clark 1998) and 34° in Japan (Kaner and Taniguchi 2017), as confirmed by the habitat area of the extinct bamboo rat (Tong 2007); so bamboo JHs could follow proto-JHs only after the latter would have crossed the northern border of the "bamboo zone" – not earlier than c. 11 kya | Nivkh *tyf-kanga* (Mamcheva 2005) and Ainu *mukkuri* (Nobuhiko 2008) – the lamella is framed from all sides, is elongated, and is bottle-shaped; non-disposable | Framed plate JH with a rope, tied through the hole(s) and an optional handle, idioglot plucking idiophone | Abrupt pulling of a rope by the right hand while freely holding the frame with the first and second fingers of the left hand, slightly touching the JH with the lips and pressing the JH with the left thumb – JH responds to a greater diversity of motions | The loudest and lowest sound of all organic materials, with the strongest rattling that is relatively homogeneous in timbre yet hollow (it isolates spectral components well, plays the pentatonic harmonic scale, and generates the bass drone) | Stage (1): more diverse technique allows the imitation of many environmental sounds and the invention of new effects, promoting the creation of program music and its use for animistic magic – to control the behavior of the imitated targets |

(continued)

**Table 8.4** (continued)

| JH material and making time | Material's availability in the JH's area of use | Traditional JH of this type | Instrument class | Sound production technique | Salient timbral quality | The principal method of TO |
|---|---|---|---|---|---|---|
| | | | | than the proto-JHs | | |
| (5) Wood from tree's trunk (*Larix*, *Pícea*, *Populus*, *Betula*, *Alnus*, *Salix*, *Cedrus*) would require about 2 hours to make with stone tools (Crabtree and Davis 1968), much expertise | Although *Larix*, *Betula*, *Alnus*, and *Salix* were present in the areas of the use of the proto-JHs during the Pleistocene, the technology of woodwork with the aid of only flint stone tools (Crabtree and Davis 1968) is more complex and time-consuming than bamboo work with similar tools (Bar-Yosef et al. 2012), which suggests that bamboo technology was first to emerge | Nivkh *khar-kongon* (Mamcheva 2005) and Evenk *kongiipkavun* (Sheikin 1986a) – the construction is the same as above | The same as above | The same as above | Loud, dense, dry, and harsh sound, the brightest of all organic materials, but imperfect blending of the attack and the aftersound – the higher and shorter "clang" stands out over the darker FF (it isolates components well, plays heptatonic scale, and uses the bass drone) | Stage (2): brighter sound and poor blending direct attention to formants, leading to the formation of clear distinctions between various vowel articulations and allowing users to establish their semantic values, promoting the rise of phonetic symbolism and "musilanguage" expression |
| (6) Bone – it would require about 25 hours to make without metallic tools (Sidéra 2011), plus a week or two of soaking the material, no | despite the omnipresence of bone, the technology of its cutting is so complex that it requires elaborate instruction for | *Tumran* of the Ugric ethnicities (Alekseyenko 1988) – the construction is | The same as above | The same as above | Loud, hollow sound with a pronounced clucking attack (well blended with the aftersound), although less | The same as above |

| | | | | | | |
|---|---|---|---|---|---|---|
| success without knowing a special technique | soaking and cutting the material (Schibler 2001) and knowledge of which bone to use (Rho et al. 1993); finalizing a small artifact takes over a day of work (Sidéra 2011), which suggests that bone technology followed wood technology | the same as above | | | harmonious and refined in isolating spectral components than the wood and bamboo instruments | |
| (7) Copper – its smelting requires several hours of heating the ore, making a furnace and accessing coal and ore, as well as expertise in fueling and extracting technologies (Coghlan 1939), requiring professional occupation (Bronson 1996) | The earliest extraction of copper occurred in Anatolia and Balkans 9–8 kya, spreading over Europe and Central Asia by c. 6 kya, reaching India (Tylecote 2002) and Siberia c. 4 kya (Sergeyeva 1981), delayed in Indochina (Hung and Chao 2016) and Korea by 2 kya (Park and Gordon 2007); copper smelting technology greatly exceeds the | Nivkh *pasla vyt' kanga* (Mamcheva 2005) and Buryat *aman khuur* made of local "red copper" (Simukhin 2006) – the lamella is bent into a hook and attached to the frame, shaped like a horseshoe – long-lasting valuable JH | Bow-shaped JH, idioglot plucking idiophone | Hitting or strumming the lamella in a variety of ways – forward, backward, pressing, or just holding the lamella with a finger, firm or loose holding of the ring by the teeth or lips, right or left shift of the frame | Prolonged ringing or buzzing sound, harmonically richer and louder than JHs made from organic materials, especially in high range; JH can play chromatic and microchromatic melodies; superb isolation of spectral components (it can use very narrow bands, permitting selective use of chords, including their use as drones) | Stage (3): longer duration and greater range allow playing several vowel and consonant articulation changes on the same stroke, establishing the JH phonemic system and putting in place rules for phonemic grouping; JH syllables merge into JH "words" which acquire their semantic values to form the glossary of expressive means to |

(continued)

**Table 8.4** (continued)

| JH material and making time | Material's availability in the JH's area of use | Traditional JH of this type | Instrument class | Sound production technique | Salient timbral quality | The principal method of TO |
|---|---|---|---|---|---|---|
| | complexity of bone work and woodwork technologies | | | (Sheikin 1986b) – overall, more diversity than in JHs made from organic materials | | convey affective states |
| (8) Iron – its smelting requires even longer heating, about 8 hours, to generate the needed t° of 1200 °C (Tylecote and Owles 1960) – twice higher than copper (Coghlan 1939), so that iron production typically calls for a 24-hour run, more efficient furnace and fuel | The earliest extraction of iron occurred in Anatolia and Iran 3 kya, swiftly spreading to Europe (Tylecote 2002), in parallel to India (Prakash and Tripathi 1986) and China (Mei et al. 2015), advancing to northeast, and reaching Siberia (Chernykh 1992) and Korea c. 2 kya (Taylor 1989); iron smelting technology exceeds the complexity of copper | Nivkh *zakanga* or *vychranga* (Mamcheva 2005) – the same construction as above | The same as above | The same as above | Very similar to copper but more homogeneous in timbre throughout the JH range and perhaps less bright – promoting homophonic texture, where melody can be placed in any part of the texture except the drone in the bass | Stage (4): greater registral homogeneity promotes melodic style, where melodic components share the same tonal quality, while other components (extra melodies or accompanying figurations) contrast it; elaboration of great variety of textures |

| | | | | | | |
|---|---|---|---|---|---|---|
| | technology and depends on the knowledge of copper smelting, therefore following the path of global distribution of a single invention (Roberts et al. 2009) | | | | | |
| (9) Steel | Steel technology was accidentally discovered in the Middle East c. 3 kya but was systemically explored much later – the Wootz steel from Southern India became the first internationally acknowledged steel in the sixth century BC, exported from there under the title of "Damascus" steel (Tylecote 2002) | Mass produced and sold generic Western JHs – the same basic construction as above but lower quality than the indigenous JHs hand-made by blacksmiths (Morgan 2017) | The same as above | The same as above | Similar to iron but brighter in the upper range – promoting the placement of melody on top of the texture and avoiding any voices or parts above it, to prevent its masking, which creates the "orthodox" homophony | Stage (5): upper range brightness promotes the use of a single broad and diverse expressive melody sustained in a self-sufficient melodic mode, causing the shift from timbral to frequency-based melopoeia |

Nine typical JH constructions are characterized in terms of the material of their makeup, time of the first availability for manufacturing, most common representative type of JH, organological classification, procedure of generating JH's sound, characteristic sound quality, and the principal model of TO

between the hygric tundra and the xerophytic steppes, of a major importance for sustenance of herbivores (as large as mammoth) in Beringia (Hibbert 1982). Its key feature is the co-dominance of the steppe herbs and the tundra dwarf shrubs, lichen, and mosses in subneutral arid soils, with high density of roots, supporting the hydro-thermic balance that enables graminoid vegetation even in the subarctic landscapes (Yurtsev 1982). The maximal diversity of the steppe plants within the Arctic is observed in the mega-Beringian sector, stretching from Russian Kolyma to Alaska (Yurtsev 1986). This area had even drier and slightly warmer steppic climate during and before the Last Glacial Maximum (LGM), especially East and West of Central Beringia (Brigham-Grette et al. 2004) – with ash-buried remnants of herbs well preserved at the Kitluk surface (Höfle et al. 2000). Reliable evidence of the first human presence at the Beringian Pacific coast dates back to 34 kya (Ikawa-Smith 2004). By 15 kya, humans left extensive traces of cultural activity in Beringia, originated perhaps in the Dyuktain culture of Lena Basin (Hoffecker and Elias 2007). This might have been the time of invention of grass JH.

*Leymus* must have attracted attention of humans as a source of food – it is edible by many animals (*Leymus racemosus* is even called “mammoth wild rye”) and is extensively used for fodder and grazing of livestock in Russia (Tsvelyov 1976, 177) and China (Wang et al. 2012b). The grinding stones, burned pebbles, and storage pits that suggest the preparation of vegetable food were found in the Tachikiri site, dated to 32 kya (Ikawa-Smith 2004). If *Leymus* was not used by Paleolithic foragers, then it must have been used by Neolithic livestock breeders and agriculturalists for strengthening the sand-based soil, as it is done today. The domesticates flourished in Eastern Asia between 10–6 kya (Dodson and Dong 2016), suggesting the timeframe for the invention of grass JH some time around 32–6 kya.

The other proto-JHs are made from trees and shrubs (Fig. 8.3). The chip/splinter construction is made of *Larix* and *Betula* in Yakutia (Dyakonova 2017) and Tuva (Suzukei 1989, 65–66). This construction was definitely used prior to the third to thefifth centuries AD in Altai – an unframed bone JH with arms and lamella of equal size, clearly resembling modern Tuvan charty-komus, was excavated in Tcheremshanka (Borodovskii 2017). The forestation of Far East occurred only after the LGM. At the LGM, c. 21 kya, North Asia, presently occupied by boreal forests, had <20% of their present woody cover, increasing after 15 kya (Tarasov et al. 2007). By 12–5 kya, *Larix* and *Betula* grew at the Arctic coast near the Lena delta – and these northwestern lands served as refugium for the forest species that spread south during the warming (Kremenetski et al. 1998). However, trees were absent in the East; in the Kolyma basin, along the Pacific coastline, until 7 kya (Anderson et al. 2002); and in Kamchatka (Klimaschewski et al. 2015). In Beringia, the abrupt climate change occurred ≈13,500 BP and resulted in proliferation of *Betula* and *Populus* after the earlier elimination of forests due to severe cold – especially *Betula* became abundant, forming the wide “birch zone” (Hopkins et al. 1982). This suggests the timeframe of 13–5 kya for the hypothetical transformation of the older Pacific grass proto-JH technology into the continental chip/splinter technology somewhere at the southern vicinities of Beringia, westside bordering with Yakutia – where mas-khomus survived until the early twentieth century (Tchakhov 2012).

The twig proto-JH construction, most likely, emerged from the grass construction (very similar geometry, Fig. 8.3), transmitted from Russian Far East westward to Tuva. *Lonicera* and *Salix*, commonly used today (along with *Filipendula* and *Caragana*) to make twig JHs in Tuva and Altai (Suzukei 2006, 159–161), were part of the taiga and cold/cool mixed forest biomes. They quickly replaced former tundra communities 11 kya over the Primorye and Amur basin, where forestation was lusher than today until 2.5 kya (Mokhova et al. 2009). *Salix*, *Filipendula*, and Poaceae were part of an emerging forest biota southwest of Amur, in Northeast China, even earlier, by 14,250 BP (Stebich et al. 2009). Around 11 kya, similar combination of *Salix* and *Filipendula* with grasses spread North of Sakhalin, to southern Kamchatka, where *Filipendula* became abundant after 6300 BP (Klimaschewski et al. 2015). *Caragana*, together with patches of *Salix* and *Populus* in riparian areas, constitutes part of the desert vegetation common in innermost Central Asia (Tarasov et al. 2007). However, cold-tolerant *Caragana jubata* reaches the north to the lower Lena River and the east to the Okhotsk Sea coast (Yurtsev 2001). In Mongolian and Northern Chinese deserts, *Caragana*-containing biota emerged during the early Holocene (Yang et al. 2004). In the Orkhon Valley of Mongolia, sand started transforming into soil, enabling a dense vegetation only during the mid-Holocene, c. 6 kya (Schwanghart et al. 2009). All of this suggests that the twig proto-JH was invented no earlier than 11 kya for Primorye and 6 kya for Mongolia.

The transition from grass to twig may have occurred directly in the Central Primorye, where temperature and humidity markedly raised by 9–6 kya, increasing the forestation, including *Salix* and *Lonicera* – the pastoralist Iman culture ca. 5–4 kya was in a position to use them (Chlachula et al. 2015). In Sakhalin itself, the early Holocene biota, 16–12 kya, included *Salix*, *Betula*, and Poaceae – replaced by treeless vegetation in the Boreal period, ≈9300 BP (Rudaya et al. 2013). It is possible that the Sakhalin inhabitants of Ogonki-5 developed the twig construction prior to 9300 BP and transmitted it to the neighboring Primorye.

Similarly, the proto-JH constructions might have evolved into a mukkuri-type JH in Sakhalin, as Mamcheva suggests (2012), sometime during the early Holocene, during temporary warming. Bamboo is confined to tropical and subtropical climate, growing lavishly in China (626 species) and Japan (84 species), modestly in Korea (6 species), and virtually absent in Mongolia and Russia (except Sakhalin), as well as in Europe (Akinlabi et al. 2017).

Bamboo has been the principal material for tool-making across the entire Southeast Asia (Pope 1989) and may have substituted stone tool technology there (Watanabe 1985). This "bamboo use hypothesis" has recently received experimental support and recognition (Bar-Yosef and Belfer-Cohen 2013). The commonality of bamboo artefacts might explain proliferation of bamboo JHs all over Southeast Asia and Oceania. However, the "bamboo zone" has remained distant from the cold arid landscape associated with proto-JHs. Its northern border advanced the farthest in Paleolithic China, approaching the Beijing latitude (Clark 1998). The Pleistocene northern border of the bamboo zone is marked by the habitat area of the extinct "bamboo rat," *Rhizomys*, that could consume nothing but bamboo, and reliable remnants of which are found no farther than 36°N, at Henan, Shaanxi, and Anyang provinces (Tong 2007). Impressions of bamboo nodes were found in sites of the Yangshao culture (7–5 kya), 35°N, far from the modern bamboo habitat (Chu 1973). By about 6 kya, the northern bamboo forest line must have been situated at least 3° north of the present latitudinal limit (J. Huang 1984). In Japan, bamboo was used to build houses at Kuzuharazawa IV, Shizuoka Prefecture (34°N) ca. 13 kya (Kaner and Taniguchi 2017). However, the subtropical climate, beneficial for bamboo, became established over Japan only ≈6 kya (Winkler and Wang 1993), making bamboo widely available.

Hence, the earliest bamboo JHs were likely to appear as the proto-JHs spread south, to wetter and warmer regions, away from the zone with dry and cold conditions required for the growth of proto-JH plants. This event could have occurred no earlier than 11 kya and more likely by 6 kya.

The wooden JH constructions, like Nivkh *khar-kongon* (Mamcheva 2005), probably succeeded the bamboo constructions – despite the availability of *Larix*, *Betula*, *Alnus*, and *Salix* across Russian Far East during the Pleistocene. The technology of manufacturing a JH from wood with the help of flint stone tools is considerably more complex and time-consuming compared to that of bamboo. Experimentally, it was demonstrated that making a pottery paddle from oak by stone tools takes 2.5 h and involves 7 operations (Crabtree and Davis 1968). Making a bamboo knife takes only 15 min and 4 operations – and making of a big spear (5–6 cm in diameter and 6–10 m long) takes 32–49 min (Bar-Yosef et al. 2012). Strong anisotropy, orthotropy, and hygroscopy (Bucur 2016) make wood a highly "capricious" material: not only its mechanical properties vary enormously depending on the direction of the cut (e.g., elasticity along the grain is 10–20 times smaller than across the grain), it easily deteriorates from exposure to high moisture and heat (Esteves and Pereira 2009) and requires special treatment of knots and spiral grains (Tsoumis 1991). Therefore, making a wooden JH requires a special knowledge without which an instrument will not sound "right," unlike making a proto-JH, and to a lesser extent than making a bamboo JH, which suggests that wooden technologies were derived from bamboo technologies (Sheikin 2002, 132).

Although bone differs from wood and bamboo by being available practically anywhere humans live, it is even more finicky for making a JH. Bone is also strongly anisotropic and hygroscopic (Spatz et al. 1996) and additionally difficult because of significant variations in mechanical properties of different cortical and trabecular bones (Rho et al. 1993). Even a small object made from antler (more consistent in properties than bone), such as a fishing hook, takes 1.5–2.5 h to make with stone tools (Robinson 1942). The procedure of bone cutting is by far more demanding than any woodwork. Without soaking the antler or the bone for at least a week or two, no cutting with a flint blade is possible at all, and a single cut requires an hour of work (Schibler 2001). Bone is even a more time-consuming material than antler: manufacturing a bone spoon that emulates early Neolithic Anatolian and Balkan artefacts takes 25 h total and is not possible without any expertise (Sidéra 2011). This makes it unlikely that the bone JH technology preceded the wood JH technology.

Metal extraction and metalwork are even more demanding than bone work. An experimental archeological study reproduced a typical procedure (digging and drying of a pit, surrounding it with stone slabs, filling it with charcoal, and placing the ore under an inverted pottery vessel) of smelting copper from the simplest copper ores and found out that no metal is produced unless the procedure is meticulously followed – the heating of ore up to 600–700° alone takes several hours and requires hermitization (Coghlan 1939). This experiment was reproduced and confirmed the results (Böhne 1968). Evidently, the complexity of smelting by far exceeds bone work and woodwork, demanding professional expertise (Bronson 1996). It took a long time before the necessary technological knowledge was accumulated in Anatolia and Balkans 9–8 kya, from where the technology quickly spread over Eurasia (Tylecote 2002). An independent discovery might have occurred in Fennoscandia (Herva et al. 2009) and Karelia (Zhuravlev 1977). The Northern Far East, where the genesis of grass proto-JH and its transition to wood and bamboo must have occurred, obtained its local metallurgy with a significant delay (Wan 2011). The Russian Far East (Sergeyeva 1981) and Korea (Park and Gordon 2007) were reached only by 2 kya. Copper and iron technologies spread to the Eastern end of Eurasia, where tool technology remained dominated by bamboo and other organic materials, at about the same time (Habu et al. 2017). Subsequently, metallic JH formed an entirely different technological and cultural tradition, bound to West, that contrasted and succeeded local wood, bone, and bamboo JH technologies in Ural area, Siberia, and Northern Far East (Li 1956; Golubkova and Ivanov 1997; Pegg 2001; Yakovlev 2001; Sheikin 2002; Mamcheva 2012; Dyakonova 2017).

Within this perspective on the evolution and availability of materials for JH manufacturing, the earliest archeological finds of JH fall somewhere around the tertiary transition. At first, the grass technology was transferred to the more solid frameless chip/splinter and twig constructions in the forested regions of northern Pacific coastline and Yakutia sometime between 32 and 5 kya. In the region of Tuva/Altai, this transition probably occurred ≈5–4 kya, when the open-steppe landscape changed into the forest-steppe one (Zhilich et al. 2017). Then, the proto-JH constructions may have evolved into the bamboo JH. In order for the wooden JH to emerge, the bamboo JH had to cross the northern "bamboo zone" border at 34°–36°N once again; spread further to the north, where trees were rare and bone would have presented a better alternative; and come back to the "bamboo zone." It was at this point that the Lower Xiajiadian bone JH was manufactured 4–3 kya. Noteworthy, Lower Xiajiadian culture was characterized by active contacts with remote northwestern Andronovo (55°N) and Seima-Turbino (56°N) cultures (which influenced local metal production) and had a mixed population, partly of a Yellow River type and partly of east- and north-Asian Mongoloid type (Liu and Chen 2012, 312). Such crossroad position made bronze, bone, and bamboo (and probably wood) JH technologies all meet at Northeast China ≈2000 BC. What makes this territory uniquely fit in the timeline of JH evolution is the climactic changes from cold to hot and back to cold throughout the Holocene (Wang 1984). Northeastern Inner

Mongolia, Heilongjiang, and northern Jilin were covered by tundra 18 kya (southern border at 44°N) and Liaoning by arid steppe; by 12 kya the tundra receded to 46°–48°N, changing Liaoning into forest-grassland; and since 9 kya no tundra survived in Northeast Chinese, transforming entirely into forest-grassland, by 6 kya supplanted by temperate forest (Winkler and Wang 1993). Specifically, *Larix* grew in Heilongjiang, Huma (51°N) c. 5200 BP; *Betula* grew there c. 8000 and again 2000 BP, also in neighboring Sanjiang Plains (46–48°N) c. 8000 BP and also about the same time in Dunhua Swamps (43.2°N), Pulandian (39.3°N), Mount Yenshang, and Beijing (39.5°N). Poaceae proto-JHs of tundra likely naturally evolved into *Larix* and *Betula* chip/splinter proto-JH and thenceforward into the bamboo JH as the local population adapted to climate changes.

## 8.7 JH Tradition and Genesis of Languages in Northern Eurasia

Northern Inner Mongolia, Liaoning, Jilin, Heilongjiang, Primorye, Sakhalin, Amur, and Khabarovsk Krai make up the area where the transition from proto-JH to bamboo JH most likely took place between 11 and 6 kya, with the epicenter near the oldest JH find (Fig. 8.2), in Manchuria – "a meeting place of diverse peoples and cultures" (P. Huang 1990). Chinese specialists have four views on inhabitants of the northern Lower Xiajiadian, to which the uncovered JH belonged: to Sushen people, to pre-Shang Chinese people, to Northern tribes (Guzhu, Shanrong, and Tufang), and to pre-Yan Chinese (Da-Shun 2002). Guzhu was a vassal state at Lulong; Shanrong was a confederation of northern mountain tribes, often fighting with the Yan kingdom; and Tufang was another hostile confederation of the north of Beijing – all belonging to "the Northern Zone" that ran along the Great Wall demarcating China from the steppe "barbarians" (Sun 2006). Sushen people are identified as the earliest ancestors of the Manchus, most probably of Tungus, but possibly of Paleo-Asiatic, origin, recognized by ancient Chinese as a political entity that unified non-Chinese ethnoses at the Amur/Ussuri region 4–3 kya (P. Huang 1990). Chinese historians kept applying terms like "Sushen" or "Shanrong" to tribal conglomerations based on what they saw as important political and territorial issues rather than ethnic identifiers, which explains why the same name was often applied to numerous different ethnicities over time. Based on matching Chinese historiographies with the Russian historic and ethnographic materials, one of the leading Russian sinologists, Vsevolod Taskin, identified three historic Northern Chinese tribal confederations as follows: Sushen, Tungusic-speaking; Donghu, Mongolic-speaking; and Xiongnu, Turkic-speaking (Taskin 1984). Their languages, with all likelihood, prevailed over Chinese at that time (Janhunen 2010).

The morphological analysis of the remains at the Dadianzi cemetery confirms that the Lower Xiajiadian population was ethnically mixed and culturally segregated (Liu and Chen 2012). The mitochondrial DNA and Y chromosome study of 14 human remains from Dadianzi indicate that the Lower Xiajiadian people descended from the Central Plain and from the north (Li et al. 2011). The wider comparison of the DNA samples from six Lower Xiajiadian sites has confirmed significant genetic differences between populations of the West Liao River Valley (Cui et al. 2013). The archeological, phylogenetic, and mitochondrial DNA analysis of 42 remains from the Jinggouzi Cemetery, c. 2485 BC, near Chifeng, Inner Mongolia, identified their closest relation to ancient Xianbei people and to modern Oroqens, confirming their association with the Donghu culture and indicating their nomadic tribal lifestyle (Wang et al. 2012a).

Better researched is the ethnic identity of people of the succeeding culture of Northeast Chinese – the Upper Xiajiadian – to which the bone JH from Chifeng, dated 1200–600 BC, belonged (Honeychurch 2015). Despite the cultural discontinuity between the Lower and Upper Xiajiadian societies, following the demise of agriculturalism, a few hereditary lines in civic planning and arts were carried through (Da-Shun 2002). The common anthropological interpretation is that the Upper Xiajiadian culture was brought by the northern militant newcomers with deep knowledge of metallurgy, who replaced the sedentary Lower Xiajiadian people and thereafter kept advancing southward (Di Cosmo 1994). However, recently this picture of cultural replacement has been overviewed in favor of more gradual transformation (Sun 2017). Also, the paleogenetic study of temporal continuity of Y chromosome lineages among the West Liao River populations during the last 5000 years did not find evidence of any population replacement (Cui et al. 2013). No significant genetic difference was found between the Lower and Upper Xiajiadian populations in the Liao River Valley (Li et al. 2011). The comparison of anatomical remains and images on archeological artefacts from Upper Xiajiadian monuments points to the Mongolic and Turkic origin of their population (Barinova 2014). The Upper Xiajiadian culture is credited with forming the earliest confederation of Donghu, Wuhuan, and Xianbei ethnic groups, who probably spoke proto-Mongolic and evolved into the state of civilization approximately eighth to fifth centuries BC (Hu 2010). Its influence stretched from Amur (Girchenko 2018) to Altai, mediated by the Munkh-Khairkhan culture in Khovd province, Mongolia (Kovalyov and Erdenebaatar 2014). The animalistic style of ornamentation that characterizes this civilization is of Scythian descent and should be regarded as the cultural interaction between the east and the west which was initiated some 4500 BP (Wang 2018).

Five bamboo JHs from the Yuhuangmiao burial, Sichuan, seventh to sixth centuries BC (Table 8.2) most likely belonged to the Shanrong people that invaded Yan in the seventh century BC and maintained contacts with northern nomadic tribes (Shulga and Girchenko 2013). All Rong tribes are believed to have been part of the southern Tungusic intrusion (Barnes 1993, 165). The spread of bamboo JHs from Liaoning to Sichuan and Shaanxi reflects the general cultural transition from the Liaodong Peninsula to the Ordos during the first millennium BC (ibid.). The southern expansion of the Upper Xiajiadian culture formed the most remote branch of the Scythian-Siberian sphere of influence, most evident in the integrity of the animalistic ornamentation from the Korean Peninsula to Central Asian steppe (Kang 2011). The Upper Xiajiadian burials show common cultural traits with the burials of Xiongnu commoners of the imperial period, indicating their shared origin (Miniayev 1991). And the similarity of the economy types and food production across Northern China points to the Donghu influence and Mongol descent, including cultural contacts, political unions, as well as migration – fueled by the competition in military technologies (Komissarov 1988). The entire area from Transbaikal to Manchuria has been the melting pot of ethnogenesis of numerous ethnicities of Indo-European, Mongolian, and Manchurian stalk throughout the first millennium BC – along with the shared pastoral nomadic lifestyle in contradistinction to the Chinese sedentary agriculturalism (Botalov 2007). The geographic border between these lifestyles coincides with the border between two general types of JH music. The indigenous pan-Siberian tradition is based on an intuitive use of timbre classes of *timbre-oriented* music. The non-indigenous Chinese and Russian/Ukrainian JH traditions that enclose the "indigenous" area at its southeast and western borders are based on the formally taught music theory of *frequency-oriented* music (Nikolsky et al. 2020). Both of these "formal" traditions constitute a certain kind of "tonality" – diatonic heptatonic for Russia/Ukraine and diatonic pentatonic for China (Nikolsky 2015).

Limited natural resources compelled nomadic tribes to keep moving in search of pastures suitable for their livestock, unleashing conflicts, forming alliances, and sustaining intense competition in food and military technologies. The ongoing chain of borrowings and intermarriages within the ever-changing tribal confederations undoubtfully affected language formation. The population of the entire Siberia is characterized by extremely low gene diversity, low density, relative scarcity of indigenous ethnoses (only 31), and languages (35 versus over 100 in Europe), mostly of Altaic and Uralic phyla, and common economy of nomadic and semi-nomadic type with common social features, e.g., clan structure, endogamy, and the levirate (Karafet et al. 2008).

The totality of the information presented above suggests that the *genesis of pan-Eurasian JH tradition occurred probably among people speaking proto-Tungusic* ≈4 kya. By 3 kya, the tribes speaking *Mongolic* languages entered the scene and dominated in the Northeast Chinese (Blench 2008). Closer to 2 kya, the *Turkic* tribes made their appearance from the west. The admixture of Tungus, Mongolic, and Turkic languages has been conceptualized as the "Altaic family." Its concept was coined in 1845 by the Finnish linguist, Matthias Castren, who grouped these languages together with the Finno-Ugric ones based on the commonality of their vowel harmony and agglutination (Bogoras 1927). Castren's colleague and compatriot, Gustaf Ramstedt, excluded the Uralic languages from the "Altaic family," replacing them with Korean (Ramstedt 1957). Nikolai Poppe also supported the inclusion of Korean, renaming Castren's family into "Uralo-Altaic" to distinguish it from the extended "Altaic" family (Poppe 1965), which introduced the distinction between the extended "macro-Altaic" and the basic "micro-Altaic" families (Georg et al. 1999). Miller (1967) extended the family further by adding Japanese to it.

However, Gerard Clauson presented arguments for discarding both extended and basic family concepts. He conducted the lexicostatistical analysis of the glossaries of "Altaic" languages, disclosing considerable discrepancy between Turkic and Mongolic languages, especially in regard to their archaic words, and the lack of genetic ties between Turkic and Tungus languages (Clauson 1956). Clauson's lexic and glottochronological approach, in turn, was criticized by Lajos Ligeti (1971) who demonstrated the inability of Clauson's analysis to disclose the ties between those languages that are already known to be genetically related. Nikolai Syromyatnikov (1971) formulated the set of principles for comparative lexic and morphological analysis and suggested to apply them to the most distant languages of the expanded "Altaic family." Sergei Yakhontov (1971) emphasized that the methodology created for the analysis of Indo-European languages is inapplicable to the agglutinative Altaic languages, and their comparative lexic analysis should include semantic gradations favoring those words that are less connected with cultural changes. Gerhard Doerfer (1963) enhanced Clauson's approach by separating the glossaries of "peripheral" and "nuclear" basic words (e.g., words denoting the body parts) and holding the latter for a trans-historic phenomenon, not affected by cultural contacts. However, Doerfer's approach also failed to confirm the existing relationship between the related languages (Andreyev and Sunik 1982). Doerfer's choice of "nuclear" words was also criticized for inadequate reduction of actual glossaries (Kolesnikova 1972). Alexander Shcherbak (1994) attempted to combine the lexic and cultural analysis in relation to Mongolic and Turkic languages, which revealed that the Mongolic glossary is based on a forest lifestyle, whereas the Turkic glossary on a steppe lifestyle. This implies the cultural rather than the genetic ties uniting Mongolic and Turkic languages, supporting Doerfer's conclusion.

On the other hand, Sergei Starostin (1991) implemented the Yakhontov methodology for the lexico-phonetic comparison of "Altaic" languages and established that Turkic, Mongolic, and Tungusic languages share about 20% of Yakhontov's basic words, whereas Japanese and Korean languages share only 13–14%. Starostin explained such low percentage as an indicator of the antiquity of the proto-Altaic language. Later, he expanded the scope of lexic analysis to 2000 words, breaking Yakhontov's list in 2 groups, of 65 and 35 words, the latter of which was considered diachronically stable, so that if the number of common words from this list between the two languages exceeded the first list, such two languages were pronounced genetically related (Starostin et al. 2003). This approach has caused extensive criticism, establishing the consensus that the common traits between "Altaic" languages are caused by language contact rather than shared ancestry (Kortlandt 2003; Georg 2004; Vovin 2005; Róna-Tas 2011; Jankowski 2013). Nevertheless, the issue of similarity between these languages continues to receive attention of researchers, promising a new synthesis. As Roy Miller noticed (1996, 15–17), the objections to the Altaic unity rest exclusively on the lexicon, whereas the arguments for Altaic unity rely on the features of verbal morphology. The *solution to the "Altaic" controversy, therefore, lies in tracking the correspondences between the morphological typologies of the languages in question*, since different ethnoses readily borrow words but stick to the same morphological principles (Johanson 2010). The proof of this is the remarkable conservation of the agglutinative languages, withstanding external influences from non-agglutinative languages.

The new methodology for lexicostatistic analysis within the morphological context was formulated by Robbeets and seems to be gaining acceptance in the field (Gözaydin 2006; Rozycki 2006; Büyükmavi 2007; Décsy 2007; Kara 2007; Dybo 2019). She was able to identify 19 verbal suffixes that reflect formal, functional, combinatory, typological, and paradigmatic co-dependencies and 14 criteria for distinguishing language contact from common ancestry (Robbeets 2015). Comparison of the common innovations of verbal morphology in combination with the phonological and lexic innovations enabled Robbeets to date the breaking of the initial protolanguage into the constituent languages and families. Like Joseph Greenberg (2000), Johanson and Robbeets (2010) replaced the controversial concept of "Altaic" with more accurate "Transeurasian" family. According to Robbeets, proto-Transeurasian language split into proto-Altaic and proto-Manchurian (Koreo-Japonic) language about 5000 BC, which in turn split the former into proto-Turkic and Tunguso-Mongolic and the latter into proto-Japonic and proto-Koreanic languages by 3000 BC (p. 506). Proto-Tungusic and proto-Mongolic languages branched out shortly before 2000 BC. Alternative application of lexicostatistic method brought dates quite close to Robbeets' methodology: 4750 BC for formation of proto-Altaic (equivalent to Robbeets' "Transeurasian"), 4350 BC for its split into proto-Turkic and Tunguso-Mongolic, and the latter's split by 3700 BC (Blažek 2009). These dates are slightly earlier than Starostin's estimates of disintegration of proto-Altaic language by 6000 BC and formation of proto-Japonic and proto-Koreanic by 3000–4000 BC (Starostin 1991). Extended etymological analysis, conducted by Starostin's team later, confirmed the 6000 BC date but moved the separation of Turkic-Mongolian and Korean-Japanese branches earlier – about 4000 BC – while delaying the formation of the Old Turkic language to the first century BC (Starostin et al. 2003, 235–6).

Even an earlier date of 10,000 BC for the breakup of the proto-Altaic was suggested by Sunik, based on his estimates of the evolution of the morphological structures (Sunik 1978). Combining the computerized lexicostatistic analysis with the cultural reconstruction moved the date for the breakup of the proto-Transeurasian language in Robbeets' estimation earlier to 5700 BC, postponing the formation of the proto-Altaic to 4600 BC, and even more so for the Koreo-Japonic – to 3300 BC – while increasing the age of the Mongolic/Turkic split to 2800 BC (Robbeets 2017). The latest update of Altaic chronology that implemented the Bayesian phylogenetic inference generally confirmed the Russian Altaists' claim of a closer relation between the Turkic and Mongolic branches while siding with the Western Altaists' claim that Tungusic, Mongolic, and Turkic families each constitute separate entities (Robbeets and Bouckaert 2018). The Mongolic and Turkic split is valued at 100% posterior clade support for the families: Korean and Japonic (at 98.7%) and Turkic (90.3%).

All in all, *the timeframe of genesis of the Far Eastern metacultural "organic" JH tradition coincides with the breakup of the proto-Transeurasian ≈8 kya* – when in Northeast China and Russian Primorye tundra gave way to a more humid forest-grassland, enabling the neighborhood of grasses, trees, and bamboo, necessary for transition from proto-JHs to bamboo and wooden JHs (Table 8.4). Talismanic use of proto-JHs corresponds to the stage of formation of numerous "dialect continuums" in low-density populations of constantly migrating nomadic and seminomadic tribes with limited use of agriculture (Robb 1993). *Emergence of mukkuri- and kanga-like JHs and onomatopoeic JH music in itself was likely to execute the role of* lingua franca *in interactions of the neighboring regional tribes, leading to the crystallization of proto-Transeurasian.*

It is not accidental that *the map of distribution of Transeurasian languages* (Robbeets and Bouckaert 2018, 146) basically *coincides with the map of modern distribution of the indigenous JH traditions* (Fig. 8.2) – with the sole exception of Turkey that currently bears no indigenous JH tradition. However, this may have been a recent omission in light of the presence of such tradition among the neighboring speakers of very close Azerbaijani (Yesipova et al. 2008) and Gagauz (Dallais et al. 2002).[21] The plausibility of this scenario is supported by the fact that the Turkish word "qopuz" (related to Mongolian "quyur"), used in reference to JH, constitutes one of the "nuclear words" for the entire Turkic language family, testifying to the cardinality of JH for nomadic Turkic tribes (Róna-Tas 1986, 25:222–3).

## 8.8 Nivkh Language's Ties to the Emergence of Jew's Harp's Eurasian Tradition

Sheikin (2002, 125–132) implemented the "nuclear" epistemological approach to the names of JH used by Siberians and established five pan-Siberian "nuclear" roots: (1) Nivkh "*kongon*" for east, (2) Ainu "mukkuri" for southeast, (3) Yakut "*khomus*" for central and southern Siberia (along with the related Manchurian "kumun" and Buryat "khur"), (4) Selkup "pynkyr" for northwestern Siberia, and (5) Mansi "tumre" for the Urals. The geographic distribution of these "nuclei" marks five

[21]Iron JH was listed in the catalog of Turkish national instrumentarium compiled by Evliya (aka Ewliya) Chelebi in the seventeenth century, with the mentioning "invented in Danzig" – however, many other instruments on that list are supplied with mythological provenances (e.g., Pythagoras, Solomon, Queen of Sheba) (Farmer 1936). It is highly probable that JH was initially brought to Anatolia around the sixth century by nomadic Turkic tribes but lost its importance and became forgotten as the traditional nomadic clan-based lifestyle gave way to the sedentary lifestyle with a centralized government, and as Turkey evolved into a regional power, absorbing the cultural achievements of its southern neighbors, famous for their cultural pedigree. At this point the exuberance of sophisticated musical instruments and ancient orchestral tradition of Mesopotamia (Krispijn 2010) and Egypt (Sachs 2008) would have completely out-shadowed the modest melodic capacities of a quiet JH, designed primarily for personal use (Alekseyev 1991b).

areas (Fig. 8.4), while their etymological connections,[22] along with the historic-ethnographic data, point to the direction they spread, namely:

1. Pacific coastline (Primorye, Khabarovsk Krai) and Sakhalin (according to Sheikin, a homeland)
2. Japan and nearby islands, Korea and Northeast Chinese (southern expansion)
3. The "belt" from the Amur basin to Mongolia and Altai plus the northern and western vicinities (northern expansion to Yakutia and westward to Middle Asia)
4. The Ob' and Yenisei River basins (further northern expansion)
5. Kama/Volga area (further western and northwestern expansion)

Not much is known about the languages of Area-1 spoken during the early Holocene. It is likely that JH inventors spoke a language ancestral to Nivkh due to the Nivkhs' unique combined use of grass, bamboo and wooden JHs. The origin of the Nivkh language presents an unresolved puzzle. Nivkh shares many lexical and grammatical features with Tungusic languages and Ainu (not to speak of all that it has borrowed from the Russian, Chinese, Mongolian, Manchu and Japanese languages), and shows typological resemblances to the Chukchi, Ainu and Amerindian languages (Mattissen 2003). Despite its multiple connections, Nivkh is an isolated Paleo-Siberian language which should be taken as an indication of the old age of its "pedigree" since all of its family relatives have gone extinct (Pevnov 2009). This "age difference" likely explains why the indigenous neighbors of Nivkhs speak all of one another's tongues except that of the Nivkh (Gruzdeva 1998). Janhunen regards the Nivkh language as part of the **Amuric family**, originated in Manchuria and

[22]The Ugric word "*tumran*" does not carry any onomatopoeic connections, unlike many other local names for JH, but descends from "*tombyra*" of the Siberian Tatars and Turks of Middle Asia and Kazakhstan – used to refer to lute-like plucking musical instruments (Alekseyenko 1988). The common feature in this derivation must have been the plucking hand action that produces a characteristic sound with a strong attack, followed by a quick decay of the sound. "*Pymel*'" of the Kets and Yughs does not have indigenous etymological connections either, probably descending from the Selkup "pynkyr" ("buzzing") and relating to shamanic tambourines ("pendir" and "fendir") of the northern Samodic peoples. The Yakut "khomus" (and its multiple Turkic varieties) descends from Old Turkic "khobus" (Nadeliayev et al. 1969, 451), which in turn, must be related to the Manchurian "kumun" – the name of the official 8-tone "state music" (Tsintsyus 1977, 2:431) and the pentatonic system theory borrowed from Ancient China (Zakharov 1875, 36). "Kumun" might have descended from the Old Korean "komungo" – "the lute of a black stork" (Tsintsyus 1977, 2:431). Sheikin lists the arguments that explain the sacred meaning of the concepts of "lute," "stork," and "black" within Ancient Manchurian cultures, and connects them to Japanese "koukin" (2002, 130). Like the word "tumran," "khobus" is related to words used in reference to plucking string instruments – which is indicative of the later origin of "khobus" in comparison to "kongon." Nivkh "kongon" has onomatopoeic origin (Mamcheva 2005). It relates to Udege "kunkai," Nanai and Ulchi "kunkai-konkai," Orok "kunka" and Negidal "konkikhi," as well as, possibly, Even "konkukan" and Evenk "kongiipkavun" (Sheikin 2002, 131) – used in reference to the frame JHs made of organic materials. Bow-shaped metallic JHs are called by the Nanai, Ulchi, Oroch, Negidal, and Orok peoples "muene," "mene," "mukhele," "mughene," "mukhene" and "mukhane" – related to Manchurian "mekheni" and Ainu "mukkuri" (ibid.). This term must have been originally formed in reference to the bamboo JH, but was appropriated by neighboring ethnic cultures in reference to the metallic instruments.

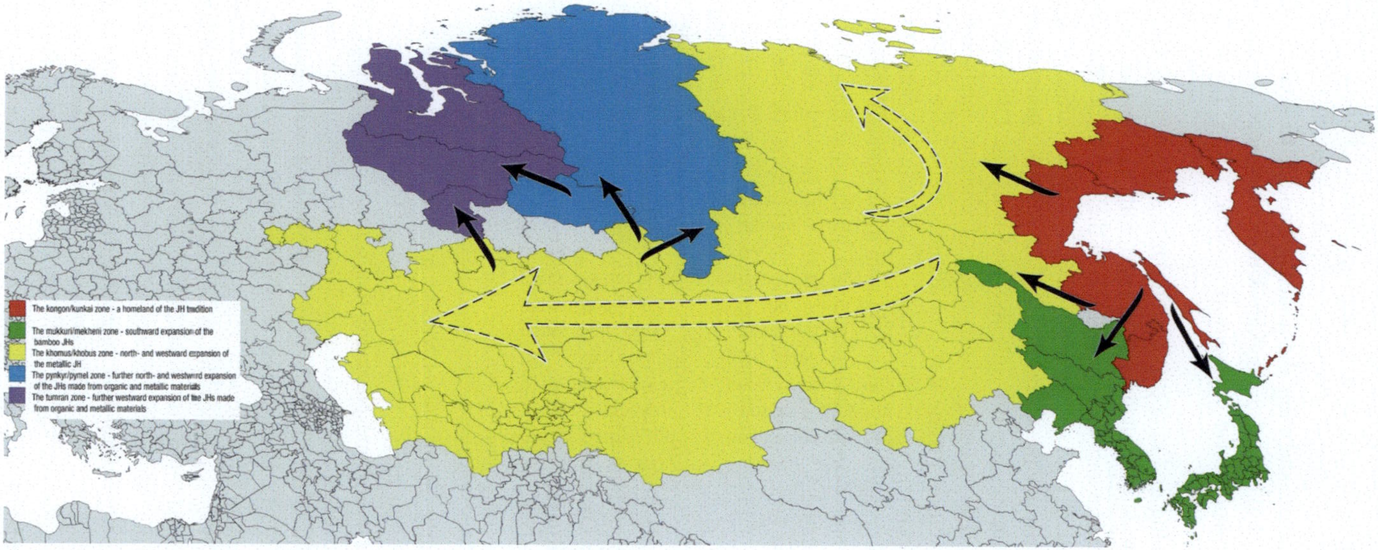


**Fig. 8.4** *Five zones of the spread of pan-Siberian JH tradition*. This map represents the geographic distribution of the indigenous JH music after Sheikin (2002, 125–132), based on his comparative etymological analysis of words for JH of local languages and available ethno-historical information – integrated with the data on trading routes from the previous figure. Red color marks the *kongon/kunkai* Pacific zone, the cradle of proto-JH music. Green color marks the *mukkuri/mekheni* southern zone of the first fully-fledged JHs made of bamboo and later of bone and other materials. Yellow color marks the *khomus/khobus* zone that originally was confined to central and southern Siberia, but expanded to west and north - indicated by two big dashed yellow arrows. This is by far the most massive zone, chiefly responsible for inter-continental spread of JH music over Eurasia. Throughout its spread, the western branch at some point met the eastward wave of the spread of metallurgy, causing metallic bow-shaped JHs to become prevalent in Turkic tradition. Blue color marks the *pyrky/pymel* west Siberian zone of JHs made mostly from organic materials, but at a later point, from metal. Purple color marks the Uralic *tumran* zone that used JHs made from organic materials as well as metals. Black arrows indicate the routes of inter-zonal cultural contacts

spread to the north, where it gave rise to Chukotko-Kamchatkan languages (Janhunen 1996, 71). The latter form a distinct language family (Fortescue 2005). Fortescue identifies the morphological and typological features that the Nivkh and the Chukotko-Kamchatkan share, deeming them strong enough to consider their common ancestry in proto-Chukotko-Kamchatkan-Amuric language, based on a *symmetrical 6-vowel system* with *vowel harmony* (Fortescue 2011). The archaeological data supports this affiliation.

Maxim Levin (1963) was the first to notice the similarities between the Okhotsk culture of Sakhalin-Hokkaido-Kurile Islands, associated with the Nivkhs, and the old Koryak culture of Kamchatka and Taigonos, tracking them to a common origin from "northeastern Paleoasiatics" (283). Ruslan Vasilevskiy (1969) furthered the list of cultural commonalities between the Okhotsk and Old Koryak cultures, tracing their past to the Neolithic Ancient Koryak culture situated by the Okhotsk coast, from Amur to Tausik Bay. The amazing consistency of climate, landscape and economy of the coastal population for the last 10 millennia must have contributed to the preservation of the Nivkh language and culture in Sakhalin as opposed to the continental Okhotsk coast, geographically open to human migrations. A missing link between the ancient Nivkh and Koryak cultures could be the maritime Tokareva culture (Lebedintsev 1998), dated 5.5 kya (Kuzmin 2000), and possibly constituting an extinct branch of the Amuric family (Fortescue 2011). One of the leading specialists in Nivkh culture, Chuner Taksami, regarded Nivkhs as *direct heirs of the Neolithic population of Beringia, holding that the comparison of the mythologies and cults of the regional cultures confirmed their common ancestry rather than later co-borrowings* (Taksami 2009). Sergei Ivanov (1963) arrived at the same conclusion in his fundamental work on ornamentation in traditional fine art of indigenous Siberian ethnicities. Aleksey Okladnikov (1955), perhaps, the biggest authority on archaeology of Siberia and Russian Far East, also considered the Amuric ancestors of Nivkhs as the prime colonizers of the entire Okhotsk coastal area, whose culture drastically differed from the neighboring continental cultures already during the Neolithic.

The genetic evidence clearly distinguishes Nivkhs from the neighboring ethnicities while *tying them to the Amur area*. Of the 8 ethnoses of the Russian coastal Far East, it is only the Amuric Nivkhs and Nanaians that do not carry haplogroup Z, typical for the Tungus, Manchurian and Turkic populations. However, they do carry the haplogroup Y (the highest in Nivkhs – 47%), associated with Korea and Kamchatka (Gubina et al. 2013). Haplogroup H is present in Nivkhs, Nanaians and Evenks (Amur) as well as in Chukchi and Yukaghirs (Chukotka and Yakutia). Haplotype S13, present in 28% of the Nivkhs and in 14% of the overall Siberian haplogroup D, probably constitutes the founding haplotype for Siberia and America – its affinity with the American haplotype AM88 indicates that Nivkhs are related to the Beringian population that migrated to America (Torroni et al. 1993). mtDNA haplogroups G1b and Y1a, common amongst the Kamchatka Koryaks, are also found amongst the Sakhalin Nivkhs (Volodko et al. 2008). And G1b is common amongst the Chukchi as well (Schurr et al. 1999). The comparative analysis of the genetic relationships of the Siberian populations confirms *the genetic connection between the Chukchi, Koryaks and Nivkhs* (Fedorova et al. 2013). The distribution of haplogroup Y, restricted to the Lower Amur region and Kamchatka, reaches its maximum frequency in Sakhalin Nivkhs' mtDNAs (66%) in combination with low intra-group diversity, which testifies to the relatively recent age of this haplogroup and of the long isolation of the Nivkh population (Starikovskaya et al. 2005). Nivkhs have the least diverse mtDNA amongst the Okhotsk coastal ethnicities and cluster together with Koreans, distant from Koryaks and Itel'mens, in the neighbor-joining tree that estimates genetic distances from the nucleotide diversity values (Schurr et al. 1999).

Interestingly enough, Nivkhs, Nanaians and Koryaks, all share the haplogroup U5 – most prevalent in the exceedingly distant Saamis (53%) and common for Western Europe (Gubina et al. 2013). This goes to substantiate Janhunen's claim that Uralic language family descends from remote Manchuria (Janhunen 2009). Yet another indicator of possible Western ties of the Amuric peoples is subhaplogroup D4 that is common among the Nivkhs (5.4%), Negidals (6.1%) and Ulchi (10.3%) of Amur as well as the Tubulars (1.4%) and Nganasans (5.2%) of Taimyr, separated by 5000 km (Starikovskaya et al. 2005). Altogether, the available genetic data suggests that the *Nivkhs came to the Okhotsk Sea from Western Siberia before the separation of America and eventually became pushed out to Sakhalin by newer incoming tribes.*

The Nivkh language bears a number of traits that can be attributed to the influence of the JH tradition.

1. **Vowel harmony**. In the past, the Nivkh vocal system was characterized by vowel harmony (**VH**) (Kreinovich 1937, 26) that regulated the alternation of high ([i], [y], [u]) and middle/low ([e], [a], [o]) vowels (Gruzdeva 1998).[23] And, as we shall see, VH very much lives up to the musical reputation of "harmony"

Leaving aside the controversies in defining VH, in the most general sense, VH can be structurally described as *the systematic agreement of vowels in adjacent moras or syllables within the phonological and morphological boundaries of a word (or its base) in regard to one or more articulatory features* (Krämer 2005).[24] Perceptually, VH prolongs the duration of a given tonal quality – especially where the latter is at risk of being misidentified (Kaun 2004). In essence, this is not different from *increasing the frequency of occurrence of a particular pitch class in TO of music*

[23]Gruzdeva's qualification of this VH as "height" was challenged by Ko, Joseph, and Whitman who interpreted it as based on the opposition of the advanced and retracted tongue root rather than the opposition in height (Ko et al. 2014). Shiraishi and Botma seem to pose a middle ground by arguing for "a synchronic pattern of co-occurrence restrictions that is based on height" which, in their opinion, nevertheless might have developed from an earlier tongue-root system (Botma and Shiraishi 2015).

[24]VH differs from other forms of assimilation by systematically triggering an alternation in "target" vowels that are positioned in direct proximity to the "trigger" vowel once a certain feature specification is met (underlyingly or in the surface form), so that the "target" vowel sounds the same as the "trigger" vowel (Krämer 2005). Such triggering usually exerts morphonological influence by highlighting the boundaries of words in a sentence.

*in order to stress its anchoring melodic function*. Frequent repetitions, especially in opening and closing of a musical phrase, ascribe tonicity (stability) to the pitch level (E. Alekseyev 1986). In contrast, the variability of the pitch value within a wider frequency "zone" characterizes "unstable" pitch classes, e.g., "the leading tone" (Garbuzov 1948). The interplay of stability and instability generates oscillations of tension and relaxation that fuel emotion-based semiosis in music (Krumhansl 2002). Tension-relaxation dialectics applies not only to the domain of frequency but also to timbre (Paraskeva and McAdams 1997) (Volodin 1972; Pressnitzer et al. 2000; Nazaikinsky 1988; Nazaikinsky and Rags 1964; Lerdahl 1987) – just involving the timbre- rather than the frequency-classes (Nikolsky et al. 2020). In JH music, vowels function as musical "timbre-classes," acting "musically" to bind adjacent sounds – unlike consonants that epitomize the tendency of speech to oppose adjacent sounds and fragment the sound stream (E. Alekseyev 1991b). Vocal music expands vowels while contracting consonants, following the paradigm "vowels sing but consonants speak" (Kolinsky et al. 2009). It is vowels that support prosodic features common to the pragmatics of both music and speech (Wennerstrom 2001).

In VH, too, it is vowels that initiate the harmony process, whereas consonants block it (Krämer 2005). "Blockers" halt harmony and usually do not themselves undergo assimilation (Rose and Walker 2011). Hence, consonants can be viewed as disharmonizers of vowels in verbal communication that is fundamentally "non-musical" – after all, natural languages have more consonants than vowels (Bonatti et al. 2007). This is because there are many more ways of fragmenting the sound-stream than there are ways of joining sounds into a single stream. VH provides one way to accomplish this joining – by making a vowel-change mark morphological boundaries so that the repetition of a vowel quality effectively binds a new syllable with the previous one into a single syntactic unit. This is not very different from binding melodic sounds into a musical phrase by making them belong to the same harmony.[25]

---

[25]This phenomenon is known in musicology in English as "harmonic rhythm" (Swain 2002) and as "harmonic pulsation" in Russian musicology (Berkov 1962). Its essence is that melodies and musical textures in Western classical tradition, starting from the XVII century, are built around the harmonic progressions that tonally fine-tune melodic structures, since the anchor tones of melody ought to constitute the "chordal tones," present in a given harmony that encompasses the entire musical texture from one moment of metric time to another (Nikolsky 2016). Harmonic changes generate the peculiar pulse whose rate is usually sustained more or less intact throughout the musical composition in a manner similar to metric pulsation. Western music, including most folk and popular music styles, abides by rather strict metric and harmonic pulsations (which often coincide). Therefore, Western musicians and listeners develop a peculiar skill of figuring out an implied harmony while listening to a melodic progression (Povel and Jansen 2001) – even in a monophonic piece of music. Restraining the melody to only the chordal tones of a certain harmony effectively binds such tones into a syntactic group (motif, phrase or sentence): e.g., the main theme of the "Blue Danube Waltz" by Johann Strauss Jr.. Harmonic pulsation is not limited to Western art music. It is as common to all forms of Western folk and popular music (e.g., "On Top of Spaghetti"). Theoretically speaking, it should be present in every form of multi-part non-Western music that affords the use of chords.

But musicality of VH goes further – involving **harmonicity**. Harmonic-to-noise ratio and subharmonic richness are known to convey affective information in animal vocal communication (Briefer 2012). Harmonicity and inharmonicity are meaningful in animal calls, used to convey, respectively, friendliness and fear (Morton 1977). There are reasons to believe that the ability to discriminate specific harmonics is not unique to humans. Tamarins were shown to discriminate signals with the upper harmonics eliminated and with the second harmonic mistuned (Weiss and Hauser 2002). This suggests that the harmonic and subharmonic analyses might not apply to the perception of music alone and that they are engaged in the discrimination of vowel-like sounds across the animal kingdom. Vowels' harmonic distinctions could contribute to semiotic distinctions.

Sound symbolism was found to occupy an important place in the pragmatics of speech, possibly constituting a cross-cultural universal (Svantesson 2017). One of the most stable associations of "high" [i] with smallness, and "low" [a] with largeness, verified for many languages (Diffloth 2006), finds a perfect match in a musical universal, where high pitches are perceived as "small" and low pitches – as "large" (Dolscheid et al. 2014). Both domains, music and phonetics, could be implementations of general cross-modality of pitch and visual size (Bien et al. 2012). JH follows suit (Nikolsky et al. 2020), so as sung vowels in lyrics (Russo et al. 2006). If Sapir (1929) and Newman (1933) were right in tying vocal symbolism to articulatory and kinesthetic visual/motoric experience, then *JH provides a par excellence tool to explore expressive capacities of vowel height, perfectly visible to the listener and sensible to the player.*

Symbolic meaning of vowels often interferes with the lexic meaning of words, but excels in purely musical applications, which suggests the *musical, prelinguistic, origin of phonetic symbolism.* VH presents a further advance in the semantic exploration of vowel height by elevating it from an elementary syntactic level to a complexity akin to the composition of a melody. Melopoeic options include maintaining the same pitch level (monotony associated with reinforcement, emotional shock, deprivation or deep mediation), ascending (increasing tension) or descending (relaxation) within the same structural syntactic unit. Within the context of a purely symbolic, non-denotational, frame of reference for the JH music, committing to one of these melopoeic options fundamentally constitutes musical semiosis. The latter, however, can obtain a formative influence on verbal communication in a language that was formed *after* the formation of a JH tradition within the same ethnos:

- Musicologically speaking, VH directs the attention of speakers and listeners to 3 musical aspects of vocal expression: *pitch*, *rhythm* and *dynamics* – manifested, accordingly, in tracking "intrinsic fundamental tone" (Lehiste and Peterson 1961), "intrinsic duration" (Lisker 1974) and "intrinsic intensity" (Möbius 2003). All three are highly reflective of the speaker's affective state. To follow the paradigm of a vowel as a specific musical instrument (Ladefoged and Disner 2012, 32), VH can be regarded as *limiting the number of "musical instruments" in the "instrumentation" of some pre-existing "piece of music,"*[26] *where a word is treated like such "orchestrated" musical piece. "Instrumentation" here is called forth to increase the intelligibility of that word.* The very idea of such "instrumentation" may have come from JH musicking.

We have already noted the importance of harmonics for JH articulations. Not only the configuration of the most salient harmonics distinguishes one JH vowel from another (Nikolsky et al. 2020) – the entire "talking khomus" style is characterized by patterns of changes in harmonic structures (Nikolsky 2017). Preliminary results of a new study[27] strongly indicate that VH of the same vowel (e.g., Yakut "*ylyym*") on JH involves matching the **harmonic signature** of the trigger-vowel by the target-vowel – unlike the rendition of the very same VH word in Yakut speech (Exs. 15–16). This suggests that in languages of people whose music features JH as a principal musical instrument, height-based VH generally emulates that of JH, but cannot exactly reproduce the harmonic signature of the repeated vowel because of the consonant assimilation. On the other hand, JH consonants are largely filtered out, ranging from complete elimination to severe attenuation, thereby protecting vowels from consonants' influence.

**Audio Examples 15–16**

Ex. 15. **The VH word "ylyym" articulated on khomus and in speech.** The harmonic signature of the first syllable "y-" is nearly perfectly matched to that of the second syllable "lyym" in the JH rendition: the harmonics Nos.1–16 retain the same dynamic configuration (the crests and troughs). The speech version does not contain such harmonic matching. In contrary, the second syllable seems to be designed to contrast the first syllable – not only by the harmonic signature, but by the intonation in pitch. This difference remains strong across multiple performances by the same person, as well as performances of other people. Demonstration by Erkin Alekseyev, used by his permission. http://chirb.it/bgDfbO

[26]Such practice, indeed, takes place in Western instrumental music. Composers, such as Praetorius, Mattheson, Telemann, Czerny, Berlioz, Glinka, Rimsky-Korsakov, Widor, and Koechlin, wrote at length about the expressive capacities of the most popular musical instruments and their appropriateness or inappropriateness for a particular musical expression. In professional music education, the study of instrumentation has become a standard course, often differentiated from orchestration – since instrumentation applies to the choice of musical instruments in composing and arranging chamber music as opposed to orchestral music. Although non-Western music lacks such formal discipline of study, nevertheless similar concerns preoccupy creation and reproduction of music in many musical traditions with rich instrumentarium: e.g., modern (Davis 2004) or historic (Krispijn 2010), as well as their combination (Pacholczyk 1993).

[27]In 2017, the International Museum of Jew's Harp at Yakutsk organized a project of comprehensive study of JH articulations on the instruments from its collection, conducted by Aleksey Nikolsky, Varvara Dyakonova, Ivan, Eduard and Erkin Alekseyevs, the report of which is currently being prepared for publication.

Ex.16. **The non-VH word "syma" articulated on khomus and in speech.** Both syllables possess different harmonic signatures in both versions, designed to emphasize syllabic contrasts. Although the consonant "m" in "syma" is not rendered any clearer than "l" in "ylyym," the vowel "a" receives completely different treatments: the performer stresses the descending leap from the fifth degree to the second degree – whereas in the example above the performer emphasized the sameness of the repeated degree. Speech version, evidently, does not proceed by such "degrees," so phonetic contrasts of adjacent consonants cause vowels to contrast whether there is a repetition of the same vowel or not. Demonstration by Erkin Alekseyev, used by his permission. http://chirb.it/xF8wAI

Nivkh JH tradition likely supplied the sonic model for Nivkh height-based VH. It is hardly possible to play JH in fast tempo without "VH" – having two or more adjacent sounds produced on exactly the same palatal position. JH performance technique[28] makes a player well aware of whether one sound reproduces the position of the previous sound, varies it or contrasts it. Each option, accordingly, generates either "perfect" VH of the same vowel (e.g., Yakut *ylyym*), "imperfect" VH of two harmonically related vowels (Yakut "*syma*"), or opposition of vowels that marks the boundary between two syntactic units (Yakut "suruk-biçige"). Nivkh might have initially featured a clear-cut VH, similar to Yakut, eventually losing it while its peculiar consonant system was taking shape.

2. **Phonotactics**. The second trait of the Nivkh language, possibly influenced by JH, is its phonotactics that affords four syllable types: vowel (V), consonant-vowel (CV), vowel-consonant (VC) and consonant-vowel-consonant (CVC) in prevailingly monosyllabic words (Gruzdeva 1998). Disyllabic roots are fewer in Nivkh, and trisyllabic roots are found in loanwords only (Shiraishi 2006). Such phonotactic structures also characterize Nivkh JH musicking (Mamcheva 2012). Fast decay limits JH in regard to the number of syllables that can be articulated on a single activation of lamella to just 2 (3 in fast tempo). And inciting lamella, in effect, generates lexically meaningless VH polysyllabic strings – e.g., Nivkh "khavr-khavr-khaf-khaf-khaf" – breaking them into "words," whenever the vowel quality is changed, as in "*khai t'ok khai-nak-nak* (Sheikin 2002, 132). In the same vein, Mansi vocable-based phrases are made: e.g., "*Gav ri ke, gav ri ke, vas py-rish, vas py-rish*!", articulated on the JH (Tchernetsov 1987, 35). Such segmentation is quite similar to syllables and words of natural languages.

[28]The hand of a player marks "syllables" by hitting, touching, or pinching the JH lamella, or pulling the rope tied to it (basically, any form of exciting the lamella), whereas the tongue position in his/her mouth controls whether the harmony is changed or retained. The change of the height along with the action on lamella usually marks the structural syntactic boundary between two musical intonations, motifs or phrases. The change of the height alone generates something akin a diphthong or even a triphthong. Performers are aware of such segmentation and usually control how many changes in articulation they make on a single act of excitement of lamella.

Another JH-like phonotactic trait is that the Nivkh language limits word-initial clusters to no more than 2 consonants, and word-final clusters to no more than 3 consonants. It also disallows plosives at the second position in an initial cluster (Gruzdeva 1998). Kreinovich (1937, 26) considered the rules for succession of consonant phonemes the most characteristic trait of Nivkh. Articulation of consonants is very difficult on JH. The most common syllables feature V, CV, and more rarely, VC and CVC (Zagretdinov 1997). Marginalization of consonants and dominance of vowels is immediately obvious in audio examples of "talking JH" (Nikolsky 2017) applicable to most traditions of Russian Far East.[29] Moreover, JH requires a performer to match consonant articulation to that of the adjacent vowels (but not the other way around!) – especially for fast music, where the movement of a performer's tongue within a syllable has to be minimized in order for the JH's "talking" to remain intelligible. This technical limitation is most obvious in the inter-syllabic transitions of the VC-CV(C) type performed on a single stroke, which causes consonants to simultaneously contrast each other while matching the preceding and the successive vowels. Peculiarity of this requirement could be responsible for the emergence of the *obstruent system of the initial consonant alternations*, determined primarily by the phonetic nature of the final segment of the preceding word – some of which have clear historic progression of development (Comrie 1981, 267–8). The practice of JH musicking must have originally put in place such phonological rules as the alternation of dental consonants after a vowel, which later acquired the syntactic significance.

3. **Pause insertions**. In Nivkh morphonology, consonant mutation is sensitive to pause insertions, which indicates the importance of temporal adjacency (Shiraishi 2006). Insertion of a pause prevents consonant mutation both for spirantization and hardening, as noticed by Kreinovich (1937, 15). Such interdependence might look strange for a language, but it is totally normative for music, where caesura generates discontinuity between musical segments, working similarly to a period between sentences in verbal syntax. For JH music, the cancellation effect of caesura is even greater, since caesura usually involves taking a breath, thereby interrupting whatever processes that were active prior to the caesura.
4. **Vowel quantity**. The fourth "JH-related" trait is the phonematic opposition of vowels by their quantity due to lengthening of a vowel in the beginning or the middle of a word before [s], [z] and [r], and more rarely, [v] and [f], especially pronounced in the main, Amuric, dialect of the Nivkh (Panfilov 1962, 1:11–12).

[29]The example of the lexically meaningful text rendered in the "talking khomus" style on JH is (in IPA): */o-huo-xaj-dɯːr o-huo-xaj e-hie-xej-diːr e-hie-xej ej-der e-re do-ɣot-tor oj-dor, oj-dor o-juo-ɣuŋ køt-tør-køt-tør kø-tyø-ɣyŋ!/* (Nikolsky 2017). The opening 13 syllables are all meaningless vocables, traditional for singing during the Yakut round-dancing and for JH playing. The following syllables (Nos.14–34) can be translated into English as: "Come on, friends! Let's jump, we'll jump, jump for a while, fly up, fly up, keep flying!" The clip can be heard here: https://chirb.it/0g6pFO. As it is evident from listening to it, most of the consonant phonemes come out as "blurred" or even "swallowed," despite the outstanding technical skills of the performer, an internationally renowned Yakut khomusist, Ivan Alekseyev.

Doubling the quantity of a vowel is rather standard in indigenous JH traditions of Russian Far East and Siberia. There, striking the JH usually occurs with metric regularity by periodic cycles where the last strike is doubled in its rhythmic value. Habituation to the same timing increment is likely to affect articulation changes within the same strike, too - as long as the decaying portion remains audible and fits in the metric pulse. The custom of estimating articulatory changes in relation to the pulse of striking can easily be transferred onto the speech medium, following the model of poetic verses through epic folklore (e.g., Nivkh *nastur* or Yakut *olonkho*). The evidence for such intuitive division of a beat pulse into 2 or 3 equal parts in Nivkh traditional music (Ex.17) is abundant in the practice of employing a repertory of rhythmic formulas, encoded as brief verbal phrases, most explicit in the tradition of playing the musical log (*tiatia chkhash*) at annual bear festivals – the most important celebration in the Nivkh calendar (Mamcheva 2012, 112–128).

**Example 17 Binary and ternary "divisions" in Nivkh traditional JH music.** This is an example of rhythmic patterning of JH articulations within the time span of the same hand stroke (Mamcheva 2012, 296). It generates rhythmic figures referenced to the regular progression of beats. The clip is selected from Mamcheva's fieldwork, used by her permission. http://chirb.it/ABn63s

All instrumental motifs of Nivkh music, no matter which musical instrument is used, tend to strictly follow the principle of correspondence between a musical sound and a verbal syllable in a dedicated formula employed for the convenience of memorization of a suitable rhythm (113). Whenever a beat is subdivided into two sounds, a corresponding syllable is broken into phonemes, and a consonant phoneme receives an additional vocable vowel, usually [a], e.g., "khurk" becomes "khur-ka" (114). The dominant metric organization in Nivkh rhythmic formulas is 8-beat heptasyllabic, where the seventh syllable receives double quantity (115). Mamcheva explains metric strictness by the long-standing tradition of reciting ritual texts in culturally important rituals.

5. **Independence of vowel lengthening from stress**. Nivkh vowels are lengthened due to the sentence prosody, as a compensation for the deletion of postvocalic fricatives in fast speech, or in songs' lyrics to assign words to the longer notes. Stress, on the other hand, is generally restricted to the *initial* syllable of a word (Shiraishi 2006). Although stress can potentially fall on any syllable, its most common position is on the first syllable (Panfilov 1962, 22) – especially for the Amur (Kreinovich 1979) and West-Sakhalin dialects (Shiraishi 2006). **Autonomy of durational and dynamic types of stressing** is typical for music in general. And JH requires even greater autonomy, since dynamic stress is mostly controlled by hand action, while durational stress to a large extent is controlled by respiration. Hand action tends to be regular, generating periodic stresses that fall on the opening of each musical phrase and motif. Patterns of breathing, on the other hand, are more volatile, changed in response to emotional state and chosen registration, causing shortening or elongation of certain sounds. This is consistent

with the different functionality of the Nivkh vowel lengthening that serves prosodic purposes – stressing that primarily supports the discreteness of words, most obvious in morphonology of compound words (Kreinovich 1979). JH might have been the prototype for such specialization that later became obscured - as stress in Nivkh language has elaborated grammatic dependencies (Mattissen 2003, 85–91).

6. **The distinctive use of two tones**. High and low tones oppose each other in Nivkh monosyllables and polysyllables (Gruzdeva 1998), which might also have originated in JH musicking. JH articulations manifest clear pitch distinctions based primarily on the vowel height, with additional influence of labialization (Nikolsky 2017) – to the extent of forming "articulation scales" (Ogotoyev 1988). *Pitch-grouping*, along with *stress-* and *pause-grouping* plays an important role in phonological and morphonological organization of the Nivkh language – each featuring *hierarchic* internal structure of its own (Mattissen 2003, 120). Their formative power may have been revealed for Nivkh speakers by the JH musicking, where pitch, dynamics and articulation style each constitute an autonomous aspect of expression that might support or compensate for another aspect. The general tendency of ascending pitch to be supported by a dynamic increase in music finds an analog in the Nivkh language. Its vowels interact in stress and tone, where the former usually supports the latter (Panfilov 1962, 1:21–2). The prototype for such interaction can be found in JH sound production, where striking JH produces the short spike of high frequency spectral content that quickly fades out, leaving the low frequency bands to ring (Adkins 1974). Perceptually, the sound produced by striking appears as the initial syllable in multi-syllabic articulations.

Based on the totality of the presented information, it seems reasonable to suggest that Nivkhs' ancestors were using grass JHs during the early Holocene and either passed this technology to the neighboring peoples of Primorye or themselves advanced to twig or wood-chip JH technology. Very much like the Altaians, the Amur and Sakhalin Nivkhs also observe the ancestor tree cult (Taksami 1977, 66).[30] In the past, every Nivkh settlement worshipped an ancestor tree of its founder. In fact, Nivkhs were even more protective of trees than Altaians, equating the cutting of any tree with murder unless the woodcutter left the *inau* (a ritual stick with splinters) on the stump. Another important celebration that symbolized the connection of the taiga with the heavenly world was the autumnal feeding of the Taiga's Master immediately after the first freezing of the closest river, which also required connecting a local ancestral tree with the transported larch tree (Ostrovsky 1997, 223). This rite could have dated back to the times of the Last Glacial Maximum. And the special cosmological importance of the larch complements the

[30]Nivkhs believe that their entire ethnos descended from the larch tree, and each of the Nivkh kins in addition venerates its own ancestral tree. During the most important annual bear festival, Nivkhs transport a dried larch tree to their ancestral tree and conjoin them to provide access for a patronizing "heaven man" to come down to Earth and protect his people (99).

fact that the larch is still commonly used by Nivkhs to make JH *khar-kongon* (Mamcheva 2005). The viability of the hypothesis of a talismanic origin of JH is supported by the long-lived tradition of the Nivkhs treating their musical instruments like live creatures: e.g., a musical log was carved to mark a symbolic "head" with roughly sketched eyes, mouth and ears, and its mouth was "fed" with lingonberry juice (Mamcheva 2008). Sounds emitted by objects were believed to have elemental power: e.g., a buzzing sound could raise wind (Kreinovich 1927). Nivkhs routinely use sound along with fire and smoke as a means to protect humans from misfortune and trouble – the custom of hanging little wooden blocks (*mil-kaunyr*) above the child's cradle, knocking each other from lulling the cradle, still survived until the twentieth century (Kreinovich 1973, 354). Talismanic JHs likely executed a similar "evil-repelling" function. All of this data completely agrees with what is known from the much better researched JH tradition of the Yakuts and Tuvans. This includes the practice of using musical instruments to conceal a natural human voice, reported by Nivkh informants and explicitly stated in epic mythology (Mamcheva 2008).

Yet another piece of evidence that ties Nivkhs and symbolic meaning of vocal articulations, employed for animistic religious practices, is the tradition of depicting so-called "**singing masks**"[31] (Okladnikov 1968, 1971; Okladnikov and Zaporozhskaya 1972; Devlet 1976, 1980; Okladnikov and Mazin 1976, 1979; Leontyev 1978; Brodianskii 1978; Arkhipov 1989; Kochmar 1994; Devlet and Devlet 2011) – petroglyphs on rocks adjacent to the river, which depict the isolated human face, stylized yet individually expressive, so that none of the uncovered images (about few hundreds) reproduce each other. Expressions are defined by detailed configuration of mouth, nose and eyes, representing a vocal articulation, charged with a certain affective state (Fig. 8.5) – in a stark contrast to the visual art of neighboring Siberian cultures, dedicated to animals and usually presenting them in groups (Okladnikov and Mazin 1979, 86). The oldest of these "singing" images (3000 BC) come from the Amur basin area (Okladnikov 1971, 83–89). Sheikin (2002, 259) was the first to compare all the available "mask" images and notice that they represent distinct articulations. Sheikin concluded that "singing masks" must have been created to perpetuate the "correct" way of articulating an ideologically important word. This could be the name of a mythological ancestor of a particular kin or ethnos – in line with the Nivkh, Chukchi, and Yukaghir traditions of always naming a presumed author of a traditional song. Masks that depict a patronizing ancestral spirit (*khambaba*) are still manufactured by Udege shamans for *kamlanie* –

[31]Russian archeologists refer to these singing masks as "*lichina*" – literally, "larva", and figuratively, "mask" in a negative sense of feigning one's look – but etymologically derived from the positive word "*lik*" (iconographic image of a holy figure). Such pictorial "masks" constitute something like a peculiar genre of Bronze Age fine art, compositionally analogous to "head portrait," but depicting a deity, spirit or ancestor rather than a concrete human being (Khlobystin 1987). Some of such images seem to represent a genuine mask by depicting, under the chin, a handle for holding the mask or laces for tying it to one's head (Devlet 1976, 6).

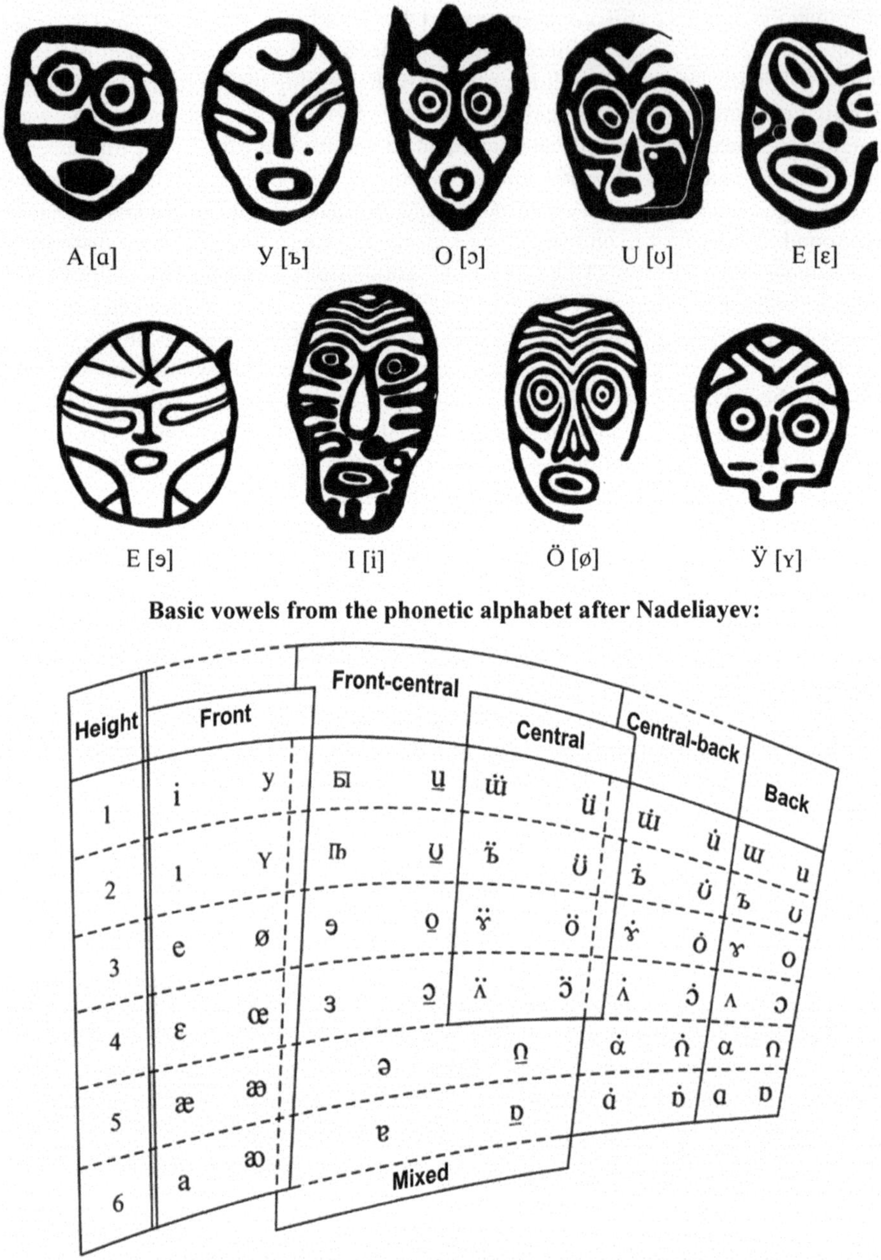


**Fig. 8.5** *Nine "singing masks" from Sikachi-Alyan, Amur, ≈3000 BC, which represent vocal articulations important to the ancestor cult of Paleo-Amuric peoples* (Sheikin 2002, 259). The images were discovered, dated and described by Okladnikov (Okladnikov 1971, 83–89). Their anthropomorphism and focus on facial expression, distinct for each of the images, strongly contrast the traditional animal-based visual art of the surrounding Siberian peoples, modern as well as

shamans use different masks for calling different spirits (Zabolotskaya 2011).[32] Alternatively, petroglyph masks could hint a "magic" password required to evoke an important spirit-master of a local landmark (Okladnikov 1971; Gemuyev and Sagalayev 1986; Alekseyev 1992; Kliashtornyi and Savinov 2005; Surazakov 2014) for personal protection or healing, as it is still practiced today (Voldina 2017). Or, the image could represent a kin's military call. Such calls (*uran)* were still used in the nineteenth century by Turkic Siberian tribes (e.g., Yakut "*urui*!") and were believed to have supernatural power (Sagalayev and Oktiabr'skaya 1990, 21). If such word consisted of more than one syllable, vowel harmony, common for most of local languages, would have made both syllables similar, representable by a single image.

The "singing mask" images are reproduced from Sheikin's book "History of music culture of Siberian ethnicities" and are used with the permission of Sheikin. The phonetic table was kindly provided by I. Seliutina and her research group, and used with their permission.

Okladnikov stresses that Amuric masks represent an uninterrupted tradition that ran for millennia and involved not only shamans, but virtually all local indigenous populations. Thus, the modern Nanai ritual sculptures (*seon*) of a hunting patron (Girki-Aiami) closely resemble the Neolithic "singing masks" of the Sakachi-Alian and are used for sacrifice and praying (1971, 104).[33] Even more common is the use of such masks and dolls in funeral rites, where a mask is supposed to carry the soul of

**Fig. 8.5** (continued) archaeological (Andreyeva 1987). Such "singing masks" present an uninterrupted tradition that still survives in sculpture, ceramics, burial and shamanic rites, as well as folklore, of modern Nanai population of the middle Amur basin. Similar in style images were uncovered only in few isolated sites in Tuva, Yakutia and Kamchatka, suggesting the substantial expansion of a Manchurian Neolithic culture during the Bronze-Iron Ages. Each of the images was identified as to which vowel it might represent by I. Seliutina, T. Ryzhykova, N. Urtegeshev and A. Dobrinina from the Laboratory of Experimental Phonetic Research at the Institute of Philology of the Russian Academy of Science, Siberian Department. Following the methodology of the founder of this institute, Nadeliayev (1960), this lab has collected an extensive database of articulations of basic phonemes of Siberian languages (Seliutina et al. 2012). Though the available images permit only rough estimation of the gradations in labialization with a gist of palatability, Nadeliayev's classification of labialization (Nadeliayev 1980) enables identification of vowels based on the ratio of mouth width to height. Each vowel is presented in the phonetic alphabet developed after Nadeliayev (Seliutina et al. 2012)

[32]Wooden ritual facial masks (*kambaba)* have been used by Nanai and Udegei shamans to look for the soul of a sick client or to make sure that the dead soul indeed reaches the underworld (*buni*). If a mask happened to fall off a shaman's face, his own soul was believed to perish. Nearly identical masks, *mugde*, were made by Nanais from the birch bark to contain the soul of the deceased (Okladnikov 1971, 106).

[33]Nanai hunters manufacture such *seons* before hunting and carry them along in a special container. If a hunt was unsuccessful, they hung a *seon* on a tree, pray to the patron, and make a sacrifice. If the

a newly deceased, securing reincarnation of future generations within a seok, thereby ensuring its procreation (Lipskii 1956).[34] Such funeral rituals have been widely spread from Yenisei to Amur, at least for the last 200 years, indicating earlier ties between Ugric and Manchu-Tungusic ideologies. The use of expressive shamanic masks to communicate with the dead ancestral souls by Buryat, Shor, Kumandin, Udege, Evenk, Mansi, Khanty, and Nenets shamans still existed in the early twentieth century (Devlet and Devlet 2011, 310–12). Archaeological finds of such funeral masks, dated 200 BC–400 AD, support the conclusion of Russian archaeologists that the Northern Far East became culturally isolated as early as the Bronze Age, preserving the earlier Neolithic traits until the Russian colonization in the seventeenth century (Andreyeva 1987).

The idea of "fixing" the articulation of a sacred word is not that different from the existing Ugric practice of reserving a particular melody for evoking a patronizing spirit (Tchernetsov 1987; Gippius 1988; Sheikin 1990). Similar "personalized" melodies are linked to ritual masks in Southern Siberia – Tuva, Buryatia and further south, at Tibet, Nepal and Bhutan – within the Lamaistic tradition of *tsam*, a theatric ritual dance (Devlet and Devlet 2011, 335–37). A dancer wears a mask that displays a specific emotional expression and along with a specific melody distinguishes one *tsam* protagonist from another. Such theatric representation may have originated in ancient pre-Lamaistic animistic customs that survived in hard-to-access mountain areas.

The territory of the Neolithic "singing masks" tradition stretched all the way from the continental coast of the Sea of Japan to Khakassia (Kyzlasov 1986, 185–7). Okladnikov concludes that the orbit of this tradition's influence likely spread even further south and southeast. Okladnikov supported this in his demonstration of how the ancient Amuric petroglyphs had set standards for the ornamental designs of traditional visual art of the Nivkh, Nanai, Ulchi, and even, to some extent, of the

prayer was "heard," a Nanai hunter draws on a patch of fish skin an icon of the *seon* Doonte, next to the sun, surrounded by 9 snakes, 9 birds, 9 horses, and 9 human figures (Lopatin 1922, 17:228).

[34]Lipskii provides perhaps the most detailed explanation of the rituals related to funeral masks and dolls. His rich personal ethnographic and archaeological experience allowed him to notice important details in burials and relate them to contemporary indigenous practices and beliefs. Masks and dolls were manufactured by the closest relative of the deceased immediately after their death, replicating their unique physical traits (scars, tattoos, etc.) Such relative dressed the doll with the clothes of the deceased person and put the mask on the doll. After 3–9 days of taking care of the doll (feeding, cleaning, and putting it to sleep), that relative broke or burned the mask together with the doll. This treatment was supposed to block the dead soul from the living and secure its transportation to the underworld (*buni*), where it had to stay in order to be available for reincarnation into future newborn generations of the dead soul's *seok*. That is why "seok" simultaneously means "corpse," "bones," and "kin" – dead bones of a corpse give life to the future posterity of the same *seok*. Here, the individual resemblance of the mask and the doll were intended to make sure that the soul would attain its proprietary (and not some foreign) *seok* settlement in the underworld, being successfully recognized and accepted by that soul's dead ancestors. Concern for correctness of this procedure reflected the incentive to keep the possessions of one's kin and secure its procreation. Within this framework, a picture of a particular vocal articulation could serve like a password ("sesame") to get access to a specific kin, its territory, and its ancestor/patron's spirit.

Ainu, supplying the latter with geometric elements, abstracted from their initial anthropomorphic identity and wrapped into zoomorphic ornaments whose resemblance to Amuric petroglyphs can be recognized only when compared side by side (Okladnikov 1971, 101–5). Okladnikov also disclosed cultural patterns common for Amuric and Northeast Chinese as well as Japanese and Oceanic prehistoric art (Okladnikov 1955). Taksami further explored the Ainu (Taksami and Kosarev 1990) and Austronesian (Taksami 2009) connections of ancient Russian Far Eastern cultures. This line of research could explain why JH retains its importance in indigenous cultures of Southeast Asia and Oceania. Millennia of cultivation of "magic" articulations within the ideological framework of ancestor cult would have cemented a stable tradition of timbre-based music.

## 8.9 Jew's Harp and the Origin of Vowel Harmony

The local Paleo-Asiatic population of the Amur area may have selected specific vowel articulations as an attribute of worship within a Neolithic cult that subsequently spread via migration and cultural contacts over a vast territory from Upper Yenisei at the west, the Altai/Sayan mountains and the Manchurian Plain at the south, and the Pacific coastline at the east (Fig. 8.2). The remnant of this culture must be the Nivkh language and cultural tradition that found refuge in Sakhalin. The JH tradition, formed within the ancestor and tree cults as part of pan-Siberian/Far Eastern religion, must have influenced language development in the process of ethnogenesis across this entire region. Proto-Nivkh probably captured the traits of TO promoted by the earliest JH and proto-JH constructions. One of its most likely influences, vowel harmony, must have emerged in some embryonal state in the ancestor of the proto-Nivkh and subsequently became lost due to language isolation and modernization. However, VH has evolved into the principal morphonological tool for languages that formed *after* the Nivkh ancestors were pushed away from the continent by the newer wave of migration at the onset of the Metal Age. The strongest manifestation of this function is found in Turkic languages (Gadzhiyeva 1997).

Stressing a specific timbral quality by means of VH is fundamental for **agglutination** – the derivation of words by stringing morphemes together without phonetic changes, reserving a single function for each "glued" morpheme. In all Turkic languages, the morphemic seams at the boundaries of neighboring morphs are stressed primarily by VH. According to Tcherkassky (1965), the entire system of phonological oppositions in Turkic languages must have emerged in no other way but as a paradigmatic base supporting the VH. In languages that engage VH by two different principles (e.g., palatability and labialization) the VH alone is responsible for delimiting about 75% of words in the flow of speech – which is achieved with great efficacy, so that a morpheme's boundaries are recognized by listeners practically without any mistakes (Melnikov 1966a).[35]

The Yakut language is considered to possess an exemplary VH system, representative of the VH in proto-Turkic (Levitskaya 1984). The Yakut language harmonizes vowels in 4 aspects: short/long, front/back, narrow/wide, labial/non-labial - involving also the consonant harmony of soft palatalized or hard non-palatalized consonants, depending on their proximity to the front or back vowel (Antonov 1997). Therefore, palatal and labial VH acquire systemic importance. In Yakut phonology, speech progressions are characterized by regular alternation of consonants and vowels, with very limited possibility of consonants to follow each other. The main phonological function of consonants is terminating the vowel, so that the syllable's boundaries are always marked by the succession of consonants (Dmitriyeva 2002). The delimiting function of VH makes accents morphonologically unimportant (Reformatskii and Vinogradov 1969). Accents always fall on the vowel in the last syllable of a word (adding an affix simply moves the accent) and does not affect the quality of non-accented vowels (Antonov 1997). Such a strict rule makes the function of chaining morphemes the prerogative of vowels.

Yakut VH closely resembles TO of music. Consistent short/long opposition of all vowels usually retains binary proportion akin to binary division that characterizes musical rhythm and meter (Fraisse 1982). The fixation of accents establishes a metric framework for it. VH simultaneously defines words syntagmatically by the "horizontal" interruption of the prolonged timbral quality and paradigmatically by defining a word as the continuity of "vertical"[36] morphonological predictability, according to the selected harmonizing feature (Vinogradov 1972a). As a result, a word as a morphological unit is organized by VH, and as a syntactic unit – by a fixed accent (Vinogradov 1970). This bi-axial definition, common for not only Yakut, but Turkic languages in general, where linear changes in continuity of the timbral quality are constantly quantized into known "classes" of harmonic relations by adopting a specific "harmonic quantor" (Vinogradov 1972b), *is functionally equivalent to*

[35]For the study organized by the International Museum of Jew's Harp at Yakutsk in 2017, mentioned above, it was very difficult even to find a set of native Yakut words that would be free from VH for comparative acoustic analysis. Practically all common Yakut words contain VH, and even those borrowed from non-VH languages (e.g., Russian) become "harmonized" in the pronunciation by native Yakut speakers. Yet another phenomenon, noticed while carrying out this project, is that native Yakut JH players intone the harmonized vowels differently than JH players for whom Yakut is a second language (Nikolsky 2017). This might constitute the "melodic" counterpart of the "harmonic" process triggered by habituation to VH.

[36]Vinogradov's use of the terms "vertical" and "horizontal" has nothing to do with the linguistic conventions of referring to "height harmony" and "cross-height," or "horizontal harmony." Vinogradov refers not to the physical parameters of vocal articulation but to the pattern of variation in speech production: *synchronic* for phonetic variation (hence, "horizontal" in a sense of "successive" adjustments, following the common convention of representing time by the horizontal axis) and diachronic for phonological variation ("vertical" in a sense of categorical rather than sequential adjustments).

*rhythmo-metric organization of melody*. In music, ongoing changes in rhythmic values are also quantized into metric classes within the framework of the composition's "metric grid" with the help of selecting a regular temporal increment for a beat (Boenn 2018). Moreover, the resultant metric pulse is more often than not marked by the "harmonic pulse" (Swain 2002). Similarities go even further, if to take into consideration the interaction of VH with the fixed accent: after all, musical metric stress and linguistic dynamic accent share the same domain of amplitude that interacts with the temporal domain of rhythm and meter.[37]

Besides temporal and dynamic regulation, VH also regulates pitch levels: back vowels tend to sound higher in pitch than front vowels (Möbius 2003), and labial vowels - lower than non-labial (Zsiga 2013). All in all, the production of *Yakut speech is not that different from melopoeia*. Each vowel receives its distinct characterization by *rhythm, meter, pitch, dynamics, and timbre*. All of these parameters are consciously tweaked by a JH player – but *not by a singer*. As was already discussed, singing tends to equalize vowels in their timbral qualities, adhering to a single ideal of euphony. JH music is the only form of music that systemically cultivates multiple articulatory oppositions:

1. High/low
2. Front/back
3. Labial/non-labial (all three are engaged simultaneously in a "classic" JH technique of restricting articulations to oral cavity alone)
4. Pharyngealized/non-pharyngealized (used when a chest cavity is added to sound production – very common for Siberian traditions)
5. Nasal/non-nasal opposition (Mazepus 1989).

Remarkably, the first four oppositions find their counterpart in VH of Turkic languages (Seliutina 2017), and this is hardly an accident. The vocalism of the initial syllable (usually a word's root) is defined in the same *3-dimensional system* (Vinogradov 1972a) as in "classic" JH playing (oppositions Nos.1–3 from the list above).

There are two principal acoustic traits that distinguish the Turkic vocal system from that of JH music: monotone FF and scarcity of consonants. However, they hardly come into play in JH semiosis. JH monotony causes quick habituation – one's ear simply ignores it, focusing on the phonological changes in the upper part of JH spectrum. And scarcity of consonants does not make JH music any less intelligible, because hand action takes over the delimiting function. Each excitation of the lamella marks a "JH syllable," whereas executing two vowels on one excitation

[37]The definition of a word's boundary by the regularity of changes in the opposing timbral qualities in conjunction with the fixed stress can be thought of as an act of listening to a melody accompanied by a metronome with a ringing tone set to kick in on a particular click with certain periodicity – in addition to setting a particular click to be louder than the others and having an additional pulse set by changes of harmonies within the entirety of music. The resultant periodicities, despite their complexity and diversity, usually exhibit pronounced metric (i.e., regular periodic) organization.

produces a "JH diphthong." An additional source of generating "syllables" on the same vowel (a-a-a) is provided by alternations of inspiration and expiration (Ex. 18).

**Audio Example 18 Tyynyy araastara** (breathing devices). This is the demonstration of the traditional Yakut playing technique of generating rhythmic patterns and timbral recoloring of the very same articulation on the same hand stroke – using changes in inspiration and expiration. Ivan Alekseyev, used by his permission. http://chirb.it/8B011O

Hand action operates on two levels: "syllabic" in JH "words" and "lexic" in JH "musical phrases." The latter occurs as a result of periodic stress that falls regularly on a certain "hand beat." For both Nivkh (Mamcheva 2012) and Yakut (Alekseyev 1976) traditions this is usually the seventh beat (Ex.19).

**Audio Example 19 Ohuokay**. This is the demonstration of a "talking khomus" style applied to the verse of a popular Yakut round-dance. It uses a clear heptasyllabic metric pattern, fit into 8 beats, where the seventh syllable receives a double value. http://chirb.it/0g6pFO

For "talking khomus," VH delineates words even clearer than for speech because JH clearly stresses whether the articulatory degree is repeated or changed, and in the latter case, whether it changed by a leap or by a step (Ex. 11):

- In essence, *striking of the JH is as "monotonous" as VH*: both of them limit the timbral variety of sounds in sake of highlighting the seams of syntactic units (Ex. 13). At the end, both regulate the distribution of timbral tension and resolution – which is one of the keynotes of musical TO.

The JH roots of the palatal/labial VH might explain its origins better than the existing linguistic explanations. The first theory was presented by Trubetzkoy – he saw VH as the consequence of the symmetric arrangement of the vowel repertory of a language (Trubetzkoy 1939, 104), where a harmonic rule emerged to make the word-initial syllable "strong" by assigning to it the *delimitative* (phonemic) function, while subsequent "weak" syllables were neutralized and executed the *distinctive* (phonetic) function (Trubetzkoy 2001, 28). An alternative explanation by Tcherkassky emphasized not phonemic but *lexic* categorization in the emergence of VH. In his opinion, VH emerged from the phonematic system with very few vowels and few possible oppositions, forming strong and weak vowels, which grew in variety as polysyllabic lexical genesis caused consolidation of morphemes, generating the *syncretic* obligatory (*non-functional*) multi-factorial VH and agglutination that subsequently caused reduction of VH by making it *functional* (Tcherkassky 1965). This theory was challenged by Melnikov. According to him, VH was brought to life not by *syncretic* but by *selective* assimilation of vowels, based on separation of morphological and semantic functions, where front/back opposition generated affixation, whereas wide/narrow opposition generated differentiation by meaning (Melnikov 1966b). This view interprets VH as an aid in parsing the flow of speech by the *least acoustically reliable aspect*, which for Turkic languages is palatality. Shcherbak (1970) further emphasized the *lexic* roots of VH - instrumental in redrawing prosodic seams, when *amorphous* morphological structures started changing into *agglutinative* ones, causing a large number of words to become functional words. In this model, shared by the majority of Russian specialists, VH initially was designed to delineate *syllables*, achieved by palatal harmony, but after proliferation of polysyllabic words, VH acquired the function of delineating *words*, forming labial harmony. Here, the rise of the initial palatal VH appears as a "random mutation" that spontaneously acquired functional use.

Lewis provided an alternative explanation, seeing VH as a method of minimizing articulatory action of the vocal apparatus (Lewis 1967, 15–16). This seems to be an oversimplification, since VH occurs in relatively few languages and often remains limited, blocked by various factors. More convincing is tying VH not to *motoric* but to *perceptual* ease in recognizing familiar words (Suomi 1983). Indeed, VH works like an error-correction tool in detection of "weak" vowels in the non-initial syllables under the condition of a fixed word stress. Maintaining the pitch level of F2 of the most salient initial "strong" syllable throughout all subsequent "weak" vowels in a word simplifies real-time identification of syllables and words by directing a listener's attention to the pitch aspect alone. Such mechanism would promote harmonization only for those vowels whose timbral quality is perceptually more salient (Kaun 2004).

Viewing VH as a perceptually driven error-correction aid prompted Ohala's attempt to facet a universal theory of VH that earned acclaim amongst Western specialists (Ohala 1994). Like Tcherkassky, Ohala considers VH the "fossilized" result of purely phonetic and non-distinctive assimilations between neighboring vowels. But unlike Tcherkassky, Ohala considers such assimilation to be present in most languages, equating VH with **coarticulation** (Öhman 1966). And here Ohala substantially differs from Russian researchers. He holds that it is the *paradigmatic* features of vowels (height, frontness, rounding, voice quality) that lead to VH – and *not* the *syntagmatic* features (length, diphthongization), which are "dynamic" rather than fixed. In the Russian tradition, *both* paradigmatic and syntagmatic features are considered crucial for VH, and the latter fundamentally differs from assimilation. According to Vinogradov, vocalism of the initial syllable in a word (root) is defined within a *3D system*, while vocalism of non-initial syllables (affixes) is defined by a *single dimension* – nevertheless, both subsystems interact, forming systemic relations, which distinguishes VH from other forms of assimilation (Vinogradov 1972a). VH's essence is *the selection of a single specific "prosodic quantor," capable of distinguishing vowels along the known phonemic oppositions, and being present in both root and affixes*. Such a quantor then is applied to co-variation of syllables as well as words, working on both levels, lexic (macro-) and morphemic (micro-). Within a word, a prosodic quantor executes a constant cumulative function, whereas within a sentence, the same prosodic quantor supports an incrementally varied "delimiting" function.[38] Vinogradov stresses that it is this *hierarchic co-regulation of quantor's influence on syllables of a word and on words in a sentence* that distinguishes VH languages from non-VH languages. Yet another important difference is that VH generates new types of sonance to integrate morphological units while distinguishing one unit from another, which *attracts the attention* of a speaker and listener, and involves semantics for drawing word boundaries – in opposition to coarticulation that remains *unnoticed*, determined by purely motoric function of speech organs, free of semantic distinctions and not generating new sonance types (Zubkova 1973).

As it follows from this brief overview, linguistic theories explain the emergence of VH by the need to simplify production and perception of syllables and words in the flow of speech, secure effective recognition of vowel phonemes, support vowel oppositions, and generate new polysyllabic words by means of agglutination. There is a taste of **circularity** in these explanations: the *past* origin of VH is explained by demonstrating its linguistic necessity as it is known *today*. This approach seems to be fundamentally inadequate – obviously, those ancient people who came up with the idea of harmonizing vowels had neither any idea of vowels, phonemes, nor harmony. It is doubtful that they pursued the deliberate goal of constructing new polysyllabic words. Certainly, the discovery of VH could have been accidental, but the deeply systemic nature of VH makes this unlikely. The acoustic aspect of speech is far from being arbitrary – certain types of sounds and methods of their integration are selected based on the psycho-acoustic criteria of their pronunciation, recognition, memorization, and categorization, which all are achieved by a limited number of options, each cultivating its own method of associating sound with syntax (Zubkova 1986). Thus, in languages that distinguish morphemes, function words tend to use fewer syllables, phonemes and phonemic combinations compared to meaningful words, which use more.

Of all the researchers, the closest to avoiding circular reasoning seems to be Malmberg, who recognized VH's tendency to *uniform the phonetic arrangement of speech* in a manner similar to ordinary *instinctive utterances* such as sigh or groan (Malmberg 1962). The reduction of differentiating linguistic capacities in VH, along with its prelinguistic association with utterances, makes sense in the *situation of*

---

[38]By "incremental" function (Rus., *stupenchataya funktsiya* – literally, "degree-like function") Vinogradov means the function that stays constant for certain "intervals" (spans) of text but changes abruptly for other "intervals." He borrows the concept of the "degree" (gradation) of a harmonic quality from Greenberg (Greenberg 1963), defining it as a *set of vowels, each of which can interact with another vowel from the same set, including itself, within some grammatically definable unit* (usually a word, but possibly a morpheme). The presence of the opposing degrees in a vocal system of a language constitutes a necessary condition for VH. Vinogradov gives Evenk language as an example of "incrementally" designed VH due to the presence of harmonic degrees. However, he notes that traces of such design can be found in virtually any VH language.

*acquisition* of new verbal skills, be it ontogenetic or phylogenetic. After all, simplification enhances learning. It is no accident that VH constitutes the first morphonological strategy of a prelinguistic infant in conjoining syllables (papa, pipi), carried out through extensive word-making games (Jakobson 1968, 84–6). Jakobson specifically states that this peculiarity of early children speech is no different than palatal VH of Uralo-Altaic languages. Indeed, VH comes in handy not only in acquisition of primary language in early childhood, but also in secondary language acquisition for adults (Pycha et al. 2003). In light of this, VH is likely to have *prelinguistic roots* and follow the *affective* coding model of both music and musilanguage. This is in line with the finding that tamarins can acquire dependencies between adjacent and non-adjacent syllables in their statistical learning – in essence, learning to recognize something exceedingly similar to VH (Newport et al. 2004). Like humans, non-human primates emit sounds similar to human vowels to express their emotional state, but cannot utter anything like human consonants (Owren et al. 1997). Hence, the idea of "harmonizing" vowels by their pitch level – one of the most salient acoustic attributes of vocalizations, accessible to human infants and many primates - might have occurred already in musilanguage, during prehistoric times.

Palatal VH is estimated by Russian turkologists as a proto-linguistic feature that must have emerged much earlier than the breakup of proto-Turkic (Shcherbak 1970, 121). The earliest documents of Old Turkic are the seventh to eighth centuries Orkhon stone inscriptions in Mongolia, which most likely use even more ancient alphabet, and contain palatal and labial VH of [a], [ï] and [u] (Kononov 1980, 66–76). The timeframe for proto-Altaic to break into proto-Turkic and Tunguso-Mongolic ranges from the most recent estimate of 3000 BC (Robbeets 2015) to the oldest at 10,000 BC (Sunik 1978). VH was reconstructed not only in proto-Turkic (Gadzhiyeva 1997), but in proto-Altaic as well (Tenishev 1997). Then, the "singing faces" on Amuric and Tuvan petroglyphs and Amuric ceramics coincide with the formation of a hypothetical proto-Turkic language and its split from proto-Mongolic (Robbeets 2017), while the earliest sites of JHs excavated in China, Mongolia, Tuva, and Western Siberia (Table 8.3) precede the formation of Old Turkic (Starostin et al. 2003, 235–6), Mongolic[39] (Rybatzki 2003) and Tungusic[40] (Robbeets 2015, 16–18). Janhunen emphasizes that in Northeast Asia, early traditions of writing have provided enough historical and textual information to support relatively accurate dating even in absolute terms (except the Nivkh and Ainu languages) – and testify to a time

[39]The earliest date for the emergence of Mongolic, calculated with the help of the Automated Similarity Judgment Program, gives 267 BC (Holman et al. 2011). However, estimates based on the lexicon of documented and living languages of the Mongolic family suggest a time depth of no more than 1000 years (Rybatzki 2003).

[40]Robbeets' dating by the end of the Han period, based on the name changes in Ancient Chinese historiographies, generally agrees with Janhunen's dating by the local Iron Age (500 BC–500 AD) (Janhunen 2012, 8).

depth of 500–2500 years for modern "Altaic" languages (Janhunen 2010).[41] Therefore, the formation of languages within Turkic, Mongolic, and Tungusic families must have occurred *under the influence of fully-fledged local JH traditions*, whereas the formation of the vocal system of Transeurasian/Altaic probably went in parallel with the formation of the pan-northeastern JH tradition (Fig. 8.2). This scenario fits well into the totality of the available information about the organology, archaeology, and ethnomusicology of JH music, overviewed above.

Geographic clustering of Turkic, Mongolic and Tungusic languages (Robbeets and Bouckaert 2018, Fig. 8.1) quite closely matches the clustering of indigenous JH traditions (Fig. 8.6). In fact, *each of these families may have been influenced by its own JH indigenous tradition*:

1. Metallic JH is indisputably favored by Turkic peoples.
2. Mongolic peoples use JHs made of all materials, including bamboo.
3. Tungusic peoples resemble Mongolic peoples except that they rarely, if ever, use bamboo JHs.
4. Japonic peoples favor bamboo JHs.

These regional differences in prefering certain materials for JH-making might have contributed to the differences in VH of the geographically corresponding language families. Like Nivkh, the Japanese language seems to have possessed VH in the past (during the Nara period), however, unlike Nivkh, its VH was palatal (Hattori 1982). Old Korean also had VH – but most probably, tongue root (Whitman 2015). Modern Mongolic dialects include palatal as well as tongue root VH. This is interpreted in 2 ways: as a "Mongolic Vowel Shift" from a palatal contrast in Old Mongolian to a tongue root contrast (Svantesson 1985), or on the contrary, as a shift from the retracted tongue root VH to the palatal VH in Kalmyk/Oirat (S. Ko 2012). The latter scenario was proposed as prototypical for the entire historic development of VH in proto-Tungusic, proto-Mongolic, and proto-Korean languages, and possibly even in Turkic and Japanese languages (including Ainu), respectively, at the Western and Eastern borders of the Transeurasian family (S. Ko et al. 2014). Then, the retracted tongue root and/or height VH can be considered an "eastern" trait, while palatal VH – a "western" trait (Janhunen 2017). Such specialization finds an accompanying trait in indigenous JH music: Turkic speakers (West) prefer playing bow-shaped metallic constructions, whereas speakers of Japanese, Korean, and Ainu (East) generally prefer frame-shaped bamboo constructions (Fig. 8.6) – as do speakers of Chinese and Austronesian languages.

Each of the JH constructions promotes its own timbral model and performance technique (Table 8.4). JH constructions made of different materials promote

[41] Janhunen describes a simple yet effective method for dating languages and language families based on the calculation of the diagnostic differences (isoglosses) between geographically adjacent languages in question. The greater the diversification, the more ancient the origin, pointing to the linguistic homeland of the language family. A similar approach can be used to date and localize the JH traditions by cross-examining JH vowels, syllables, and words specific for affective expression within common genres (e.g., calming lullaby or affirmative hunting prayer) of indigenous JH music.

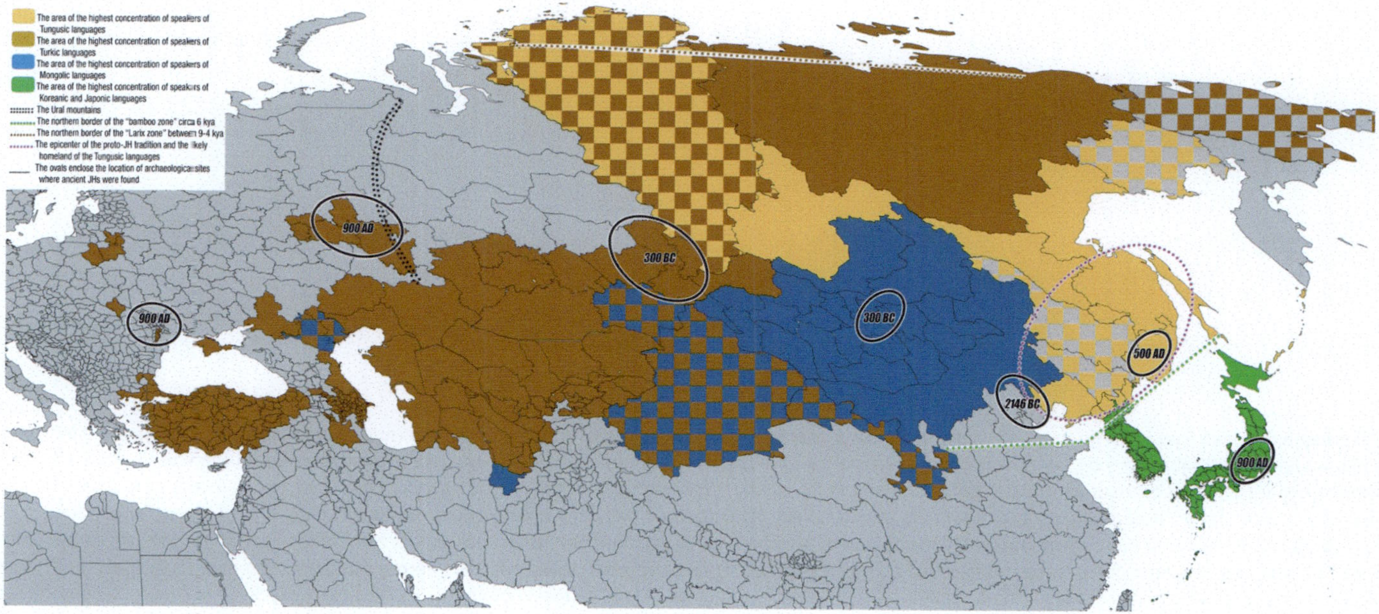


**Fig. 8.6** *Correlation of archaeological, ethnographic and paleo-ecological data on distribution of traditional indigenous JH music in Eurasia with geographic locations of the greatest language communities for constituent families of the Transeurasian (Altaic) languages*. This figure visualizes the geographic distribution of the Transeurasian languages (Robbeets and Bouckaert 2018) from Turkic (brown color), Tungusic (orange), Mongolic (blue) and Koreanic/Japonic language families (green). The areas of their admixture are indicated by the checker pattern that mixes the related colors. Gray color indicates non-Transeurasian languages. A number of cues are taken from Fig. 8.2 to corroborate the linguistic and JH demographics. Black ovals enclose the area where ancient JHs have been discovered by archaeologists. The date in an oval shows the earliest estimation of the archaeological finds in that area. The pink dotted oval surrounds the area where proto-JHs were most likely invented and put into use, and where the Tungusic languages have emerged (Janhunen 2012). The dotted brown line marks the northmost boundary of forestation (*Larix*) between 9 and 4 kya, and the dotted green line – the northmost boundary of the bamboo zone circa 6 kya – extended even further to north toward Japan to account for its warmer maritime climate. The overall distribution of language families matches

the pattern of clustering of the archaeological sites in such a way that each of the families receives at least one dedicated JH cluster. Widely dispersed along the path from northeast to southwest (from Chukotka to Bulgaria and Lithuania), patches of Turkic languages indicate the enormous spread of Turkic conquests following mastering of horse breeding and metallurgy by Turkic tribes (Chernykh 1992) – in agreement with their clear preference of metallic JHs (Sheikin 2002). There are 3 JH clusters associated with Turkic territories, the dating of which suggests the epicenter in Altai/Sayan mountains, followed by western, southwestern, northern and northeastern expansion. However, it is possible that the epicenter was somewhere in southern Yakutia, and the absence of archaeological data is explained by hostile Yakut climate and lack of exploration. As wide and patchy is the distribution of Mongolic languages – from Kalmykia and Afghanistan to Inner Mongolia – testifying of military expansion of Mongolic peoples during the late antiquity (Honeychurch 2015) and the Middle Ages (Botalov 2007). Unlike Turkic prevalence of uniformed bow-shaped metallic JHs, Mongolic JHs greatly vary in their making: they are commonly made of bamboo, wood, bone and metal (Chuluunbaatar 2016). Tungusic peoples are also dispersed over a sizeable area – inverting the distribution of Turkic peoples, from northwest to southeast (from the Kara Sea to the Okhotsk Sea). Tungusic peoples resemble Mongolic peoples in using diverse materials to make JHs. The only difference is that Tungusic JHs hardly ever use bamboo that does not grow in northern latitudes. Tungusic archaeological sites are located in Primorye and dated significantly later than Mongolic and Turkic sites. Taking in consideration its proximity to the Lower Xiajiadian JH in Northeast China and the above-mentioned studies on Tungusic ethnogenesis, it is reasonable to conclude that proto-Tungusic population cultivated proto-JHs and borrowed more advanced JH technologies from Mongolic neighbors at a later point. Finally, Japonic and Koreanic languages are confined to a narrow region on the eastern outskirt of Eurasia, where the earliest JH sites are contemporary to the Western Turkic ones. However, in this part of the world, the favorite material for JH-making is bamboo. Hence, both sides of Transeurasian region are marked by strong JH preferences: for bow-shaped iron/steel by Turkic speakers in the West and for frame-shaped bamboo by Japanese (and Ainu) speakers. At the center, Mongolic and Tungusic speakers make JHs from all materials, except bamboo by the Tungus

different musical textures[42]: JH drone can sound like a "tone," a divisible "double-note" (discrete vertical dyad) or an indivisible "chord," whereas the part of JH's spectrum above the drone can be filled by different sonic material - a discrete melodic "tone," a "double-note," a "chord" or a special effect of indefinite pitch (Nikolsky et al. 2020, Appendix). Metallic and bamboo JHs substantially differ in TO:

- Metallic bow-shaped constructions specialize in producing *homophonic* textures based on a single expressive melody,
- Bamboo and wooden frame constructions hardly ever use such textures and instead cultivate *polyphonic* textures with multiple simultaneous melodies.

Homophonic textures tend to promote greater "horizontal" melodic diversity (consecutive contrasts), while polyphonic – "vertical" (simultaneous contrasts). The former is likely to generate finer categorization of gradients - homophonic JH melodies usually engage considerably more modal degrees and wider ambitus than polyphonic melodies. "Western" palatal/labial VH and JH "articulatory scales" of the bow-shaped constructions that support purely frequency-based TO generally oppose "Eastern" height/tongue root VH and JH "articulatory scales" of the frame-shaped constructions that support only rough contrasts between a few marginal "articulatory degrees." These frame-shaped scales tend to combine poorly differentiated multiple variation parameters. Metallic JHs seem to exceed bamboo JHs in harmonic clarity, so that 8-degree articulation scales come out clearly *only* on metallic instruments. There is experimental evidence that indigenous JH players can only reliably recognize *gradations of vowels in height* by ear, struggling with the recognition of variations in palatality and labiality (Nikolsky et al. 2020). On the other hand, in cultural practice, reproduction of one player's sounds by another player is routinely used within the pan-Eurasian genre of romantic serenading duets (Haid 1999), which must include variations in palatality and labiality. The matter of what exactly JH players are capable of detecting in JH music by ear requires thorough experimental research.

[42]Typological differences in musical texture are recognized in *multi-part music* of Western classical tradition (Ratner 1980, 108–180) and non-Western traditional music (Agamennone 1996). But "single-part" (i.e., music conceptualized as a single melody) monodic and heterophonic forms of music also can be texturally analyzed (Swan 1943) – since textural changes, as a rule, accompany intra-sectional transitions in a musical form (Berry 1987, 184–300). In such cases, texture is comprised not by multiple parts but by different patterning of a single stream of sounds into such components as melody, passages, embellishments, repeated figurations of various melodic contours, rhythms, and registers (Skrebkov 1973, 136). In essence, any music that has "form" (i.e., contains distinct structural changes) also has texture. In this most general sense, texture should be defined as a *specific arrangement of the totality of musical sounds engaged in production of a musical work or its autonomous portion in frequency/time to comprise a particular type of presentation of music* (Frayonov 1981). The same applies to timbre-based music, including solo JH that can be broken in parts based on their musical functionality: e.g., melody, drone, accompaniment (Dobzhanskaya 1991).

Within this perspective, the Nivkh, Tuvan, and Yakut proto-JHs, characterized by the most indistinct spectral textures, appear to represent the earliest stage of JH's influence on VH of local languages. The Nivkh language could constitute a vestige of proto-Amuric that originated in central Manchuria (Janhunen 2010), possibly related to "proto-Manchurian." In this scenario, Tuvan and Telengit twigs might be the Western "proto-JH" satellites of proto-Altaic, while Nivkh grass VH – the Eastern satellite of proto-Manchurian in the first split of proto-Transeurasian. Invention of mukkuri-like bamboo instruments must have accompanied the formation of Tunguso-Mongolic at the northeastern outskirt of the "bamboo zone," while in the West, proto-Turkic branched off into Turkic languages as their westward spread met the eastward spread of metallurgy along the northern border of the Eurasian Steppe Belt (Great Steppe). Janhunen argues that the homeland of proto-Turkic lied somewhere near Mongolia, whereas proto-Tungusic originated in Manchuria (Janhunen 2010). This distinction corresponds to the availability of bamboo and trees in well-forested Manchuria in contrast to the overall scarcity of vegetation in Mongolia, which was likely to promote the development of bone and metal JH technologies. The breakup of proto-Tungusic, proto-Japonic/Koreanic and proto-Mongolic into language families probably was influenced by the extent of availability of bamboo. While spreading further north, Tungusic languages must have developed under the influence of wooden and bone JHs. Koreanic/Japonic languages most likely originated in the Korean peninsula (Janhunen 2010), where bamboo's abundance turned the mukkuri type into the reference model for JH construction. And Mongolic languages occupied an interim position – both regarding the complexity of their VH and the diversity of Mongolian JH materials and constructions.

It seems that the determining factor in the similarity of TO between the JH traditions of principal Transeurasian families is primarily the material of manufacture, followed by the geometric shape, e.g., all bamboo mukkuri-like instruments produce more or less similar textures that differ from all metallic bow-shaped instruments, no matter which part of Eurasia these instruments were manufactured in.[43]

Yet another language family closely related to the "Western" palatal VH and the metallic bow-shaped JH tradition is Uralic. If Janhunen is right in his estimation of the Northeast Asiatic descent of proto-Uralic (Janhunen 2009), the VH similarity of Uralic and Turkic languages should be attributed to their shared ecological base of

[43]The indigenous traditional music performed on *mukkuri* in Japan shares remarkable similarity of TO and spectral textures with bamboo-made Mongolian *khulsan khuur* and neighboring Tuvan *kuluzun* and *cheler komus*. These textures are analyzed in the Appendix ("Spectral texture and the influence different materials have on it on the example of Jaw Harp") to the article "The overlooked tradition of 'personal music' and its place in the evolution of music" (Nikolsky et al. 2020). Culturally, Tuva has been under strong Mongolian influence (until 1912 Tuva was administered as part of Outer Mongolia). So, it is possible that the Mongolian *khulsan khuur* influenced the Tuvan *cheler komus*. To establish whether their similarity owes to the material of making or cultural contacts and whether Mongolian and Tuvan traditions are related to Ainu, it would be necessary to cross-analyze other remote indigenous JH bamboo traditions.

the JH tradition – both families expanded in parallel westward toward Europe, from southeast to northwest, crossing their paths in northern latitudes and absorbing metallurgical know-how along their way. So, all in all, almost the entire Northern Asia constitutes the domain of likely formative influence of strong indigenous JH traditions on local languages. To ultimately confirm or deny this, it would be necessary to thoroughly define the vocal systems of local JH traditions and compare them to vocal systems of corresponding languages.

## 8.10 Conclusion

This paper examines the formation of languages and music systems stemming from a hypothetical common ancestor musilanguage in northeastern Eurasia, seeking to disclose common traits between tonal organization (TO) of music and the vocal system of languages currently used by indigenous people in this region. Vowel harmony (VH), characteristic for most indigenous languages of this part of the world, seems to constitute one of the features of TO that originated in the pan-Siberian tradition of Jew's harp (JH) music. The peculiarity of palatal, labial, and height forms of VH found in northeastern Eurasian languages parallels the vocal systems of indigenous JH traditions of people speaking these languages. The entire Transeurasian (formerly "Altaic") language family nearly perfectly matches the geographic distribution of indigenous JH music. The timeframe of the emergence of the JH tradition seems to coincide with the estimated time of the breakup of the proto-Transeurasian language, and the formation of modern languages of Transeurasian constituent families must have occurred *after* the JH tradition became established over the Urals, Siberia, Mongolia, and the Far East. Some other VH languages (e.g., Chamorro and Xinaliq) might also belong to the sphere of influence of the pan-Siberian JH tradition.

The solid archaeological date of the oldest found JH (Lower Xiajiadian culture, XXII-XI centuries BC) concurred with the breakup of the proto-Tungusic and proto-Mongolic languages, according to the latest lexicostatistic studies. Earlier development of the JH tradition can be inferred from 4 main sources: 1) ethnographic research of the surviving traditions of manufacturing JH by indigenous people who maintain a traditional lifestyle, 2) paleo-ecological data on the availability of the traditional materials for JH-making in different parts of northeastern Asia, 3) relative complexity of the manufacturing process, and 4) comparative analysis of the etymological relations between the indigenous names for JHs. This interdisciplinary approach allows us to define the most likely order of the succession of the most typical JH constructions and their corresponding technologies. An additional source of information, related to JH, is the petroglyph tradition of depicting "singing faces" that render a specific vocal articulation. This tradition existed in the 3rd millennium BC in the Amuric basin and reached Tuva by the 2nd millennium

BC. The possible use of such images can be inferred from the surviving rituals and fine art of the Amuric peoples. Similar insight into the probable prehistoric use of so-called "proto-JHs," prepared from readily available organic materials (such as grass, wood-chips, and twigs) is provided by the surviving traditions of playing such musical instruments and the ramifications of the belief systems that underlie the surviving cult of "ancestral trees" in the Altai and Amur/Sakhalin areas.

The specific traits of VH prevailing in the main Transeurasian families as well as in the Uralic languages and in the reconstructed Transeurasian protolanguages might be explained by the geographic distribution of JH constructions made of different materials and therefore favoring different geometric forms. In particular, metallic bow-shaped and bamboo frame JH models oppose each other by their sonic properties, TO, and playing techniques, supporting a number of local indigenous traditions that are practiced, respectively, at the western (bow-shape prevalent) and the eastern (frame-shape prevalent) borders of the vast JH-dominant area. Their geographic disposition matches that of Turkic (plus Uralic) versus Japonic/Korean language families (plus Ainu and Nivkh languages). The acoustic differences between JH constructions made of different materials contribute to the differences in the TO of JH music, which in turn may have contributed to the phonemic repertories of indigenous languages that emerged after the local JH traditions had established their ideological importance and social status. VH here appears as the adaptation of the JH techniques of sound production for verbal communication. VH is not the only linguistic feature trackable to JH performance. This paper enumerates a number of other phonological, prosodic, and phonotactic traits common for both local languages and local JH traditions. In general, the "timbre-class" makeup of JH music closely resembles the phonemic makeup of natural languages – to the extent that the phenomenon of agglutination, commonly accompanying VH, can be regarded as the application of the rules of stacking "timbre-classes" together in a stream of JH music. JH music appears to have elaborated and conserved the repertory of devices of "intonational prosody" ancestral to both musical and linguistic intonations, within the prehistoric musilanguage that preceded the proto-Transeurasian language.

## References

Adkins CJ (1974) Investigation of the sound-producing mechanism of the jew's harp. J Acoust Soc Am 55:667–670. https://doi.org/10.1121/1.1914580

Agamennone M (1996) Polifonie. Procedimenti, tassonomie e forme: una riflessione "a più voci.". Edizioni Il Cardo, Venice

Agapitov NN, Khangalov MN (1885) Data for research on shamanism in Siberia. Shamanism amongst Buryats from the Irkutsk Province [Материалы для изучения шаманства в Сибири. Шаманство у бурят Иркутской Губернии]. The proceedings of the East Siberian section of the Imperial Russian Geographic Society 14:1–61

Akinlabi ET, Anane-Fenin K, Akwada DR (2017) Bamboo: the multipurpose plant. Springer, Berlin

Aksyonov AN (1964) Tuvan folk music: materials and studies [Тувинская народная музыка: материалы и исследования]. Muzyka [Музыка], Moscow

Aleksandrova N (2017) Frame-shaped Jew's harps in archeological monuments at the Kama and Vyatka interfluve [Пластинчатые варганы в археологических памятниках междуречья Камы и Вятки]. In: Matsiyevskii IV (ed) *Problems of organology*, vol 10. Russian Institute of History of Arts [Российский институт истории искусств], Sankt-Petersburg, pp 83–91

Alekseyenko AM (1988) Musical instruments of the peoples of North East Siberia [Музыкальные инструменты народов севера Западной Сибири]. In: Taksami CM (ed) Material and spiritual cultures of Siberian peoples [Материальная и духовная культура народов Сибири]. Nauka, The Institute of Ethnography of the USSR Academy of Science, Leningrad, pp 5–23

Alekseyev EY (1976) Problems in the modal genesis (on the example of Yakut folksong): analysis [Проблемы формирования лада (на материале якутской народной песни): исследование]. Muzyka [Музыка], Moscow

Alekseyev EY (1986) Early folkloric intonation. Pitch aspect [Раннефольклорное интонирование: звуковысотный аспект]. Sovetskii Kompozitor [Сов. композитор], Moscow

Alekseyev IY (1988) The art of playing Yakut khomus [Искусство игры на якутском хомусе]. Preprint, Yakutsk

Alekseyev EY (1990) Folkloric notation. Theory and practice [Фольклористическая нотация. Теория и практика]. Soviet Composer [Советский композитор], Moscow

Alekseyev EY (1991a) Jew's Harp: undisclosed perspectives of research... [Варган: нераскрытые исследовательские перспективы]. In: Alekseyev E (ed) Jew's Harp (khomus) and its music. Proceedings of the first All-Union conference, 1988 [Варган (хомус) и его музыка. Материалы I Всесоюзной конференции 1988]. Yakut Institute of language, literature and history of the USSR Academy of Science, Yakutsk, pp 6–14

Alekseyev EY (1991b) Jew's Harp that is playing, singing, talking... [Варган играющий, поющий, говорящий]. In: 2nd International Jew's Harp Congress-Festival at Yakutsk, 19–25 June 1991. International Jew's Harp Society, Yakutsk

Alekseyev NA (1992) In: Gurvich IS (ed) Traditional religious beliefs of Turkic-speaking peoples of Siberia [Традиционные религиозные верования тюркоязычных народов Сибири]. Nauka, Novosibirsk

Alekseyev EY, Levin TC (1990) Tuva: Voices from the Center of Asia. Smithsonian Folkways Recordings, New York

Alekseyeva GG (1986) Khomus, play your song for them! [Заиграй им, хомус, свою песню]. In: Grigoryeva A (ed) Music of Russia: musical creativity and musical life in republics of Russian Federation [Музыка России: музыкальное творчество и музыкальная жизнь республик Российской Федерации], vol 8. Sovetskii Kompozitor [Советский композитор], Moscow, pp 321–332

Alexeyev I, Shishigin SS (2004) The Music of the Yakut Traditional Khomus. J Int Jew's Harp Soc 1:88–94

Anderson PM, Lozhkin AV, Brubaker LB (2002) Implications of a 24,000-Yr palynological record for a Younger Dryas cooling and for boreal forest development in Northeastern Siberia. Quatern Res 57:325–333. https://doi.org/10.1006/qres.2002.2321. Cambridge University Press

Andreyev ND, Sunik OP (1982) On the problem of relationship between Altaic languages and methods of its solution [О проблеме родства алтайских языков и методах ее решения]. Voprosy Jazykoznanija (Topics in the study of language) 2:26–36

Andreyeva JV (1987) Bronze age of far East [Бронзовый век Дальнего Востока]. In: Rybakov BA (ed) Archaeology of the USSR. Bronze Age of the forest zone of the USSR [Археология СССР. Эпоха бронзы лесной полосы СССР]. Nauka, Moscow, pp 351–357

Antonov NK (1997) Yakut language [Якутский язык]. In: Tenishev ER (ed) Languages of the world. Turkic languages [Языки мира: Тюркские языки]. Russian Academy of Science, Kyrkyzstan, Bishkek, pp 17–34

Arbib MA (2012) How the brain got language: the mirror system hypothesis. Oxford University Press, New York

Arbib MA, Iriki A (2012) Evolving the language-and music-ready brain. In: Arbib MA (ed) Language, music, and the brain: a mysterious relationship. The MIT Press, Cambridge, MA, pp 359–375

Arkhipov ND (1989) Ancient cultures of Yakutia [Древние культуры Якутии]. Yakut Book Publishing, Yakutsk

Balzer MM (2016) Shamanic worlds: rituals and Lore of Siberia and Central Asia. Shamanic Worlds: rituals and Lore of Siberia and Central Asia, 2nd edn. Routledge, Abingdon-on-Thames. https://doi.org/10.4324/9781315487335

Barinova E (2014) The problem of interaction of China with Central Asia during the Bronze Age based on the available material culture [Проблема взаимодействия Китая с Центральной Азией в бронзовом веке (по данным материальной культуры)]. Tomsk State Univ J 380:67–79

Barnes GL (1993) China, Korea and Japan: the rise of civilization in East Asia

Barr W (1994) The eighteenth century trade between the ships of the Hudson's Bay Company and the Hudson Strait Inuit. Arctic 47:236–246. https://doi.org/10.2307/40511572. Arctic Institute of North America

Bar-Yosef O, Belfer-Cohen A (2013) Following Pleistocene road signs of human dispersals across Eurasia. Quat Int 285:30–43. https://doi.org/10.1016/J.QUAINT.2011.07.043. Pergamon

Bar-Yosef O, Eren MI, Yuan J, Cohen DJ, Li Y (2012) Were bamboo tools made in prehistoric Southeast Asia? An experimental view from South China. Quat Int 269:9–21. https://doi.org/10.1016/j.quaint.2011.03.026

Beck CM, Deeds EE, Pozorski S, Pozorski T (1983) Pajatambo: an 18th century roadside structure in Peru. Hist Archaeol 17:55–68. https://doi.org/10.1007/BF03374031. Springer

Beliayev VM (1933) Musical Instruments of Uzbekistan [Музыкальные инструменты Узбекистана]. Gos Muz Izdat [Гос. музыкальное изд-во], Moscow

Berkov V (1962) Harmony and musical form [Гармония и музыкальная форма]. Soviet Composer [Советский композитор], Moscow

Berndt A, Hähnel T (2010) Modelling musical dynamics. In: Brandenburg K (ed) Proceedings of the 5th audio mostly conference: a conference on interaction with sound, Piteå, Sweden — September 15–17, 2010. Fraunhofer Institute for Digital Media Technology, Ilmenau, pp 1–8. https://doi.org/10.1145/1859799.1859817

Berry W (1987) Structural functions in music. Dover Publications, New York

Bettini A (2017) A course in classical physics 4 – waves and light, Undergraduate lecture notes in physics. Springer, Berlin. https://doi.org/10.1007/978-3-319-48329-0

Bien N, ten Oever S, Goebel R, Sack AT (2012) The sound of size: crossmodal binding in pitch-size synesthesia: a combined TMS, EEG and psychophysics study. NeuroImage 59:663–672. https://doi.org/10.1016/j.neuroimage.2011.06.095. Elsevier Inc

Blažek V (2009) Koguryo and Altaic. On the role of Koguryo and other Old Korean idioms in the Altaic etymology. Ural-Altaic Stud 1:13–25. https://doi.org/10.31826/9781463234188-002 In-t inostrannykhîâzykov

Blench R (2004) Musical aspects of Austronesian culture. In: Bacus L, Glover I (eds) Proceedings of the Xth Association of Southeast Asian Archaeologists Conference, London 2004. Kay Williamson Educational Foundation, Cambridge, pp 1–10

Blench R (2008) Stratification in the peopling of China: how far does the linguistic evidence match genetics and archaeology? In: Sanchez-Mazas A, Blench R, Ross M, Peiros I, Lin M (eds) Past human migrations in East Asia: matching archaeology, linguistics and genetics. Routledge, London, pp 105–132. https://doi.org/10.4324/9780203926789

Boenn G (2018) Computational models of rhythm and meter. Computational models of rhythm and meter. Springer, Cham. https://doi.org/10.1007/978-3-319-76285-2

Bogoras W (1927) In memory of M.A. Castren: to the 75th passing anniversary [Памяти М.А. Кастрена (к 75-летию дня смерти)]. Academy of Science of USSR, Leningrad

Böhne C (1968) Über die Kupferverhüttung der Bronzezeit: Schmelzversuche mit Kupferkieserzen. Archaeol Aust 44:49–60. F. Deuticke

Bonatti LL, Peña M, Nespor M, Mehler J (2007) On consonants, vowels, chickens, and eggs. Psychol Sci 18:924–925. https://doi.org/10.1111/j.1467-9280.2007.02002.x

Borodovskii AP (2007) Ancient carved horn from Southern Siberia of the Paleometallic Age [Древний резной рог Южной Сибири эпохи палеометалла]. The Novosibirsk State University, Novosibirsk

Borodovskii AP (2017) Bone Jaw Harps and their workpieces of the Hunno-Sarmatian period in the North Altai region [Костяные варганы и их заготовки гунно-сарматского времени на территории Северного Алтая]. In: Derevianko AP, Molodin VI (eds) Problems of archaeology, ethnography, anthropology of Siberia and adjacent territories [Проблемы археологии, этнографии, антропологии Сибири и сопредельных территорий], vol 23. The Institute of Archaeology and Ethnography of Russian Academy of Science, Novosibirsk, pp 279–283

Botalov SG (2007) Hunnus and Turks: historic-archaeological reconstruction [Гунны и тюрки (историко-археологическая реконструкция)]. The Ural department of the Russian Academy of Science, Cheliabinsk

Botma B, Shiraishi H (2015) Phonetic (non-)explanation in historical phonology: Duration, harmony, and dissimilation. In: White A (ed) The second Edinburgh symposium on historical phonology, The University of Edinburgh, Scotland, 3/12/15–4/12/15. University of Edinburgh, Edinburg, pp 1–15

Bowling DL (2013) A vocal basis for the affective character of musical mode in melody. Front Psychol 4:464. https://doi.org/10.3389/fpsyg.2013.00464

Briefer EF (2012) Vocal expression of emotions in mammals: mechanisms of production and evidence. J Zool 288:1–20. https://doi.org/10.1111/j.1469-7998.2012.00920.x

Brigham-Grette J, Lozhkin AV, Anderson PM, Glushkova OY (2004) Paleoenvironmental Conditions in Western Beringia before and during the Last Glacial Maximum. In: Madsen DB (ed) Entering America: Northeast Asia and Beringia Before the Last Glacial Maximum. The University of Utah Press, Salt Lake City, pp 29–62

Brodianskii DL (1978) Yet another field of neolithic art in Far East [Еще одна область неолитического искусства на Дальнем Востоке]. In: Vasilyevskii RS (ed) At the origin of creativity. Primitive art [У истоков творчества. Первобытное искусство]. Nauka, Novosibirsk, pp 133–141

Bronson B (1996) Metals, specialization, and development in early Eastern and Southern Asia. In: Wailes B (ed) Craft Specialization and Social Evolution: In Memory of V. Gordon Childe. The University Museum of Archaeology and Anthropology, University of Pennsylvania, Philadelphia, pp 177–186

Brown S (2000a) The "Musilanguage" model of language evolution. In: Brown S, Merker B, Wallin NL (eds) The origins of music. MIT Press, Cambridge, MA, pp 271–300. https://doi.org/10.1037/e533412004-001

Brown S (2000b) Evolutionary models of music: from sexual selection to group selection. In: Tonneau F, Thompson NS (eds) Perspectives in ethology, vol 13. Springer, Boston, MA, pp 231–281. https://doi.org/10.1007/978-1-4615-1221-9_9

Brown S (2001) Are music and language homologues? Ann N Y Acad Sci 930:372–374. https://doi.org/10.1111/j.1749-6632.2001.tb05745.x

Brown S (2017) A joint prosodic origin of language and music. Front Psychol 8.. Frontiers:1894. https://doi.org/10.3389/fpsyg.2017.01894

Bucur V (2016) Handbook of materials for string musical instruments. Springer, Berlin. https://doi.org/10.1007/978-3-319-32080-9

Bulgakova TD (2001) Music in traditional Nanai culture [Музыка в традиционной нанайской культуре]. Inform Bull Russian Found Basic Res 3:213–218

Büyükmavi M (2007) Review of: is Japanese related to Korean, Tungusic, Mongolic and Turkic? by Martine Irma Robbeets. Oriens Extremus Harrassowitz Verlag. https://doi.org/10.2307/24047680

Canave-Dioquino C, Santos RP, Maceda J (2008) The Philippines. In: Miller T, Williams S (eds) The Garland Handbook of Southeast Asian Music. Routledge, London, pp 415–445
Chernykh EN (1992) Ancient metallurgy in the USSR: the early metal age (trans: Wright S). Cambridge University Press, Cambridge
Chlachula J, Lynsha VA, Kolaczek P, Tarasenko VN (2015) Neolithic and aeneolithic environments in the Central Primor'ye Region (the Bol'shaya Ussurka Valley), the Russian Far East. Quat Int 370:127–144. https://doi.org/10.1016/J.QUAINT.2014.12.065. Pergamon
Christian D (2000) Silk roads or steppe roads? The silk roads in world history. J World Hist 11:1–26. https://doi.org/10.1353/jwh.2000.0004
Chu K-C (1973) A preliminary study on the climatic fluctuations during the last 5,000 years in China. Sci Sinica XVI. Science China Press:226–256. https://doi.org/10.1360/YA1973-16-2-226
Chuluunbaatar O (2016) Rare archaeological musical artefacts from Ancient Tombs in Mongolia. In: Jähnichen G (ed) Studia instrumentorum musicae popularis, New series, vol 4. MV Wissenschaft, Münster, pp 225–250
Clark JD (1998) The Early Palaeolithic of the eastern region of the Old World in comparison to the West. In: Petraglia MD, Korisettar R (eds) Early human behaviour in global context: the rise and diversity of the Lower Palaeolithic record. Routledge, London/New York, pp 425–438. https://doi.org/10.4324/9780203203279
Clauson G (1956) The case against the Altaic theory. Central Asiatic J II:181–187. https://doi.org/10.2307/41926354. Harrassowitz Verlag
Clayton MRL (1996) Free rhythm: ethnomusicology and the study of music without metre. Bull School Orient Afr Stud 59:323–332. https://doi.org/10.1017/S0041977X00031608. Cambridge University Press
Clayton MRL (2000) Time in Indian music: rhythm, metre, and form in North Indian rag performance. Oxford University Press, Oxford
Coghlan HH (1939) Prehistoric copper and some experiments in smelting. Trans Newcomen Soc 20:49–65. https://doi.org/10.1179/tns.1939.005. Routledge
Comrie B (1981) The languages of the Soviet Union. Cambridge University Press, Cambridge
Crabtree DE, Davis EL (1968) Experimental manufacture of wooden implements with tools of flaked stone. Science (New York, N.Y.) 159:426–428. https://doi.org/10.1126/science.159.3813.426. American Association for the Advancement of Science
Crane F (1968) The Jew's Harp as aerophone. Galpin Soc J 21:66–69. https://doi.org/10.2307/841429. Galpin Society
Crane F (2007) Review of Jew's Harps in European archaeology. J Int ew's Harp Soc 4:68–69
Cross, Ian, and Iain Morley. 2009. The evolution of music: theories, definitions and the nature of the evidence. In Communicative musicality: exploring the basis of human companionship, ed. Trevarthen C, Malloch S. London: Oxford University Press, pp 61–82
Cross I, Tecumseh Fitch W, Aboitiz F, Iriki A, Jarvis ED, Lewis J, Liebal K, Merker B, Stout D, Trehub SE (2013) Culture and Evolution. In: Arbib MA (ed) Language, music, and the brain: a mysterious relationship. MIT Press, Cambridge, MA, pp 541–562
Cui Y, Li H, Ning C, Zhang Y, Lu C, Zhao X, Hagelberg E, Zhou H (2013) Y Chromosome analysis of prehistoric human populations in the West Liao River Valley, Northeast China. BMC Evol Biol 13:216. https://doi.org/10.1186/1471-2148-13-216. BioMed Central
Czaplicka M (1914) Aboriginal Siberia, a study in social anthropology. Clarendon Press, Oxford
Davis RF (2004) Ma'lūf: reflections on the Arab Andalusian music of Tunisia. Scarecrow Press, Lanham, MD
Dallais P, Weber S, Briner C, Liengme J (2002) The drymba among the Hutsul in the Ukrainian Carpathians: a recent ethnomusicological survey. VIM Vierundzwanzigsteljahrsschrift der Internationalen Maultrommelvirtuosengenossenschaft 10:8–29
Da-Shun G (2002) Lower Xiajiadian culture. In: The archaeology of Northeast China: beyond the Great Wall, Nelson SM (ed), tran: Shan M. Routledge, London, pp 161–195
de Ramón IA, Rivera IA (1982) Indigenous music of Venezuela. World Music 24:22–37

Décsy G (2007) Review of Robbeets, Martine 2005. Is Japanese related to Korean, Tungusic, Mongolic and Turkic? Eurasian Stud Yearb 79:157

Devlet MA (1976) Petroglyphs of Ulug-Khem [Петроглифы Улуг-Хема]. Nauka, Moscow

Devlet MA (1980) Petroglyphs of Mugur-Sargol [Петроглифы Мугур-Саргола]. Nauka, Moscow

Devlet YG, Devlet MA (2011) The treasures of rock art of North and Central Asia [Сокровища наскального искусства Северной и Центральной Азии]. Institute of Archaeology at Russian Academy of Science, Moscow

Di Cosmo N (1994) Ancient inner Asian Nomads: their economic basis and its significance in Chinese history. J Asian Stud 53:1092. https://doi.org/10.2307/2059235. Cambridge University Press

Dickinson K (2001) 'Believe'? Vocoders, digitalised female identity and camp. Popular Music 20:333–347. https://doi.org/10.1017/S0261143001001532. Cambridge University Press

Diffloth G (2006) I: big, a: small. In: Hinton L, Nichols J, Ohala JJ (eds) Sound symbolism. Cambridge University Press, Cambridge, pp 107–113

Dmitriyeva YN (2002) Comparative grammar of Russian and Yakut languages [Сопоставительная грамматика русского и якутского языков]. State Yakut University named after M.K. Ammosov, Yakutsk

Dobzhanskaya O (1991) Chukchi playing patterns on frame Jew's Harps [Чукотские наигрыши на рамном варгане]. In: Alekseyev EY (ed) Jew's Harp (khomus) and its music. Proceedings of the first All-Union conference, 1988 [Варган (хомус) и его музыка. Материалы I Всесоюзной конференции 1988]. Yakut Institute of language, literature and history of the USSR Academy of Science, Yakutsk, pp 52–58

Dodson J, Dong G (2016) What do we know about domestication in eastern Asia? Quat Int 426:2–9. https://doi.org/10.1016/j.quaint.2016.04.005

Doerfer G (1963) Bemerkungen zur Verwandtschaft der sog. altaische Sprachen. In: Doerfer G (ed) Türkische und mongolische Elemente im Neupersischen, vol 1. Franz Steiner Verlag, Wiesbaden, pp 51–105

Dolscheid S, Hunnius S, Casasanto D, Majid A (2014) Prelinguistic infants are sensitive to space-pitch associations found across cultures. Psychol Sci 25:1256–1261. https://doi.org/10.1177/0956797614528521

Dournon-Taurelle G (1975) La guimbarde. Paris Nanterre University, Paris

Dournon-Taurelle G, Wright J (1978) Les Guimbardes du Musée de l'Homme. Institut d'Ethnologie, Muséum national d'Histoire naturelle, Paris

Drabkin W (2001) Theme. In: Sadie S, Tyrrell J (eds) The new grove dictionary of music and musicians. Macmillan Publishers, London. https://doi.org/10.1093/gmo/9781561592630.article.27789

Duvan ND (2000) Personality of Ulch Shaman and his ritual [Личность ульчского шамана и его ритуал]. In: Sheikin Y (ed) Musical Ethnography of Tungus-Manchu peoples. International Conference at Yakutsk, 17–23 Aug., 2000 [Музыкальная этнография тунгусо-маньчжурских народов]. The Institute of Problems of Exploration of North, Yakutsk, pp 51–53

Duvan ND (2003) Musical instruments of Ulch people [Музыкальные инструменты ульчей]. In: Ruban N, Melnikova T (eds) Papers of Grodekovsky Museum, vol 6. Khabarovsk regional folk museum named after Grodekovsky, Khabarovsk, pp 57–72

Dyakonova VP (1981) Tuvan shamans and their social role in society [Тувинские шаманы и их социальная роль в обществе]. In: Vdovin IS (ed) Problems of history of social consciousness of Siberian Aborigenes [Проблемы истории общественного сознания аборигенов Сибири]. Nauka, Leningrad, pp 129–164

Dyakonova V (2017) Musical instruments of Sakha people in light of classic typologies [Музыкальные инструменты народа саха в свете классических типологий]. The Arctic State Institute of Art and Culture, Yakutsk

Dyakonova VE, Grigoryeva VG (2017) On the healing properties of sound in traditional Sakha culture [О целительных свойствах звука в традиционной культуре саха]. In: Kharitonova

VI (ed) Medical ethnography: contemporary approaches and conceptions [Медицинская этнография: современные подходы и концепции]. Institute of Ethnography & Anthropology of Russian Academy of Science, OOO Publicity, Moscow, pp 160–172

Dybo A (2019) New trends in European studies on the Altaic problem. J Lang Relationship 14:71–106. https://doi.org/10.31826/9781463237288-007

Emsheimer E (1986) Jew's harps in Siberia and Middle Asia [Варганы в Сибири и Средней Азии]. In: Matsiyevskii I (ed) Problems of traditional instrumental music of peoples of USSR: instrument-performer-music [Проблемы традиционной инструментальной музыки народов СССР: Инструмент-исполнитель-музыка]. LGITMiK, Leningrad, pp 80–94

Esteves BM, Pereira HM (2009) Wood modification by heat treatment: a review. Bioresources 4:370–404

Farmer HG (1936) Turkish instruments of music in the seventeenth century. J R Asia Soc Great Britain & Ireland 68:1–43. https://doi.org/10.1017/S0035869X00076322. Cambridge University Press

Fedorova SA, Reidla M, Metspalu E, Metspalu M, Rootsi S, Tambets K, Trofimova N et al (2013) Autosomal and uniparental portraits of the native populations of Sakha (Yakutia): implications for the peopling of Northeast Eurasia. BMC Evol Biol 13:127. https://doi.org/10.1186/1471-2148-13-127. BioMed Central

Feodorov GB (1954) The results of 3-year work in Moldovia in Slavic-Russian archaeology [Итоги трёхлетних работ в Молдавии в области славяно-русской археологии]. Brief Commun Inst Archaeol 56:8–23

Fernyhough C (2009) Dialogic thinking. In: Adam W, Ignacio M (eds) Private speech, executive functioning, and the development of verbal self-regulation. Cambridge University Press, Cambridge, pp 42–52

Fitch WT (2010) The evolution of language. Cambridge University Press, Cambridge

Fletcher NH, Rossing TD (1998) The physics of musical instruments. Springer, New York/London

Fortescue M (2005) Comparative Chukotko-Kamchatkan Dictionary. Mouton de Gruyter, Berlin

Fortescue M (2011) The relationship of Nivkh to Chukotko-Kamchatkan revisited. Lingua 121:1359–1376. https://doi.org/10.1016/j.lingua.2011.03.001

Fox L (1988) The Jew's Harp: a comprehensive anthology, 2nd edn. Associated University Presses/ Bucknell University Press, London/Lewisburg

Fraisse P (1982) Rhythm and tempo. In: Deutsch D (ed) Psychology of music. Academic Press, New York, pp 149–180

Frayonov VP (1981) Texture [фактура]. In: Keldysh Y (ed) Encyclopedia of Music [Музыкальная энциклопедия]. Soviet Encyclopedia [Советская энциклопедия], Moscow

Gabrielsson A (1999) The performance of music. In: Psychology of music. Academic Press, New York, pp 501–602

Gadzhiyeva NE (1997) Turkic languages [Тюркские языки]. In: Tenishev ER (ed) Languages of the world. Turkic languages [Языки мира: Тюркские языки]. Russian Academy of Science, Kyrkyzstan, Bishkek, pp 17–34

Garbuzov N (1948) Zoonal nature of pitch hearing [Зонная природа звуковысотного слуха]. Russian Academy of Science, Moscow

Garbuzov N (1950) Zonal nature of tempo and rhythm [Зонная природа темпа и ритма]. Academy of Science of USSR, Moscow

Garbuzov N (1955) Zonal nature of hearing of dynamics [Зонная природа динамического слуха]. Gos Muz Izdat, Moscow

Gemuyev IN, Sagalayev AM (1986) Religion of Mansi people. Cult locations of the 19th-early 20th centuries [Религия народа манси. Культовые места XIX — начала XX в.]. Nauka, Novosibirsk

Georg S (2004) Sergei Starostin, Anna Dybo, and Oleg Mudrak (eds.): etymological dictionary of the Altaic languages (2003). Diachronica 21:445–450. https://doi.org/10.1075/dia.21.2.12geo

Georg S, Michalove PA, Ramer AM, Sidwell PJ (1999) Telling general linguists about Altaic. J Linguis 35:65–98. https://doi.org/10.1017/S0022226798007312. Cambridge University Press

Gippius YV (1988) Ritual instrumental melodies of the Bear festival of the Ob' Ugric peoples [Ритуальные инструментальные наигрыши Медвежьего праздника обских угров]. In: Gippius YV (ed) Folk musical instruments and instrumental music: the collection of articles and materials [Народные музыкальные инструменты и инструментальная музыка: сборник статей и материалов в двух частях], vol 2. Sovetskii Kompozitor, Moscow, pp 164–175

Girchenko EA (2018) Ceramics of Baijinbao – the Basic Site of the Bronze Age in the Heilongjiang Province [Керамика стоянки байцзиньбао - опорного памятника эпохи бронзы в провинции Хэйлунцзян]. Vestnik NSU. Series: History and Philology 17:9–15. https://doi.org/10.25205/1818-7919-2018-17-4-9-15

Godøy RI (2004) Gestural imagery in the service of musical imagery. In: Gesture-based communication in human-computer interaction: 5th International Gesture Workshop, GW, vol 2003, pp 55–62. https://doi.org/10.1007/978-3-540-24598-8

Golubkova AN, Ivanov AG (1997) Ancient Jaw Harps of Kama Area [Древние Варганы Прикамья]. The courier of the Udmurt University 8:93–107

Gözaydin N (2006) Dergi ve kitap Dünyasından. Türk Dili 653:467–471

Greenberg JH (1963) Vowel harmony in African languages. In: Houise M (ed) Actes du II Colloque international de linguistique négro-africaine, Dakar 12–16 avril 1962. Université de Dakar, Dakar, pp 33–38

Greenberg JH (2000) Indo-European and its closest relatives: the Eurasiatic language family. Stanford University Press, Stanford

Grigoryan GA (1957) Music culture of Yakut ASSR [Музыкальная культура Якутской АССР]. In: Litinskii GI (ed) Music culture of the autonomous republics of the Russian Federation [Музыкальная культура автономных республик РСФСР]. Muzgiz, Moscow, pp 331–348

Grigoryev SS (1981) The Theoretic Course of Harmony [Теоретический Курс Гармонии]. Muzyka [Музыка], Moscow

Gruzdeva E (1998) Nivkh. Lincom Europa, Munich

Gubina MA, Girgol'kau LA, Babenko VN, Damba LD, Maksimov VN, Voevoda MI (2013) Mitochondrial DNA polymorphism in populations of aboriginal residents of the Far East. Russian J Genet 49:751–764. https://doi.org/10.1134/S1022795413070065. Springer US

Habu J, Lape PV, Olsen JW (2017) Introduction. In: Habu J, Lape PV, Olsen JW, Eastep AM (eds) Handbook of East and Southeast Asian archaeology. Springer, New York, pp 1–10

Haid G (1999) The trump and eroticism. Vie rundzwanzigsteljahrsschrift der Internationalen Maultrommelvirtuosengenossenschaft 8:60–80

Hansen V (2012) The Silk Road: a new history. Oxford University Press, Oxford

Hargreaves DJ (1986) The developmental psychology of music. Cambridge University Press, Cambridge

Hattori S (1982) Vowel harmonies of the Altaic languages, Korean and Japanese. Acta Orientalia Academiae Scientiarum Hungaricae 36:207–214

Hauser-Schäublin B (1995) Puberty rites, women's Naven, and initiation: women's rituals of transition in Abelam and Iatmul culture. In: Lutkehaus NC, Roscoe PB (eds) Gender rituals: Female initiation in Melanesia. Routledge, London, pp 33–53

Herva V-P, Ikäheimo J, Kuusela J-M, Nordqvist K (2009) Close encounters of the copper kind. In: Berge R, Jasinski ME, Sognnes K (eds) Proceedings of the 10th Nordic TAG conference: Stiklestad, Norway. Britsh Archaeological Reports, London, pp 109–115

Hibbert D (1982) History of the Steppe: Tundra concept. In: Hopkins DM, Matthews JV, Schweger CE, Young SB (eds) Paleoecology of Beringia. Academic, New York, pp 153–156

Hoffecker JF, Elias SA (2007) Human ecology of Beringia. Columbia University Press, New York

Höfle C, Edwards ME, Hopkins DM, Mann DH, Ping C-L (2000) The full-glacial environment of the Northern Seward Peninsula, Alaska, reconstructed from the 21,500-Year-Old Kitluk Paleosol. Quat Res 53:143–153. https://doi.org/10.1006/qres.1999.2097. Cambridge University Press

Holman EW, Brown CH, Wichmann S, Müller A, Velupillai V, Hammarström H, Sauppe S et al (2011) Automated dating of the world's language families based on lexical similarity. Current

Anthropol 52:841–875. https://doi.org/10.1086/662127. University of Chicago Press, Chicago, IL

Honeychurch W (2015) Inner Asia and the spatial politics of empire: Archaeology, mobility, and culture contact. Inner Asia and the Spatial Politics of Empire: archaeology, mobility, and culture contact. Springer, New York. https://doi.org/10.1007/978-1-4939-1815-7

Hopkins DM, Matthews JV, Schweger CE, Young SB (1982) Paleoecology of Beringia—a synthesis. In: Hopkins DM, Matthews JV, Schweger CE, Young SB (eds) Paleoecology of Beringia. Academic, New York, pp 425–444

Houle G (1987) Meter in music, 1600–1800: performance, perception, and notation. Indiana University Press, Bloomington

Hsu T-H (2001) Taiwan: music of the Taiwan aborigines. In: Provine RC, Tokumaru Y, Witzleben L (eds) The Garland encyclopedia of world music, Volume 7. East Asia: China, Japan and Korea, vol 7. Routledge, London, pp 523–529

Hu AJ (2010) An overview of the history and culture of the Xianbei (‘Monguor’/‘Tu’). Asian Ethnicity 11:95–164. https://doi.org/10.1080/14631360903531958

Huang J (1984) Changes of sea-level since the Late Pleistocene in China. In: Whyte RO (ed) The evolution of the East Asian environment, vol I. Centre of Asian Studies, University of Hong Kong, Hong Kong, pp 309–319

Huang P (1990) New light on the origins of the Manchus. Harv J Asiat Stud 50:239. https://doi.org/10.2307/2719229

Hughes DA (1954) Music in fixed rhythm. In: Hughes DA (ed) New Oxford history of music: early medieval music up to 1300, vol 2. Oxford University Press, Oxford, pp 311–352

Hung H-c, Chao C-y (2016) Taiwan's Early Metal Age and Southeast Asian trading systems. Antiquity 90:1537–1551. https://doi.org/10.15184/aqy.2016.184. Cambridge University Press

Huron D (2006) Sweet Anticipation: Music and the Psychology of Expectation. MIT Press, Cambridge, MA

Ikawa-Smith F (2004) Humans along the Pacific Margin of Northeast Asia before the Last Glacial Maximum: evidence for their presence and adaptations. In: Madsen DB (ed) Entering America: Northeast Asia and Beringia before the Last Glacial Maximum. The University of Utah Press, Salt Lake City, pp 285–310

Ikhtisamov KS (1988) On the problem of comparative investigation of two-part throat singing and instrumental music of Turkic and Mongol peoples [К проблеме сравнительного изучения двухголосного гортанного пения и инструментальной музыки у тюркских и монгольских народов]. In: Gippius Y (ed) Folk musical instruments and instrumental music [Народные музыкальные инструменты и инструментальная музыка], vol 2. Sovetskii Kompozitor [Советский композитор], Moscow, pp 197–216

Ivanov SV (1963) Ornaments of Siberian peoples as a historico-ethnographic source (based on materials of the 19–20th centuries). Peoples of the North and Far East [Орнамент народов Сибири как историко-этнографический источник (по материалам (XIX начало XX в). Народы Север]. Academy of Science of USSR Publishing, Moscow/Leningrad

Ivanov SV (1975) Masks of Siberian peoples [Маски народов Сибири]. Avrora, Leningrad

Ivanov SI (1999) The lyrical poetry of the music of the “Talking” Khomus [Лирика музыки “говорящего” хомуса]. Vierundzwanzigsteljahrsschrift 8:89–92

Jaffe-Berg E (2001) Forays into Grammelot the language of nonsense. J Dram Theor Crit 2:3–16

Jakobson R (1968) Child language, aphasia and phonological universals. Mouton, The Hague

Janhunen J (1996) Manchuria: an ethnic history. Helsinki, Finno-Ugrian Society

Janhunen J (2009) Proto-Uralic–what, where, and when. The quasquicentennial of the Finno-Ugrian society:57–78

Janhunen J (2010) Reconstructing the language map of prehistorical Northeast Asia. Stud Orient Electron 108:281–304

Janhunen J (2012) The expansion of Tungusic as an ethnic and linguistic process. In: Whaley LJ, Mal'chukov A L'v (eds) Recent advances in Tungusic linguistics, vol 89. Harrassowitz Verlag, Wiesbaden, pp 5–16

Janhunen J (2017) Korean vowel system in North Asian perspective. In: Robbeets M (ed) Transeurasian linguistics, vol 1. Routledge, London, pp 181–193
Jankowski H (2013) Altaic languages and historical contact. In: Ko T-h (ed) Current trends in Altaic linguistics. Altaic Society of Korea, Seoul, pp 523–546
Jensen TW (2014) Emotion in languaging: languaging as affective, adaptive, and flexible behavior in social interaction. Front Psychol 5:720. https://doi.org/10.3389/fpsyg.2014.00720. Frontiers
Johanson L (2010) The high and low spirits of Transeurasian language studies. In: Johanson L, Robbeets M (eds) Transeurasian verbal morphology in a comparative perspective: genealogy, contact, chance, Wiesbaden, pp 7–20
Johanson L, Robbeets M (2010) Introduction. In: Johanson L, Robbeets MI (eds) Transeurasian verbal morphology in a comparative perspective: genealogy, contact, chance, vol 78. Harrassowitz, Wiesbaden, pp 1–5
Joukowsky AM (1965) The teaching of ethnic dance: Macedonia, Greece, Serbia, Bulgaria, Rumania, Czechoslovakia, France, Poland, Russia, Ukraine. J. Lowell Pratt, New York
Kaner S, Taniguchi Y (2017) The development of pottery and associated technological developments in Japan, Korea, and the Russian Far East. In: Habu J, Lape PV, Olsen JW, Eastep AM (eds) Handbook of East and Southeast Asian archaeology. Springer, New York, pp 321–346. https://doi.org/10.1007/978-1-4939-6521-2
Kang I (2011) The spread of the Animal Style in the I millenium BC over the territory of Liaoning and Korean Peninsula in relation to the culture of the bronze violin-shaped daggers [Распространение звериного стиля в I тыс. до н.э. на территории провинции Ляонин (КНР)]. In: Molodin VI, Hansen S (eds) «Terra Scythica» Materialien des internationalen Symposiums «Terra Scythica» (17.–23. August 2011, Denisov-Höhle, Altai). Verlag des Instituts für Archäologie und Ethnographie der SA RAW, Novosibirsk, pp 82–96
Kara G (2007) Review of Robbeets, Martine 2005. Is Japanese related to Korean, Tungusic, Mongolic and Turkic? Anthropol Linguist 49:95–98
Karafet TM, Osipova LP, Hammer MF (2008) The effect of history and life-style on genetic structure of North Asian populations. In: Sanchez-Mazas A, Blench R, Ross M, Peiros I, Lin M (eds) Past human migrations in East Asia: matching archaeology, linguistics and genetics. Routledge, London, pp 395–415
Kartomi MJ (2012) Musical journeys in Sumatra. University of Illinois Press, Champaign
Kartomi MJ, Anderson Sutton R, Suanda E, Williams S, Harnish D (2008) Indonesia. In: Miller T, Williams S (eds) The Garland handbook of Southeast Asian Music. Routledge, London, pp 334–405
Kaun AR (2004) The typology of rounding harmony. In: Hayes B, Kirchner R, Steriade D (eds) Phonetically based phonology. Cambridge University Press, Cambridge, pp 87–116. https://doi.org/10.1017/CBO9780511486401.004
Kharlap M (1978) The Tactus system of musical rhythmics [Тактовая система музыкальной ритмики]. In: Kholopova V (ed) The problems of musical rhythm: collection of essays [Проблемы музыкального ритма: Сборник статей]. Muzyka [Музыка], Moscow, pp 48–104
Khlobystin LP (1987) Bronze age of Eastern Siberia [Бронзовый век Восточной Сибири]. In: Rybakov BA (ed) Archaeology of the USSR. Bronze Age of the forest zone of the USSR [Археология СССР. Эпоха бронзы лесной полосы СССР]. Nauka, Moscow, pp 327–350
Kholopov Y (1976) Mode [Лад]. In: Keldysh Y (ed) Encyclopedia of music [Музыкальная энциклопедия]. Soviet Encyclopedia [Советская энциклопедия], Moscow
Khudiakov IA (1969) In: Bazanov VG (ed) A brief description of Varkhoyanskii District [Краткое описание Верхоянского округа]. Nauka, Leningrad
Kita S (2003) Pointing: Where language, culture, and cognition meet. Lawrence Erlbaum Associates, Mahwah. https://doi.org/10.4324/9781410607744
Kiyokbayev DG (1958) Phonetics of Bashkir language: the attempt of descriptive and comparative historic research [Фонетика башкирского языка: опыт описательного и сравнительно-исторического исследования]. Bashkir Book Publishing, Ufa

Kliashtornyi SG, Savinov DG (2005) The steppe empires of ancient Eurasia [Степные империи древней Евразии]. State Sankt-Petersburg University, Sankt-Petersburg

Klimaschewski A, Barnekow L, Bennett KD, Andreev AA, Andrén E, Bobrov AA, Hammarlund D (2015) Holocene environmental changes in southern Kamchatka, Far Eastern Russia, inferred from a pollen and testate amoebae peat succession record. Glob Planet Change 134:142–154. https://doi.org/10.1016/J.GLOPLACHA.2015.09.010. Elsevier

Ko S (2012) Tongue root harmony and vowel contrast in Northeast Asian languages. Cornell University, Ithaca. https://doi.org/10.2307/j.ctvckq52c

Ko S, Joseph A, Whitman JB (2014) Comparative consequences of the tongue root harmony analysis for proto-Tungusic, proto-Mongolic, and proto-Korean. In: Robbeets M, Bisang W (eds) Paradigm change: in the transeurasian languages and beyond, vol 161. John Benjamins, Amsterdam, pp 141–176. https://doi.org/10.1075/slcs.161.13ko

Kochmar NN (1994) Petroglyphs of Yakutia. History and culture of the East of Asia [Писаницы Якутии. История и культура Востока Азии]. The Institute of Archaeology and Ethnography of Russian Academy of Science, Novosibirsk

Koizumi F, Tokumaru Y, Yamaguchi O (1977) Asian musics in an Asian perspective, Report of Asian traditional performing arts. Heibonsha Limited Publishers, Tokyo

Kolesnikova VD (1972) The names of human body parts in Altaic languages [Названия частей тела человека в алтайских языках]. In: Tsintsius VI (ed) Sketches of comparative lexicology of Altaic languages [Очерки сравнительной лексикологии алтайских языков]. Nauka, Leningrad, pp 71–103

Kolinsky R, Lidji P, Peretz I, Besson M, Morais J (2009) Processing interactions between phonology and melody: vowels sing but consonants speak. Cognition 112:1–20. https://doi.org/10.1016/j.cognition.2009.02.014

Kolltveit G (2006) Jew's harps in European archaeology. Archaeopress, Oxford

Kolltveit G (2009) The Jew's Harp in Western Europe: trade, communication, and innovation, 1150–1500. Yearb Tradit Music 41:42–61. https://doi.org/10.2307/25735478. International Council for Traditional Music

Kolltveit G (2016) Jew's Harps of bone, wood and metal how to understand construction, classification and chronology. In: Eichmann R, Fang J, Koch LC (eds) Studien zur Musikarchäologie, vol 10. Verlag Marie Leidorf GmbH, Rahden, pp 63–73

Kolosovsky AS (1986) On musical instruments of Sakhalin Nivkhs [О музыкальных инструментах сахалинских нивхов]. In: Shubina OA, Vysokov MV (eds) Ethnographic research of the Sakhalin Regional History Museum [Этнографические исследования Сахалинского областного краеведческого музея]. Academy of Sciences of USSR, Yuzhno-Sakhalinsk, pp 56–70

Komissarov SA (1988) The complex of Ancient Chinese armament [Комплекс вооружения древнего Китая]. Nauka, Novosibirsk

Konchev VY (2004) The school of playing Altai Khomus [Школа игры на Алтайском Комусе]. Ministry of Culture and Cinema of Altai Republic, Gorno-Altaisk

Kononov AN (1980) The grammar of the language of Turkic Runic monuments of the 7–9th centuries AD [Грамматика языка тюркских рунических памятников VII-IX вв.]. Nauka, Leningrad

Kortlandt F (2003) Indo-Uralic and Altaic revisited. In: Johanson L, Robbeets MI (eds) Transeurasian verbal morphology in a comparative perspective: genealogy. Harrassowitz Verlag, Wiesbaden, pp 153–164

Kovalyov AA, Erdenebaatar D (2014) The discovery of a new culture of the advanced Bronze Age in central Eurasia: Munkh-Khairkhan culture [Открытие в центре Евразии новой культуры эпохи развитой бронзы (мунх-хайрханская культура)]. Russian Archaeol Yearb 4:194–225

Krämer M (2005) Vowel harmony and correspondence theory. Language, vol 81. Mouton de Gruyter, Berlin. https://doi.org/10.1353/lan.2005.0185

Kraxenberger M, Menninghaus W (2016) Emotional effects of poetic phonology, word positioning and dominant stress peaks in poetry reading. Sci Stud Lit 6:298–313. https://doi.org/10.1075/ssol.6.2.06kra. John Benjamins

Kreinovich EA (1927) A diary [Дневник]. Sakhalin Regional Museum of Natural Science, Yuzhno-Sakhalinsk. doi:6473-122

Kreinovich EA (1928) A diary [Дневник]. Sakhalin Regional Museum of Natural Science, Yuzhno-Sakhalinsk. doi:6473-123

Kreinovich EA (1937) Phonetics of Nivkh (Gilyak) language [Фонетика нивхского (гиляцкого) языка]. Uchpedgiz [Учпедгиз], Moscow

Kreinovich EA (1973) Nivkhgu: the enigmatic inhabitants of Sakhalin and Amur [Нивхгу: загадочные обитатели Сахалина и Амура]. Nauka, Moscow

Kreinovich EA (1979) Nivkh language [Нивхский язык]. In: Sanzheyev GD (ed) Languages of Asia and Africa [Языки Азии и Африки], vol 3. Nauka, Moscow, pp 295–392

Kremenetski CV, Sulerzhitsky LD, Hantemirov R (1998) Holocene history of the northern range limits of some trees and shrubs in Russia. Arct Alp Res 30:317–333

Krispijn T (2010) Musical ensembles in ancient Mesopotamia. In: Dumbrill R, Finkel I (eds) Proceedings of the international conference of near Eastern Archaeomusicology, held at the British Museum, December 4–6, 2008. Iconea Publications, London, pp 125–150

Krumhansl CL (2002) Music: a link between cognition and emotion. Curr Dir Psychol Sci 11:45–50. https://doi.org/10.1111/1467-8721.00165

Kungurov SN (1994) Udmurt traditional musical instruments [Удмуртские традиционные музыкальные инструменты]. Ministry of Culture of Udmurt Republic, Izhevsk

Kuz'mina EE (2008) In: Mair VH (ed) The prehistory of the Silk Road. University of Pennsylvania Press, Philadelphia

Kuzmin YV (2000) Radiocarbon chronology of the Stone Age cultures on the Pacific Coast of Northeastern Siberia. Arctic Anthropol 37:120–131. https://doi.org/10.2307/40316521. University of Wisconsin Press

Kypchakova L'V (2006) On the matter of the Tree Cult of Altaians [К вопросу о культе деревьев у алтайцев]. Siberian Pedagogical J 3:130–136

Kyzlasov LR (1986) The most ancient Khakassia [Древнейшая Хакасия]. Moscow State University named after Lomonosov, Moscow

L'vova EE, Oktiabr'skaya IV, Sagalayev AM (1989) The traditional worldview of Turks of Southern Siberia. Humanbeing. Society [Традиционное мировоззрение тюрков Южной Сибири. Человек. Общество]. Nauka, Novosibirsk

Ladefoged P, Disner SF (2012) Vowels and consonants, 3rd edn. Wiley-Blackwell, Chichester, West Sussex

Lebedintsev AI (1998) Maritime cultures of the north coast of the Sea of Okhotsk. Arctic Anthropol 35:296–320. https://doi.org/10.2307/40316471. University of Wisconsin Press

Ledang OK (1972) On the acoustics and the systematic classification of the Jaw's Harp. Yearb Int Folk Music Counc 4:95–103

Lehiste I, Peterson GE (1961) Some basic considerations in the analysis of intonation. J Acoust Soc Am 33:419–425. https://doi.org/10.1121/1.1908681

Leipp É (1963) Un «vocoder» mécanique: La guimbarde. Annales des Télécommunications 18:82–87. https://doi.org/10.1007/bf03011324. Springer-Verlag

Lerdahl F (1987) Timbral hierarchies. Contemp Music Rev 2:135–160. Taylor & Francis Group. https://doi.org/10.1080/07494468708567056

Leontyev NV (1978) Anthropomorphic images of Okunevskaya culture [Антропоморфные изображения окуневской культуры]. In: Siberia, Central and Eastern Asia in antiquity: neolithic and Age of Metal [Сибирь, Центральная и Восточная Азия в древности. Неолит и эпоха металла]. Nauka, Novosibirsk, pp 88–118

Leontyev AE (1986) Volgo-Baltic trading route during the 9th century AD [Волжско-Балтийский торговый путь в IX в.]. Brief Commun Inst Archaeol 183:3–9

Leshchenko NV, Prokopets SP (2015) Medieval musical instruments (based on data of Primorye artifacts [Средневековые музыкальные инструменты (по материалам памятников Приморья)]. Russia Pacific 4:222–235

Levin MG (1963) Ethnic origins of the people of Northeastern Asia (trans: Michael HN). University of Toronto Press, Toronto

Levin TC, Suzukei V (2006) Where rivers and mountains sing: sound, music, and nomadism in Tuva and beyond. Indiana University Press, Bloomington

Levitskaya LS (1984) Proto-Turkic vocalism [Пратюркский вокализм]. In: Tenishev ER (ed) Comparative historic grammar of Turkic languages. Phonetics [Сравнительн-о-историческая грамматика тюркских языков. Фонетика], Moscow, pp 67–82

Lewis G (1967) Turkish grammar. Oxford University Press, Oxford

Li H (1956) A comparative study of the Jew's Harps among the Aborigines of Formosa and East Asia. Bull Inst Ethnol Acad Sin 1:85–140

Li H, Zhao X, Zhao Y, Li C, Si D, Zhou H, Cui Y (2011) Genetic characteristics and migration history of a bronze culture population in the West Liao-River valley revealed by ancient DNA. J Hum Genet 56:815–822. https://doi.org/10.1038/jhg.2011.102

Liberman AM, Mattingly IG (1989) A specialization for speech perception. Science 243:489–494. https://doi.org/10.1126/science.2643163

Lieberman P (2008) A wild 50,000-year ride. In: Bengtson JD (ed) In Hot Pursuit of language in prehistory. John Benjamins Publishing Company, Amsterdam, pp 359–371. https://doi.org/10.1075/z.145.27lie

Ligeti L (1971) Altaic theory and lexicostatistics [Алтайская теория и лексикостатистика]. Voprosy Jazykoznanija (Topics in the Study of Language) 15:21–33

Lipskii AN (1956) Some problems of Tashtyk culture in light of Siberian ethnography of 200 BC-400 AD [Некоторые вопросы таштыкской культуры в свете сибирской этнографии (II-й в. до н.э. - IV в. н.э.)]. In: Mishurov AM (ed) Local history collection [Краеведческий сборник], vol 1. Khakass Book Publisheers, Abakan, pp 9–92

Lisker L (1974) On "explaining" vowel duration. Glossa 8:233–246

Liu L, Chen X (2012) The archaeology of China: from the late Paleolithic to the early bronze age. Cambridge University Press, Cambridge

Loewy JV (1995) The Musical Stages of Speech: A Developmental Model of Pre-Verbal Sound Making. Music Therapy 13:47–73. https://doi.org/10.1093/mt/13.1.47. Oxford University Press

London J (2004) Hearing in time: psychological aspects of musical meter. Oxford University Press, Oxford, NY

Lopatin IA (1922) Amuric, Ussurian and Sungarian Goldi. An attempt of ethnographic investigation [Гольды амурские, уссурийские, сунгарийские. Опыт этнографического исследования], vol 17. ZOIAK VRGO, Vladivostok

Louhivuori A (2006) Tonal development of a child's song improvisations: a case study. In: Paananen P, Fredrikson M (eds) The proceedings of the first European conference on developmental psychology of music, 17–19 November 2005, University of Jyväskylä, Finland. University of Jyväskylä, Jyväskylä, pp 287–290

Madison G, Paulin J (2010) Ratings of speed in real music as a function of both original and manipulated beat tempo. J Acoust Soc Am 128:3032–3040. https://doi.org/10.1121/1.3493462

Malmberg B (1962) Problem of a method in synchronic phonetics [Проблема метода в синхронной фонетике]. In: The new in linguistics [Новое в лингвистике], Zvegintsev VA (ed), tran: Shevoroshkina VV, vol 2. Foregin Literature Publishing, Moscow, pp 340–389

Mamcheva NA (2005) Nivkh Jew's Harps [Нивхские варганы]. Courier of Sakhalin Museum 12:271–284

Mamcheva NA (2008) Sacral significance of Nivkh musical instruments [Сакральное значение музыкальных инструментов нивхов]. Izvestia: Herzen Univ J Human Sci 86:132–135

Mamcheva NA (2012) Musical instruments in Nivkh traditional culture [Музыкальные инструменты в традиционной культуре нивхов]. GUP Sakhalinskaya Regional Press, Yuzhno-Sakhalinsk

Maslov AL (1911) Illustrated description of musical instruments from Dashkovskii Ethnographic Museum in Moscow [Иллюстрированное описание музыкальных инструментов, хранящихся в Дашковском Этнографическом Музее в Москве], vol 2. The society of connoisseurs of Natural Sciences, Anthropology and Ethnography, Moscow

Mattissen J (2003) Dependent-head synthesis in Nivkh: a contribution to a typology of polysynthesis. John Benjamins Publishing, Amsterdam

Matusky P (2008) Borneo: Sabah, Sarawak, Brunei, and Kalimantan. In: Miller T, Williams S (eds) The Garland handbook of Southeast Asian music. Routledge, London, pp 406–414

Maynard-Smith J (1976) Evolution and the theory of games. Am Sci 64:41–45

Mazel L (1972) Problems of Classical Harmony [Проблемы классической гармонии]. Muzyka, Moscow

Mazepus VV (1989) On physical grounds of sound generation on Jew's Harp [О физических основах звукообразования при игре на варганах]. In: Sheikin Y (ed) Musical ethnography of North Asia [Музыкальная этнография Северной Азии], vol 10. Novosibirsk State Conservatory named after Glinka, Novosibirsk, pp 155–161

Mazepus VV, Galitskaya SP (1997) Musical culture of Siberia. Traditional culture of indigenous people of Siberia [Музыкальная культура Сибири. Традиционная культура коренных народов Сибири], Shindin BA (ed), vol 1. Novosibirsk State Conservatory named after Glinka, Novosibirsk

Mazhitov NA (1981) Burial mounds of Southern Urals VIII–XII centuries [Курганы Южного Урала VIII–XII вв.]. Nauka, Moscow

McLean M (1999) Weavers of Song: polynesian music and dance. University of Hawai'i Press, Honolulu

McPhee C (1955) Children and music in Bali. In: Mead M, Wolfenstein M (eds) Childhood in contemporary cultures. University of Chicago Press, Chicago, pp 70–98

Mei J, Wang P, Chen K, Lu W, Wang Y, Liu Y (2015) Archaeometallurgical studies in China: some recent developments and challenging issues. J Archaeol Sci 56:221–232. https://doi.org/10.1016/J.JAS.2015.02.026. Academic Press

Melnikov GP (1966a) Morphological structure of language and means of word differentiation [Морфологический строй языка и средства словоразграничения]. In: Studies in phonology [Исследования по фонологии]. Nauka, Moscow, pp 263–284

Melnikov GP (1966b) M.A. Tcherkassky. Turkic vocalism and vowel harmony [М. А. Черкасский. Тюркский вокализм и сингармонизм]. Voprosy Jazykoznanija (Topics in the Study of Language) 15:129–137

Mersenne M (1957) Harmonie universelle: the books on instruments. Martin Nijhoff, Leiden

Miller RA (1967) The Japanese language. University of Chicago Press, Chicago

Miller RA (1996) Languages and history: Japanese, Korean and Altaic. Institute for Comparative Research in Human Culture, Oslo

Miniayev SS (1991) Zone of the Scythian world at the northeast of Chinese People's Republic: finds and problems [Зона скифского мира на северо-востоке КНР: находки и проблемы]. Soc State China [Общество и государство в Китае] 22:171–175. Moscow

Möbius B (2003) Gestalt psychology meets phonetics – an early experimental study of intrinsic F0 and intensity. In: Solé MJ, Recasens D, Romero J (eds) Proceedings of the 15th international congress of phonetic sciences: Barcelona 3–9 August 2003. Causal Productions Pty, Barcelona, pp 2677–2680

Mokhova L, Tarasov P, Bazarova V, Klimin M (2009) Quantitative biome reconstruction using modern and late Quaternary pollen data from the southern part of the Russian Far East. Quat Sci Rev 28:2913–2926. https://doi.org/10.1016/J.QUASCIREV.2009.07.018. Pergamon

Monelle R (2006) The musical topic: hunt, military and pastoral. Indiana University Press, Indianapolis

Montagu J, Burton J (1971) A proposed new classification system for musical instruments. University of Illinois Press on behalf of Society for Ethnomusicology 15:49–70. https://doi.org/10.2307/850387. University of Illinois Press Society for Ethnomusicology

Morgan DA (2008) Organs and bodies: the Jew's Harp and the anthropology of musical instruments. University of British Columbia, Vancouver. https://doi.org/10.14288/1.0066561

Morgan DA (2017) Speaking in tongues: music, identity, and representation in Jew's harp communities. SOAS University of London, London

Morton ES (1977) On the Occurrence and Significance of Motivation-Structural Rules in Some Bird and Mammal Sounds. Am Natural 111:855–869. https://doi.org/10.1086/283219. The University of Chicago Press The American Society of Naturalists

Nadeliayev VM (1960) The project of universal unified phonetic transcription (UUPT) [Проект универсальной унифицированной фонетической транскрипции (УУФТ)]. Nauka, Moscow/Leningrad

Nadeliayev VM (1980) Articulatory classification of vowels [Артикуляционная классификация гласных]. In: Nadeliayev VM (ed) Phonetic studies of Siberian languages [Фонетические исследования по сибирским языкам]. Nauka, Novosibirsk, pp 3–91

Nadeliayev VM, Nasilov DM, Tenishev ER, Shcherbak AM (1969) Old Turkic dictionary [Древнетюркский словарь]. Nauka, Leningrad

Nazaikinsky YV (1972) On psychology of human musical perception [О психологии музыкального восприятия]. Muzyka, Moscow

Nazaikinsky YV (1973) On Constants in Perception of Music [О Константности в Восприятии Музыки]. In: Nazaikinsky YV (ed) Musical Art and Science [Музыкальное искусство и наука], vol 2. Muzyka [Музыка], Moscow, pp 59–98

Nazaikinsky YV (1977) Interconnection between the intervallic-based and degree-based representation of music in the development of a musical ear [Взаимосвязи интервальных и ступеневых представлений в развитии музыкального слуха]. In: Agazhanov A (ed) Development of musical hearing [Воспитание музыкального слуха]. Muzyka, Moscow, pp 25–77

Nazaikinsky YV (1988) The sonic world of music [Звуковой мир музыки]. Muzyka [Музыка], Moscow

Nazaikinsky YV, Rags YN (1964) Perception of musical timbres and the significance of the individual harmonics in a sound [Восприятие музыкальных тембров и значение отдельных гармоник звука]. In: Skrebkov SS (ed) Application of the acoustic methods inmusicology [Применение акустических методов в музыкознании]. Muzyka, Moscow, pp 79–100

Newman SS (1933) Further experiments in phonetic symbolism. Am J Psychol 45:53–75. https://doi.org/10.2307/1414186

Newport EL, Hauser MD, Spaepen G, Aslin RN (2004) Learning at a distance II. Statistical learning of non-adjacent dependencies in a non-human primate. Cogn Psychol 49:85–117. https://doi.org/10.1016/J.COGPSYCH.2003.12.002. Academic Press

Nikolsky A (2015) Evolution of tonal organization in music mirrors symbolic representation of perceptual reality. Part-1: Prehistoric. Front Psychol 6. https://doi.org/10.3389/fpsyg.2015.01405

Nikolsky A (2016) Evolution of Tonal organization in music optimizes neural mechanisms in symbolic encoding of perceptual reality. Part-2: ancient to seventeenth century. Front Psychol. https://doi.org/10.3389/fpsyg.2016.00211

Nikolsky A (2017) On the methodology of the analysis of tonal organization of Jaw Harp music [К методам анализа тоновой организации варганной музыки]. In: Novikova OV (ed) Systemic methods of the research on musical culture, International scientific practical conference in memory of V.V. Mazepus [Системные методы изучения музыкальной культуры, Международная научно-практическая конференция памяти В.В.Мазепуса, 31/X-1/XI 2017]. Novosibirsk State Conservatory named after Glinka, Novosibirsk

Nikolsky A (2018) Commentary: the 'Musilanguage' model of language evolution. Front Psychol 9:75. https://doi.org/10.3389/fpsyg.2018.00075

Nikolsky A, Alekseyev EY, Alekseyev IY, Dyakonova V (2017) Prolegomena of modal organization of Jaw harp music on the example of the articulatory degrees of "talking khomus" [Пролегомена ладовой организации варганной музыки на примере использования артикуляционных ступеней в "говорящем" якутском хомусе]. In: Novikova OV (ed) Systemic methods of the research on musical culture, International scientific practical conference in memory of V.V. Mazepus [Системные методы изучения музыкальной культуры, Международная научно-практическая конференция памяти В.В.Мазепуса, 31/X-1/XI 2017]. Novosibirsk State Conservatory named after Glinka, Novosibirsk

Nikolsky A, Alekseyev EY, Alekseyev IY, Dyakonova V (2019a) What the “Talking Jew’s Harp” is saying: Jew’s harp and personal song as the foundation of timbre-oriented musical systems [О чем говорит «говорящий варган»: варган и личная песня как основание темброво-ориентированных музыкальных систем]. Lang Folklore Indigenous Peoples of Siberia 1:5–32. https://doi.org/10.25205/2312-6337-2019-1-5-32

Nikolsky A, Alekseyev EY, Alekseyev IY, Dyakonova V (2019b) The overlooked tradition of ‘personal music’ and its place in the evolution of music. Front Psychol

Nikolsky A, Alekseyev EY, Alekseyev IY, Dyakonova V (2020) How, where and when did the authentic Jaw Harp traditions form in Siberia and Far East [Как, где и когда складывались аутентичные варганные традиции Сибири и Дальнего Востока]. Languages and Folklore of Indigenous Peoples of Siberia 1

Nobuhiko C (2008) The music of the Ainu. In: Tokita AM, Hughes DW (eds) The Ashgate research companion to Japanese music. Ashgate Publishing, Ltd, Burlington, pp 323–344

Ogotoyev PP (1988) Problems of notation of Yakut khomus [Проблема нотации якутского хомуса]. In: Sheikin Y (ed) Musical ethnography of North Asia [Музыкальная этнография Северной Азии], vol 10. Novosibirsk State Conservatory named after Glinka, Novosibirsk, pp 150–154

Ohala JJ (1994) Towards a universal, phonetically-based, theory of vowel harmony. In: Farag AA, Yang J, Jiao F (eds) Proceedings of the 3rd international conference on spoken language processing. Yokohama, Japan September 18–22, 1994. University of California, Berkeley, Berkeley, pp 491–494

Ohala JJ, Eukel BW (1987) Explaining the intrinsic pitch of vowels. In: Channon R, Shockey L (eds) In honor of Ilse Lehiste, vol 60. Foris, Dordrecht, pp 207–215. https://doi.org/10.1121/1.2003351. Acoustical

Öhman SEG (1966) Coarticulation in VCV utterances: spectrographic measurements. J Acoust Soc Am 39:151–168. https://doi.org/10.1121/1.1909864

Okladnikov AP (1955) At the source of the culture of peoples of Far East [У истоков культуры народов Дальнего Востока]. In: Fyodorov GB (ed) По следам древних культур. От Волги до Тихого океана. The State Publishing of cultural-educational literature [Государственное издательство культурно-просветительной литературы], Moscow, pp 225–260

Okladnikov AP (1968) Representation of faces of ancient Amur [Лики древнего Амура]. West Siberian Book Publishing, Novosibirsk

Okladnikov AP (1971) Petroglyphs of Lower Amur [Петроглифы Нижнего Амура]. Nauka, Leningrad

Okladnikov AP, Mazin AI (1976) Petroglyphs of Olyokma river and Upper Amur Region [Писаницы реки Олёкмы и Верхнего Приамурья]. Nauka, Novosibirsk

Okladnikov AP, Mazin AI (1979) Petroglyphs of the Aldan river valley [Писаницы бассейна реки Алдан]. Nauka, Novosibirsk

Okladnikov AP, Zaporozhskaya VD (1972) Petroglyphs of Middle Lena [Петроглифы Средней Лены]. Nauka, Leningrad

Oleszczak Ł, Michalczewski K, Borodovskii AP, Pokutta DA (2018) Chultukov Log 9 – A settlement from the Xiongnu-Xianbei-Rouran period in the Northern Altai. Eurasian Prehistory 14:153–178

Ostrovsky AB (1997) Nivkh mythology and belief system [Мифология и верования нивхов]. Petersburg vostokovedeniye, Sankt-Petersburg

Owren MJ, Seyfarth RM, Cheney DL (1997) The acoustic features of vowel-like grunt calls in chacma baboons (Papio cynocephalus ursinus): implications for production processes and functions. J Acoust Soc Am 101:2951–2963. https://doi.org/10.1121/1.418523

Pacholczyk J (1993) Early Arab suite in Spain: an investigation of the past through the contemporary living traditions. Rev Musicol 16:358–366. Available https://www.jstor.org/stable/20795894. Accessed 21 February 2016

Pakendorf B (2007) Contact in the prehistory of the Sakha (Yakuts): linguistic and genetic perspectives. Netherlands Graduate School of Linguistics, (LOT), Utrecht

Panfilov VZ (1962) The grammar of Nivkh language [Грамматика нивхского языка], vol 1. Academy of Science of USSR, Leningrad

Paraskeva S, McAdams S (1997) Influence of timbre, presence/absence of tonal hierarchy and musical training on the perception of musical tension and relaxation schemas. In: Proceedings of the International Computer Music Conference, Thessaloniki, Greece, September 25–30, 1997. Michigan Publishing, Ann Arbor, pp 438–441

Park JS, Gordon RB (2007) Traditions and transitions in Korean bronze technology. J Archaeol Sci 34:1991–2002. https://doi.org/10.1016/j.jas.2007.01.010

Pashina OA, Vasilyeva YY, Danchenkova NY, Dorokhova YA, Lapina VA, Matsiyevsky IV (2005) Folk musical creativity [Народное музыкальное творчество]. Kompozitor [Композитор], Sankt-Petersburg

Pashkovskii O (2012) Jew's harps XIII-XVII centuries in archaeological finds [Дримби XIII – початку XVII ст. за археологічними знахідками]. New research on the monuments of the Cossack era in Ukraine [Нові дослідження пам'яток козацької доби в Україні] 12:248–253

Pegg C (2001) Mongolian music, dance and oral narrative: performing diverse identities, vol 1. University of Washington Press, Seattle

Pevnov AM (2009) On some features of Nivkh and Uilta. In: Toshirō T (ed) Saharin gengo sekai. Hokkaidō daigaku daigakuin bungaku kenkyū-ka, Sapporo, pp 113–125

Picken L (1957) The music of far Eastern Asia. In: Wellesz E (ed) New Oxford history of music: ancient and oriental music, vol 1. Oxford University Press, London/New York, pp 135–189

Pignocchi JL (2002) Los Mapuche y el trompe [The Mapuche and the Trompe]. Vierundzwanzigsteljahrsschrift der International Maultrommelvirtuosengenossenschaft 10:31–38

Pignocchi JL (2004) Trompas in the Zapata Gollan collection, found in the Ruins of Santa Fe la Vieja, from the years 1573–1660. J Int Jew's Harp Soc 2:11–21

Pressnitzer D, McAdams S, Winsberg S, Fineberg J (2000) Perception of musical tension for nontonal orchestral timbres and its relation to psychoacoustic roughness. Percept Psychophys 62:66–80. Springer-Verlag. https://doi.org/10.3758/BF03212061

Podmaskin VV (2003) Folk musical instruments of Tunguso-Manchus and Paleoasians [Народные музыкальные инструменты тунгусо-маньчжуров и палеоазиатов: проблемы типологии]. In: Bereznitsky SV (ed) Typology of culture of the indigenous peoples of Far East Russia: materials for the historic ethnographic atlas [Типология культуры коренных народов Дальнего Востока России: Материалы к историко-этнографическому атласу]. Dalnauka, Vladivostok, pp 102–119

Pope GG (1989) Bamboo and human evolution. Nat Hist 10:49–57

Popov AI (1949) Data on history of Yakut religion of the former Viliui District [Материалы по истории религии якутов бывшего Вилюйского округа]. In: Tolstov SP (ed) The anthology of the Museum of anthropology and ethnography [Сборник музея антропологии и этнографии], vol 11. Academy of Science of USSR [Изд-во Академии наук СССР], Moscow/Leningrad, pp 255–323

Poppe NN (1965) Introduction to Altaic Linguistics. Otto Harrassowitz, Wiesbaden

Porter J, Powers HS, Cowdery J, Widdess R, Davis R, Perlman M et al (2001) Mode. Modal scales and traditional music. Middle East and Asia. In: The new grove dictionary of music and musicians. Macmillan, London. https://doi.org/10.1093/gmo/9781561592630.article.43718

Poss NF (2012) Hmong music and language cognition: An interdisciplinary investigation. The Ohio State University, Columbus

Povel D-J, Jansen E (2001) Perceptual mechanisms in music processing. Music Perception 19:169–197. https://doi.org/10.1525/mp.2001.19.2.169. University of California Press

Powers HS, Wiering F (2001) Mode. The term. Medieval modal theory. Modal theories and polyphonic music. In: The new grove dictionary of music and musicians. Macmillan, London. https://doi.org/10.1093/gmo/9781561592630.article.43718

Prakash B, Tripathi V (1986) Iron technology in ancient India. Hist Metallur:568–579

Pugh-Kitingan J (1977) Huli language and instrumental performance. Ethnomusicology 21:205–232. doi:papers3://publication/uuid/6F7213F0-9DD7-41E8-811E-348C911DED84

Pycha A, Nowak P, Shin E, Shosted R (2003) Phonological rule-learning and its implications for a theory of vowel harmony. In: Tsujimura M, Garding G (eds) WCCFL proceedings. Cascadilla Press, Somerville, pp 101–114

Ramstedt G (1957) Einführung in die altaische Sprachwissenschaft, Aalto P (ed), vol 1–2. Suomalais-Ugrilainen Seura, Helsinki

Ratner LG (1980) Classic music: expression, form, and style. Schirmer Books, New York

Read G (1969) Music notation: a manual of modern practice. Allyn and Bacon, Boston

Reformatskii AA, Vinogradov VA (1969) Harmony of vowels, accent and word prosody [Сингармонизм, ударение и просодия слова]. Voprosy Jazykoznanija (Topics in the Study of Language) 18:123–125

Rho JY, Ashman RB, Turner CH (1993) Young's modulus of trabecular and cortical bone material: Ultrasonic and microtensile measurements. J Biomech 26:111–119. https://doi.org/10.1016/0021-9290(93)90042-D

Robb J (1993) A social prehistory of European languages. Antiquity 67:747–760. https://doi.org/10.1017/S0003598X00063766. Cambridge University Press

Robbeets M (2015) Diachrony of verb morphology: Japanese and the Transeurasian languages. De Gruyter Mouton, Berlin

Robbeets M (2017) The language of the Transeurasian farmers. In: Robbeets M, Savelyev A (eds) Language dispersal beyond Farming. Benjamins, Amsterdam, pp 145–162. https://doi.org/10.1093/jole/lzy007

Robbeets M, Bouckaert R (2018) Bayesian phylolinguistics reveals the internal structure of the Transeurasian family. J Lang Evol 3:145–162. https://doi.org/10.1093/jole/lzy007. Oxford University Press

Roberts BW, Thornton CP, Pigott VC (2009) Development of metallurgy in Eurasia. Antiquity 83:1012–1022. https://doi.org/10.1017/S0003598X00099312

Robinson E (1942) Shell fishhooks of the California Coast

Roesner EH (2001) Rhythmic modes (modal rhythm). In: The new grove dictionary of music and musicians. Macmillan, London. https://doi.org/10.1093/gmo/9781561592630.article.23337

Róna-Tas A (1986) In: Abraham V (ed) Language and history: contributions to comparative Altaistics, vol 25. Universitas Szegediensis de Attila Jozsef Nominata, Szeged

Róna-Tas A (2011) Recent trends in Mongolic studies, Acta Orientalia Academiae Scientiarum Hungaricae, vol 64. Magyar Tudományos Akadémia

Rose S, Walker R (2011) Harmony systems. In: Goldsmith J, Riggle J, Alan CLY (eds) The handbook of phonological theory. Wiley-Blackwell, Oxford, pp 240–290. https://doi.org/10.1002/9781444343069.ch8

Rouget G (1985) Music and trance: a theory of the relations between music and possession. University of Chicago Press

Roux L, Charles Constant FM (1950) De Bergpapoea's van Nieuw – Guinea en hun Woongevied, Koninklijk, vol 2. Brill, Leiden

Rozycki WV (2006) Review of Robbeets, Martine 2005. Is Japanese related to Korean, Tungusic, Mongolic and Turkic? Mongolian Stud 28:114–11\

Rudaya NA, Vasilevskii AA, Grishchenko VA, Mozhaev AV (2013) Environmental conditions of the late paleolithic and early neolithic sites in Southern Sakhalin. Archaeol Ethnol Anthropol Eurasia 41:73–82. https://doi.org/10.1016/j.aeae.2013.11.007

Russo FA, Vuvan DT, Thompson WF (2006) Setting words to music: effects of phoneme on the experience of interval size. In: Baroni M, Addessi AR, Caterina R, Costa M (eds) 9th

International Conference on Music Perception & Cognition, Bologna, Italy. Bolonia University Press, Bologna, pp 1246–1250. doi:88-7395-155-4
Rybatzki V (2003) Intra-Mongolic taxonomy. In: The Mongolic languages. Routledge, London, pp 364–390
Sachs C (1917) Die Maultrommel. Eine typologische Vorstudie. Zeitschrift für Ethnologie 49:185–200. https://doi.org/10.2307/23031384. Dietrich Reimer Verlag GmbH
Sachs C (1953) Rhythm and Tempo: a study in music history. Norton, New York
Sachs C (1960) Primitive and medieval music: a parallel. J Am Musicol Soc 13:43–49
Sachs C (2008) The rise of music in the ancient world, East and West. Dover Publications, New York
Sadie S (2001) Movement. In: Sadie S, Tyrrell J (eds) The New Grove dictionary of music and musicians. Macmillan Publishers, London. https://doi.org/10.1093/gmo/9781561592630.article.19258
Sagalayev AM, Oktiabr'skaya IV (1990) Traditional worldview of Turks of Southern Siberia. Sign and ritual. [Традиционное мировоззрение тюрков Южной Сибири. Знак и ритуал]. Nauka, Novosibirsk
Sam S-A (2008) The Khmer people of Cambodia. In: Miller T, Williams S (eds) The Garland handbook of Southeast Asian Music. Routledge, London, pp 85–120
Sapir E (1929) A study in phonetic symbolism. J Exp Psychol 12:225–239. https://doi.org/10.1037/h0070931
Schibler J (2001) Experimental production of Neolithic bone and antler tools. In: Choyke AM, Bartosiewicz L (eds) Crafting Bone: skeletal technologies through time and space. Proceedings of the 2nd meeting of the (ICAZ) Worked Bone Research Group Budapest, 31 August–5 September 1999. The Aquincum Museum & ICAZ, Budapest, pp 49–60
Schurr TG, Sukernik RI, Starikovskaya YB, Wallace DC (1999) Mitochondrial DNA variation in Koryaks and Itel'men: population replacement in the Okhotsk Sea-Bering Sea region during the neolithic. Am J Phys Anthropol 108:1–39. https://doi.org/10.1002/(SICI)1096-8644(199901)108:1<1::AID-AJPA1>3.0.CO;2-1
Schwanghart W, Frechen M, Kuhn NJ, Schütt B (2009) Holocene environmental changes in the Ugii Nuur basin, Mongolia. Palaeogeogr Palaeoclimatol Palaeoecol 279:160–171. https://doi.org/10.1016/J.PALAEO.2009.05.007. Elsevier
Seliutina IY (2017) Principles of organization of synharmonic systems in South Siberian Turkic languages as indicators of language complexity [Принципы организации сингармонических систем в южносибирских тюркских языках как индикаторы языковой сложности]. NSU Vestnik. Series: Linguistics and Intercultural Communication 15:5–26. https://doi.org/10.25205/1818-7935-2017-15-4-5-26
Seliutina IY, Urtegeshev AY, Letiagin AY, Shevela AI, Dobrinina AA, Esenbayeva GA, Savelov AA, Rezakova MV, Ganenko YA (2012) Articulatory databases of indigenous Turkic ethnoses of Southern Siberia (according to magnetic resonance tomography and digital rentgenography) [Артикуляторные базы коренных тюркских этносов Южной Сибири (по данным МРТ и цифровой рентгенографии)]. Russian Academy of Science, Siberian Department, Novosibirsk
Sergeyeva NF (1981) The oldest copper metallurgy in South Eastern Siberia [Древнейшая металлургия меди юга Восточной Сибири]. Nauka, Moscow
Sermier C (2002) Mongolia: empire of the steppes. Odyssey, Fyshwick
Shcherbak AM (1970) Comparative phonetics of Turkic languages [Сравнительная фонетика тюркских языков]. Nauka, Leningrad
Shcherbak AM (1994) Introduction into comparative research of Turkic languages [Введение в сравнительное изучение тюркских языков]. Nauka, Sankt-Petersburg
Shchurov V (1995) Khomus: Jew's Harp music of Turkic peoples in the Urals, Siberia, and Central Asia. PAN Records, Leiden. doi:PAN 2032CD
Sheikin YI (1986a) Musical instruments of Jurchen [Музыкальные инструменты чжурчженей]. In: Matsiyevskii I (ed) Problems of traditional instrumental music of peoples of the USSR: instrument-performer-music [Проблемы традиционной инструментальной музыки народов СССР: Инструмент-исполнитель-музыка]. LGITMiK, Leningrad, pp 100–109

Sheikin YI (1986b) Musical instruments of Ude: etimology, construction, playing patterns [Музыкальные инструменты удэ: этимология, конструкции, наигрыши]. In: Golovneva II (ed) Musical creativity of the peoples of Siberia and Far East [Музыкальное творчество народов Сибири и Дальнего Востока]. Novosibirsk State Conservatory named after Glinka, Novosibirsk, pp 38–72

Sheikin YI (1990) Instrumental music of Yugra [Инструментальная музыка Югры]. Novosibirsk State Conservatory named after Glinka, Novosibirsk

Sheikin YI (2002) History of music culture of Siberian ethnicities: a comparative historic investigation [История музыкальной культуры народов Сибири: сравнительно-историческое исследование]. Eastern Literature, Russian Academy of Science [Восточная литература РАН], Moscow

Shiraishi H (2006) Topics in Nivkh phonology. In: Groningen dissertations in linguistics (GRODIL). The University of Groningen, Groningen. https://doi.org/10.1002/anie.201611532

Shishigin SS (2015) Learn to play khomus [Хомус тарда үөрэниҥ]. RIO Media Holding, Yakutsk

Shulga PI (2015) The Yuhuangmiao Cemetery in Northern China (the 7th–6th centuries B.C.) [Могильник Юйхуанмяо в Северном Китае (VII-VI века до нашей эры)], Kozhin PM (ed). Publishing Department of the Institute of Archaeology and Ethnography, Russian Academy of Science, Novosibirsk

Shulga PI, Girchenko EG (2013) New data on the cultural connections of the Yuhuangmiao in Northern China [Новые данные о взаимосвязях культуры Юйхуанмяо в Северном Китае]. In: Derevianko AP, Molodin VI (eds) Problems of archaeology, ethnography, anthropology of Siberia and adjacent territories [Проблемы археологии, этнографии, антропологии Сибири и сопредельных территорий], vol 19. Institute of archaeology and ethnography of Russian Academy of Science, Novosibirsk, pp 387–390

Sidéra I (2011) Fabriquer des cuillers en os: L'exemple de Kovacevo. Studia Praehistorica 14:55–62

Simukhin AI (2006) Technological traditions in production of nonferrous metallic artifacts in Transbaikalia [Технологические традиции в производстве изделий из цветного металла в Забайкалье]. State Buryat University, Russian Academy of Science, Ulan-Ude

Skrebkov S (1973) Artistic principles of musical styles [Художественные принципы музыкальных стилей]. Muzyka [Музыка], Moscow

Spatz H-C, O'Leary EJ, Vincent JFV (1996) Young's Moduli and Shear Moduli in Cortical Bone. Proc Biol Sci 263:287–294. https://doi.org/10.2307/50610. Royal Society

Starikovskaya EB, Sukernik RI, Derbeneva OA, Volodko NV, Ruiz-Pesini E, Torroni A, Brown MD et al (2005) Mitochondrial DNA diversity in indigenous populations of the Southern Extent of Siberia, and the origins of Native American Haplogroups. Ann Hum Genet 69:67–89. https://doi.org/10.1046/j.1529-8817.2003.00127.x

Starostin SA (1991) The Altaic problem and origins of Japanese language [Алтайская проблема и происхождение японского языка]. Nauka, Moscow

Starostin SA, Dybo AV, Mudrak OA (2003) Etymological dictionary of the Altaic languages. Brill, Leiden

Startsev AF (2017) Ethnic beliefs of Tungus-Manchurians regarding nature and society [Этнические представления тунгусо-маньчжуров о природе и обществе]. Dal'nauka, Vladivostok

Stebich M, Mingram J, Han J, Liu J (2009) Late Pleistocene spread of (cool-)temperate forests in Northeast China and climate changes synchronous with the North Atlantic region. Glob Planet Change 65:56–70. https://doi.org/10.1016/j.gloplacha.2008.10.010. Elsevier

Suleimanova RN (1994) The remnants of shamanism amongst Bashkirs [Пережитки шаманства у башкир]. In: Gabdrafikov I'd, Kiyekbayev MD, Shakurova FA (eds) The ethnological research in Bashkortostan [Этнологические исследования в Башкортостане]. Russian Academy of Science, Ufa, pp 119–130

Sun Y (2006) Colonizing China's Northern Frontier: Yan and her neighbors during the early Western Zhou period. Int J Hist Archaeol 10:159–177. https://doi.org/10.1007/s10761-006-0005-3

Sun Y (2017) Identity and artifacts on the north central and northeastern frontier during the period of state expansion in the late second and the early first millennium BCE. In: Linduff KM, Sun Y,

Cao W, Liu Y (eds) Ancient China and its Eurasian Neighbors: Artifacts, Identity and Death in the Frontier, 3000–700 BCE. Cambridge University Press, Cambridge, pp 72–145. https://doi.org/10.1017/9781108290555.004

Sunik OP (1978) The editorial introduction [Предисловие редактора]. In: Sunik OP (ed) The sketches on comparative morphology of Altaic languages [Очерки сравнительной морфологии алтайских языков]. Nauka, Leningrad, pp 1–9

Suomi K (1983) Palatal Vowel Harmony: A Perceptually Motivated Phenomenon? Nordic J Linguist 6:1–35. https://doi.org/10.1017/S0332586500000949. Cambridge University Press

Surazakov LS (2014) Altaic shamanism [Алтайский шаманизм]. In: Yekeyeev NV (ed) Altaians: ethnic history. Traditional culture. Modern development [Алтайцы: Этническая история. Традиционная культура. Современное развитие]. The Institute of Scientific Research named after Surazakov, Gorno-Altaisk, pp 330–349

Suzukei V (1989) Tuvan traditional musical instruments [Тувинские традиционные музыкальные инструменты]. Tuvan Book Publishing [Тувинское книжное издательство], Kyzyl

Suzukei V (2006) The configuration of the development of Tuvan musical culture: dynamics of the axiological aspect [Конфигурация развития музыкальной культуры Тувы: динамика аксиологического аспекта]. Kemerovo University of Culture and Arts, Kemerovo

Suzukei V (2010) Khomus in the traditional culture of the Tuvans [Хомус в традиционной культуре тувинцев]. Tyvapoligraf, Kyzyl

Svantesson JO (1985) Vowel harmony shift in Mongolian. Lingua 67:283–327. https://doi.org/10.1016/0024-3841(85)90002-6. North-Holland

Svantesson JO (2017) Sound symbolism: the role of word sound in meaning. Wiley Interdisciplinary Rev Cogn Sci:e01441. https://doi.org/10.1002/wcs.1441

Svantesson JO, Tayanin D (2003) Sound symbolism in Kammu expressives. In: Solé M, Recasens D, Romero J (eds) Proceedings of the 15th international congress of phonetic sciences. Barcelona. Universitat Autònoma de Barcelona, Barcelona, pp 2689–2692

Swain JP (2002) Harmonic Rhythm: Analysis and Interpretation. Oxford University Press, Oxford

Swan AJ (1943) The nature of the Russian folk-song. Musical Q 29:498–516. https://doi.org/10.1093/mq/XXIX.4.498. Oxford University Press

Syromyatnikov N (1971) The methodology of comparative historic research of the common morphemes in Altaic languages [Методика сравнительно-исторического изучения общих морфем в алтайских языках]. In: Sunik OP (ed) Problems of commonality of the Altaic languages [Проблема общности алтайских языков]. Nauka, Leningrad, pp 51–64

Tadagawa L (2007) Asian Excavated Jew's Harps: a checklist. J Int Jew's Harp Soc 4:5–11

Tadagawa L (2016) Asian Excavated Jew's Harps: a checklist (1) - Lamellate Jew's Harps (1). Inst Ethnomusicol Bull Tokyo Coll Music 5:57–70

Tadagawa L (2017a) The khomus is my red deer on which I fly through the middle world (Khomus in the shamanic practice of Tuva: Research issues). New Res Tuva 2:165–176. https://doi.org/10.25178/nit.2017.2.7

Tadagawa L (2017b) Asian excavated Jew's Harps: a checklist (2) – Lamellate Jew's Harps (2). Inst Ethnomusicol Bull Tokyo Coll Music 6:57–68

Tadysheva NO (2016) Traditional etiquette of Altaians in the aspect of family relations [Традиционный этикет алтайцев в семейно-родственном аспекте]. Tomsk J Linguist Anthropol Res 2:98–105

Tadysheva NO (2018) The reflection of tree cult in traditional beliefs of Sayano-Altaic Turks [Отражение культа дерева в традиционном мировоззрении тюрков Саяно-Алтая]. Tomsk State Univ J Hist 51:118–123. https://doi.org/10.17223/19988613/51/17

Taksami CM (1977) The cult system of Nivkhs [Система культов у нивхов]. In: Ol'derogge D (ed) The monuments of culture of peoples of Siberia and North (2nd half of the 19-20th centuries) [Памятники культуры народов Сибири и Севера: (2-я половина XIX - начало XX в.)], vol 23. Nauka, Leningrad, pp 90–116

Taksami CM (2009) The common elements in the traditional culture of peoples of Pacific North [Общие элементы в традиционной культуре народов Тихоокеанского Севера]. Izvestia: Herzen Univ J Human Sci 106:15–20

Taksami CM, Kosarev VD (1990) Who are you, Ainu? The draft of history and culture [Кто вы, айны? Очерк истории и культуры]. Mysl, Moscow

Tarasov P, Williams JW, Andreev A, Nakagawa T, Bezrukova E, Herzschuh U, Igarashi Y, Müller S, Werner K, Zheng Z (2007) Satellite- and pollen-based quantitative woody cover reconstructions for northern Asia: verification and application to late-quaternary pollen data. Earth Planet Sci Lett 264:284–298. https://doi.org/10.1016/J.EPSL.2007.10.007. Elsevier

Taskin VS (1984) Materials on the history of ancient nomadic peoples of the Donghu group [Материалы по истории древних кочевых народов группы дунху]. Nauka, Moscow

Taylor S (1989) The introduction and development of iron production in Korea: a survey. World Archaeol 20:422–433. https://doi.org/10.1080/00438243.1989.9980082. Taylor & Francis Group

Tchakhov AI (2012) Ancient Yakut musical instruments: the manufacturing technology [Старинные якутские музыкальные инструменты: технология изготовления] (trans: Shamayeva MI). Alaas, Yakutsk

Tcherkassky MA (1965) Turkic vocalism and synharmonism. The attempt of historic-typological investigation [Тюркский вокализм и сингармонизм. Опыт историко-типологического исследования]. Nauka, Moscow

Tchernetsov VN (1987) Sources on ethnography of Western Siberia [Источники по этнографии Западной Сибири], Lukina NV, Ryndina OM (eds). The State Tomsk University, Tomsk

Tenishev ER (1997) Altaic languages [Алтайские языки]. In: Languages of the world. Turkic languages [Языки мира: Тюркские языки]. Russian Academy of Science, Kyrkyzstan, Bishkek, pp 7–16

Thiemel M (2001) Dynamics. In: The new grove dictionary of music and musicians. Macmillan, London. https://doi.org/10.1093/gmo/9781561592630.article.08458

Tiukhteneva SP, Khalemba A, Sherstova LI, Dobzhanskaya O (2006) Spiritual culture [Духовная культура]. In: Funk DA, Tomilov NA (eds) Turkic peoples of Siberia [Тюркские народы Сибири]. Nauka, Moscow, pp 429–462

Tompkins D (2010) How to wreck a nice beach : the vocoder from World War II to hip-hop: the machine speaks. Melville House, Brooklyn

Tong HW (2007) Occurrences of warm-adapted mammals in north China over the Quaternary Period and their paleo-environmental significance. Sci China Ser D Earth Sci 50:1327–1340. https://doi.org/10.1007/s11430-007-0096-7. Science in China Press

Torroni A, Sukernik RI, Schurr TG, Starikovskaya YB, Cabell MF, Crawford MH, Comuzzie AG, Wallace DC (1993) mtDNA variation of aboriginal Siberians reveals distinct genetic affinities with Native Americans. Am J Hum Genet 53:591–608

Touma HH (1996) The music of the Arabs. Amadeus Press, Portland, OR

Trevarthen C, Delafield-Butt JT, Schögler B (2011) Psychobiology of Musical Gesture: Innate Rhythm, Harmony and Melody in Movements of Narration. In: Gritten A, King E (eds) New Perspectives on Music and Gesture. Ashgate, Aldershot & Burlington, VT, pp 11–45

Trias S (2010) Helmholtz and coupled resonators acoustics In Jew's harp playing. Telemark University College, Notodden

Trubetzkoy NS (1939) Principles of phonology (trans: Balthaxe CAM). University of California Press, Berkeley

Trubetzkoy NS (2001) N.S. Trubetzkoy: studies in general linguistics and language structure (trans: Liberman A, Taylor M). Duke University Press, Durham

Tsintsyus VI (1977) Comparative dictionary of Tunguso-Manchuric languages. Materials for the etymological dictionary [Сравнительный словарь тунгусо-маньчжурских языков. Материалы к этимологическому словарю], vol 2. Nauka, Leningrad

Tsoumis GT (1991) Science and technology of wood: structure, properties, utilization. Van Nostrand Reinhold, New York

Tsvelyov NN (1976) The grasses of the USSR [Злаки СССР], Fyodorov AA, ed. Nauka, Leningrad

Turan FA (2006) Kinship in the Altaic World: Proceedings of the 48th Permanent International Altaistic Conference. In: Boikova EV, Rybakov RB (eds) Kinship in the Altaic World: Proceedings of the 48th Permanent International Altaistic Conference, Moscow 10–15 July, 2005. Harrasowitz Verlagh, Wiesbaden, pp 329–332

Turbat T (2017) Archaeological Jew's Harps from Ancient Nomadic Cultures of Eastern Eurasia [Евразийн зүүн хэсгийн эртний нүүдэлчдийн археологийн дурсгалаас илэрсэн хэл хуурууд]. In: Archaeological Survey (ed) Tsagaan Turbat, vol 36. Academic Council of the Institute of History and Archeology of the Ministry of Nature, Environment and Tourism, pp 169–180

Tylecote RF (2002) A history of metallurgy. Maney Pub., for the Institute of Materials, London

Tylecote RF, Owles E (1960) A second-century iron smelting site at Ashwicken, Norfolk. Norfolk Archaeol 32:142–162

Tyukhteneva SP (2015) Personality and society amongst Altaians: from clan affiliation to all-Altaic identity [Личность и общество у алтайцев: от родовой принадлежности до общеалтайской идентичности]. Bull Kalmyk Inst Humanitar Stud Russian Acad Sci 4:72–81

Uchida R, Catlin A (2008) Music of Upland Minorities in Burma, Laos, and Thailand. In: Miller T, Williams S (eds) The Garland handbook of Southeast Asian Music. Routledge, London, pp 303–316

Ushnitskiy VV (2016) The contribution of Russian explorers and researchers of the 18-early 20th centuries in the study of ethnogenesis of Sakha people [Вклад российских путешественников и исследователей XVIII - начала XX в. в изучение этногенеза народа саха]. Tomsk State Univ J:150–155. https://doi.org/10.17223/15617793/407/23

Vainshtein S (1991) The World of Nomads of the Center of Asia [Мир кочевников центра Азии]. Nauka, Moscow

Val'kova VB (1992) Musical thematicism, cognition, culture [Музыкальный тематизм, мышление, культура]. State Nizhegorodskii University, Nizhnii Novgorod

Vasilevich GM (1969) Evenki: the historic-ethnographic sketches (18th-beginning of 20th centuries) [Эвенки: историко-этнографические очерки (XVIII-начало XX в.)]. The Institute of Ethnography of the USSR Academy of Science, Leningrad

Vasilyev VY (2016) Genesis of shamanic instruments of Sakha people in light of the image of Mother-Beast of Turks, Mongols and Tungus [Генезис шаманских инструментов народа саха в свете образа матери-зверя у тюрков, монголов и тунгусов]. Northeastern Humanitar Bull 4:18–27

Vasilyevskii RS (1969) The origin of the ancient Koryak Culture on the Northern Okhotsk Coast (trans: Chard CS). Arctic Anthropol 6:150–164. https://doi.org/10.2307/40315694. University of Wisconsin Press

Verbitsky VI (1865) Seoks (bone, kin, generation) of Altaians [Сеоки (кость, род, поколение) алтайцев]. Tomsk Provincial Gazette 42:14

Vinogradov VA (1970) Vowel harmony and word phonology [Сингармонизм и фонология слова]. In: Yunusaliyev B, Yudakhin K (eds) Turkological research [Тюркологические исследования]. Ilim, Frunze, pp 106–117

Vinogradov VA (1972a) On interpretation of synharmonism as a morphonological phenomenon [К интерпретации сингармонизма как морфонологического явления]. In: Shaumian SK (ed) Problems of structural linguistics [Проблемы структурной лингвистики]. Nauka, Moscow, pp 342–353

Vinogradov VA (1972b) Typology of vowel harmoony in languages of Africa and Eurasia [Типология сингармонических тенденций в языках Африки и Евразии]. In: Zhurinsky AN (ed) Problems of African linguistics. Typology, comparative analysis, description of languages [Проблемы африканского языкознания. Типология, компаративистика, описание языков]. Nauka, Moscow, pp 125–163

Volodin AA (1972) Psychological aspects of perception of music [Психологические аспекты восприятия музыки]. The Institute of Evolutionary Physiology and Biochemistry Named After Sechenov, Moscow

Voldina TV (2017) Song as a curative remedy in traditional culture of Yugric people of Ob' [Песня как целительное средство в традиционной культуре обских угров]. In: Kharitonova VI (ed) Medical ethnography: contemporary approaches and conceptions [Медицинская этнография: современные подходы и концепции]. Institute of Ethnography & Anthropology of Russian Academy of Science, OOO Publicity, Moscow, pp 141–151

Volodko NV, Starikovskaya EB, Mazunin IO, Eltsov NP, Naidenko PV, Wallace DC, Sukernik RI (2008) Mitochondrial genome diversity in Arctic Siberians, with particular reference to the evolutionary history of Beringia and Pleistocenic Peopling of the Americas. Am J Human Genet 82:1084–1100. https://doi.org/10.1016/J.AJHG.2008.03.019. Cell Press

von Hornbostel EM (1913) Melody and Scale. C.F. Peters, Leipzig

von Hornbostel EM, Sachs C (1914) Systematik der Musikinstrumente, ein Versuch. Zeitschrift für Ethnologie, Organ der Berliner Gesellschaft für Anthropologie, Ethnologie und Urgeschichte 46:553–590. https://doi.org/10.1017/CBO9781107415324.004. Dietrich Reimer Verlag GmbH

Vovin A (2005) The end of the Altaic controversy In memory of Gerhard Doerfer. Central Asiatic J. Harrassowitz Verlag. https://doi.org/10.2307/41928378

Wallin NL (1991) Biomusicology: neurophysiological, neuropsychological, and evolutionary perspectives on the origins and purposes of music. Pendragon Press, Hillsdale

Wan X (2011) Early development of bronze metallurgy in Eastern Eurasia, Sino-Platonic papers, vol 213. University of Pennsylvania, Philadelphia

Wang PK (1984) Progress in late Cenozoic palaeoclimatology of China: a brief review. In: Whyte RO (ed) The evolution of the East Asian environment, vol 1. Center of Asian Studies, Univ. Hong Kong, Hong Kong, pp 165–187

Wang P (2018) Interaction of bronze and early iron age cultures in South Siberia, Xinjiang and Northern China [Взаимодействие культур бронзового и раннего железного века южной Сибири, Синьцзяна и северного Китая]. Vestnik NSU. Series: History and Philology 17:16–29. https://doi.org/10.25205/1818-7919-2018-17-4-16-29

Wang H, Lu C, Ge B, Zhang Y, Zhu H (2012a) Genetic characteristics of an ancient nomadic Group in Northern China. Hum Biol 84:6

Wang T, Wang G, Wang D, Tabil LG (2012b) Investigation of the Compression Characteristics of Leymus chinensis L. In: ASABE Annual International Meeting, Dallas, Texas, July 29–August 1, 2012. The American Society of Agricultural and Biological Engineers (ASABE), St. Joseph, pp 12–1341112. https://doi.org/10.13031/2013.42057

Watanabe H (1985) The chopper-chopping tool complex of eastern asia: an ethnoarchaeological-ecological reexamination. J Anthropol Archaeol 4:1–18. https://doi.org/10.1016/0278-4165(85)90011-X. Academic Press

Weiss DJ, Hauser MD (2002) Perception of harmonics in the combination long call of cottontop tamarins, Saguinus oedipus. Anim Behav 64:415–426. Academic Press

Wennerstrom AK (2001) The music of everyday speech: prosody and discourse analysis. Oxford University Press, Oxford

Wermke K, Mende W (2009) Musical elements in human infants' cries: in the beginning is the melody. Music Sci 13:151–175. https://doi.org/10.1177/1029864909013002081

Westermann DH (1937) Laut und Sinn in einigen Westafrikanischen Sprachen. Archiv für vergleichende Phonetik 1:154–172; 193–212

Whitman J (2015) Old korean. In: Brown L, Yeon J (eds) The Handbook of Korean Linguistics. Wiley Blackwell, Malden, pp 421–438. https://doi.org/10.1002/9781118371008

Whitridge P (2015) The sound of contact: historic Inuit music-making in northern Labrador. North Atlantic Archaeol 4:17–42

Wiley RH (1983) The evolution of communication: information and manipulation. In: Halliday TR, Slater PJB (eds) Animal behavior: communication, vol 2. Blackwell Publishers, London, pp 156–189

Winkler MG, Wang PK (1993) The late-quaternary vegetation and climate of China. In: Wright HE, Kutzbach JE, Webb T, Ruddiman WF, Street-Perrott FA, Bartlein PJ (eds) Global climates since the last glacial maximum. University of Minnesota Press, Minneapolis, pp 221–264

Winsler A (2009) Still talking to ourselves after all these years: a review of current research on private speech. In: Winsler A, Montero I (eds) Private speech, executive functioning, and the development of verbal self-regulation. Cambridge University Press, Cambridge, pp 3–41

Wright M (2004) The search for the origins of the Jew's Harp. Silk Road 2:49–55

Wright M (2011) The Jew's Harp trade in Colonial America. Galpin Soc J 64:209–218. https://doi.org/10.2307/23209396. Galpin Society

Wright M, Impey A (2007) The Birmingham-KwaZulu connection. J Int Jew's Harp Soc 4:44–48

Wurz S (2010) Interpreting the fossil evidence for the evolutionary origins of music. South Afr Humanit 21:395–417
Yakhontov S (1971) The lexis as a trait of genetic relation of languages [Лексика как признак родства языков]. In: Sunik OP (ed) Problems of commonality of the Altaic languages [Проблема общности алтайских языков]. Nauka, Leningrad, pp 110–120
Yakovlev VI (2001) Traditional musical instruments of Volgo-Uralic peoples: formation, development and functionality [Традиционные музыкальные инструменты народов Волго-Уралья: формирование, развитие и функционирование]. The State Kazan University, Kazan
Yang X, Rost KT, Lehmkuhl F, Zhenda Z, Dodson J (2004) The evolution of dry lands in northern China and in the Republic of Mongolia since the Last Glacial Maximum. Quat Int:69–85. https://doi.org/10.1016/S1040-6182(03)00131-9
Yesipova MV, Wright J, Sheikin YI, Gerasimov OM, Gorokhovik EM, Putilov BN, Bocharov YS, Frayonova OV, Pichugin PA (2008) Jew's Harp [Варган]. In: Yesipova MB (ed) Musical instruments [Музыкальные инструменты]. Deka-VS, Moscow
Yurtsev BA (1982) Relics of the xerophyte vegetation of Beringia in northeastern Asia. In: Hopkins DM, Matthews JV, Schweger CE, Young SB (eds) Paleoecology of Beringia. Academic, New York, pp 157–177
Yurtsev BA (1986) Megaberingia and the cryo-xeric stages of the history of its vegetation cover [Мегаберингия и криоксерические этапы истории ее растительного покрова]. In: Probatova NS (ed) Komarovsky readings [Комаровские чтения], vol 33. Far Eastern Scientific Center of the Russian Academy of Science, Vladivostok, pp 3–53
Yurtsev BA (2001) The Pleistocene "Tundra-Steppe" and the productivity paradox: the landscape approach. Quat Sci Rev 20:165–174. https://doi.org/10.1016/S0277-3791(00)00125-6
Zabolotskaya PY (2011) Shamanic masks of Siberian peoples [Шаманские маски народов Сибири]. Siberian Philol J 3:24–34
Zagoskin NP (1910) Russian waterways and the marine business in pre-Petrine Russia [Русские водные пути и судовое дело в до-петровской России]. Lito-tipografiya Kharitonova, I. N, Kazan
Zagretdinov RA (1991) The history of Bashkir kubyz [История башкирского кубыза]. In: Alekseyev EY (ed) Jew's Harp (khomus) and its music. Proceedings of the first All-Union conference, 1988 [Варган (хомус) и его музыка. Материалы I Всесоюзной конференции 1988]. Yakut Institute of language, literature and history of the USSR Academy of Science, Yakutsk, pp 67–70
Zagretdinov RA (1997) In: Zinovyeva T, Alkin M (eds) The school of playing kubyz: a practical methodological aid [Школа игры на кубызе: Учебно-Методическое Пособие]. Belaya Reka, Ufa
Zakharov II (1875) Complete Manchuro-Russian dictionary [Полный маньчжурско-русский словарь]. Typography of the Imperial Academy of Sciences, Sankt-Petersburg
Zhilich S, Rudaya N, Krivonogov S, Nazarova L, Pozdnyakov D (2017) Environmental dynamics of the Baraba forest-steppe (Siberia) over the last 8000 years and their impact on the types of economic life of the population. Quat Sci Rev 163:152–161. https://doi.org/10.1016/J.QUASCIREV.2017.03.022. Pergamon
Zhirkova RR (1991) The methodic supplement for learning to play khomus [Методическое пособие для разучивания игры на хомусе]. Yakutsk University, Yakutsk
Zhuravlev AP (1977) Eneolithic Karelia [Энеолит Карелии]. The institute of language, literature and history of Karelian chapter of USSR Academy of Sciences, Petrozavodsk
Zsiga EC (2013) The sounds of language: an introduction to phonetics and phonology. Wiley-Blackwell, Oxford
Zubkova LG (1973) Accommodation or vowel harmony? [Аккомодация или сингармонизм?]. Proceedings of the University of People's Friendship named after Lumumba 43:215–225
Zubkova LG (1986) On the relation of sound to meaning of a word in a language system (On the problem of "arbitrariness" of a language sign) [О соотношении звучания и значения слова в системе языка (К проблеме «произвольности» языкового знака)]. Voprosy Jazykoznanija (Topics in the Study of Language):55–66

# Chapter 9
# Were Musicians As Well as Artists in the Ice Age Caves Likely with Autism Spectrum Disorder? A Neurodiversity Hypothesis

**Nobuo Masataka**

**Abstract** This is an attempt to explore the nature of the activity in association with the most ancient evidence of music, bone flutes, as well as the nature of the Ice Age cave drawings and paintings with which the bone flutes were discovered and its relationship with the evolution of languages. First, the author reviews evidence for the similarities between characteristics of such drawings and paintings and those produced by individuals with autism spectrum disorder (ASD) who have failed to acquire languages unlike neurotypical individuals. Both are extremely realistic when animals are depicted whereas human figures can be only crudely created. Those characteristics are due to the possibility that the creators of the cave art produced animal images by visual realism but human figures by their own knowledge of humans (intelligence realism) as do contemporary individuals with ASD who are impaired with the cognition of interpersonal communication. The discovery of bone flutes in the caves indicates the close association of such space as that suitable to resonate produced sounds with folk culture with rudimentary form of instrumental music. On the basis of recent findings that children with ASD are highly precocious with respect to musical talent, one can reason the production of some sort of landscape music including mimicking of animal vocalizations by using bone flutes in the Ice Age. The author hypothesizes that at that time, caves were used as the audio-visual environments where ancient people, before going to hunting, mitigate their cognitive dissonance by viewing the depicted drawings and paintings, simultaneously exposed to the rudimentary form of music that is known to be effective for the mitigation, created by individuals likely to be with ASD. They can be referred to as neurodivergent from the perspective of the conceptualization of this neurodevelopmental disorder as a manifestation of neurodiversity of humans.

**Keywords** Ice Age art · Chauvet cave · Lascoux · Autism spectrum disorder · Origins of music · Bone flute · Neurodiversity · Giftedness

N. Masataka (✉)
Primate Research Institute, Kyoto University, Inuyama, Aichi, Japan
e-mail: masataka.nobuo.7r@kyoto-u.ac.jp



N. Masataka (ed.), *The Origins of Language Revisited*,
https://doi.org/10.1007/978-981-15-4250-3_9

## 9.1 Outline of the Neurodiversity Hypothesis

This chapter is devoted to the consideration about the advantages and disadvantages the people in the Ice Age enjoyed that were caused by the emergence of languages. The consideration is based upon the notion of "neurodiversity" that, seemingly, must have appeared historically first in this period with humans exclusively, according to which, there may be a trade-off relationship between the development of linguistic performance and that of artistic performance in individuals in this age. The author argues about the possibility that the developed social communication on the basis of languages had realized only at the expense of the development of other human capabilities such as music and art.

Since its first discovery in the nineteenth century, people have been surprised by the realism demonstrated by the European Ice Age art that has been found in more than 200 caves in southwestern Europe, especially in France and Spain (Marshack 1995), being created by numerous individuals, by the flickering light of oil lamps or torches using natural pigments including charcoal, kaolin, violet and black manganese oxide, red and yellow ferrous oxide, and water as a medium (Bahn 1988). The cave art images, composed in such demanding circumstances, still have the power to move the viewer with their sheer virtuosity. In particular, the paintings of Chauvet, dating to approximately 32,500 years ago, have prompted many individuals to marvel at this early flowering of the modern human mind (Humphrey 1998) (Fig. 9.1). They appear to be robust evidence of a new type of mind at work (Chauvet et al. 1996).

It has been claimed that these and other examples of Ice Age art (e.g., paintings in Lascoux) demonstrate the presence of high-level conceptual thought (Mithen 2007). The Chauvet cave is testimony that modern humans were capable of the type of symbolic thought and sophisticated visual representation that was beyond Neanderthals or each of these painted animals (Mithen 2007). Their creators might have specifically intended to represent and communicate information, and they might have had a long tradition of artistry behind them; for example, it has been noted that "We now know that more than 30,000 years ago artists had acquired a complete mastery of their technical means."

**Fig. 9.1** Representative images in the Chauvet cave

This review addresses the hypothesis that some of the people of the ice ages (32,500 years or so ago) who were involved in cave drawings and bone flute playing were likely to have been on the autism spectrum and failed to acquire languages unlike others. This represents a significant revision of a similar proposal from 20 years ago that now takes into account the advent of new knowledge and understanding of autism spectrum disorder (ASD). The author places ASD more in the light of a "strength-oriented view" of ASD in terms of the utility of ASD in neurodiversity, which had survival advantages, especially for hunter-gatherer groups. While autism has been categorized as a disorder, some selective advantages have been claimed for it recently that may represent a balance toward "folk physics" at the expense of "folk psychology". These enhanced abilities extend beyond technical skills and include heightened sensory sensitivities. However, such potential trade-off skills remain to be investigated with regard to advantageous or disadvantageous behaviors in the distant past.

Given the fact that autism is for the most part highly heritable and subject to some elements of positive selection, autism is considered to be more likely to affect human culture in a community as the size of the community decreases, as in the Ice Age period, by exerting so-called "bottle-neck effects" upon the population in the community. Here the author attempts to argue for how autism spectrum conditions become human diversity (neurodiversity) on the basis of generalization of the condition, such as that individuals with ASD (neurodivergent individuals) will be particularly focused on activities of drawing and painting of particular objects with extraordinary realism as well as musical activities. The author proposes the hypothesis that owing to this, art and music co-emerged in the Ice Age period.

## 9.2 How Has the Mind of the Most Ancient Artists Been Understood, So Far?

While such paintings and drawings may strike most people as wondrous, two articles that were published after that of Chauvet et al. (1996) drew attention to evidence indicating that the miracle they represent may not be at all of the kind most people think (Humphrey 1998). These two articles were written independently of one another, but interestingly, both of them reported similarities between the Ice Age art and drawings created by children who were determined to be savants (profoundly impaired intellectually but showing striking ability in a particular context) among children who were diagnosed as having autism according to the definition of Kanner (1943) and who were observed by the respective authors.

One of them was a 7-year-old boy named Jamie, born in the United States. He exhibited typical characteristics of the developmental disorder, e.g., severe social impairment defined as the inability to engage in two-way interaction, severe linguistic impairment, and the absence of imaginative pursuits, with instead the substitution of repetitive behavior. Parental reports described that his drawing ability came to be recognized suddenly when he became 5 years old without previous experience. He has continued to draw since then, sometimes for 5 h at a time, churning out endless, detailed images, often of buildings. Infrequently and as an afterthought, he may add color to his drawings, using crayon to his unusual images of houses, interiors, auto wrecks, fights, and police vehicles. The Towering Inferno, a skyscraper from the movie by the same name, is one of his favorite subjects, particularly the fight scene in the Promenade Room and a sequence from the film showing flames licking windows, paper flattering like confetti, and an external glass elevator breaking free. During the first 8 years, humans rarely appeared in his drawings.

Jamie's imagery as a young child was based upon the concrete world, in adventure movies such as *Towering Inferno* and *Home Alone*, as well as upon the models he had constructed using his huge collection of Legos. Nevertheless, Kellman claimed striking commonalities between his keen eye for observed reality, solid ability to utilize foreshortening, perspective, three-dimensionality, and his use of structure as the primary compositional focus and those features seen in Ice Age art.

The author of the other report (Humphrey 1998) made a similar claim, referring to the drawings made a young girl, Nadia, who was profoundly mentally impaired and had virtually no language, despite her graphic skills. She was born in the United Kingdom in 1967 and failed to develop any language until the age of 6 and had been physically clumsy. At the age of six, her vocabulary consisted of only 10 one-word utterances, which she used rarely (Selfe 1977). When she was only 3 years old, she started to show extraordinary drawing ability and suddenly produced line drawing of animals from memory, with quite uncanny photographic accuracy and graphic fluency. At present one can view published versions of them (Selfe 1977).

Humphrey (1998) mentioned some similarities between Nadia's drawing and the Ice Age art: (1) striking naturalism and realism of the individual animals and (2) the fact that such realism generally applies exclusively to animals. In the Ice Age art, animals of all sorts engage in their vigorous lives. They include horses, reindeer, roe deer, giant sloths, wooly mammoths, aurochs or wild cattle, rhinoceros, ibex, cave bears, bison, lions, birds, insects, and fish (including sturgeon and salmon) that are carved, incised, or painted on cave walls with natural pigments or molded with mud on cave floors (Humphrey 1998). These images still cause one to catch one's breath at the freshness and vitality of their beautifully rendered forms. Although some of these animals, such as the wooly mammoth, the giant sloth, and the cave bear, are extinct, many of the other animals have not changed a great deal in form since the time when sheets of ice ground across Europe thousands of years ago. Nevertheless, virtually nothing except these animals was depicted in any Ice Age cave.

Having referred to the fact that no human figure was discovered from the Chauvet or the Lascoux cave (Leroi-Gourham et al. 1979) except for a few crudely drawn figures (Fig. 9.2), he found that Nadia also drew human figures extremely rarely and, when she did draw them, she could only could do it poorly as compared to her drawing of animals (Fig. 9.3). This led Humphrey to reason that the creators of the Ice Age art actually had distinctively pre-modern minds, were little given to symbolic thought, and were essentially self-taught and untrained. Both of these essays

**Fig. 9.2** Representative images of animals and human figure drawn in the Lascoux cave

**Fig. 9.3** Representative images of a horse and a girl drawn by a 5-year-old girl with ASD, Nadia

regarded the similarities between the Ice Age art and drawings of the "savants" as indications that modern human intelligence was lacking in the ancient "artists."

## 9.3 Similarities of the Ice Age Art and that Produced by Contemporary Individuals with ASD

It has taken two decades since the above-mentioned argument occurred. During the period, our knowledge about developmental disorders has been expanded dramatically. That was enough to revise the notion about this developmental disorder, on which Kellman and Humphrey (1998) documented their claims. Indeed, even the term "ASD" did not exist in the last century. At present, it is regarded as developmental disorders that are characterized by social communicative difficulties and restricted behaviors and interests (American Psychiatric Association 2013).

**Fig. 9.4** Representative pieces of drawings produced by a 9-year-old boy with ASD, N.H

According to a report published in 2012, ASD affects approximately 1 in every 88 individuals in the United States and is believed to be lifelong, congenital, and highly heritable (Constantino et al. 2013). Accumulated evidence suggests that as many as 300 to 500 distinct genes are involved in the etiology, with no single locus accounting for more than 1% of cases (State and Sestan 2012). Consequently, ASD must be a diagnosis made purely on the basis of behavior (Lord et al. 2011). The genetic heterogeneity of ASD poses a stark challenge for understanding the condition biologically: with so many different "causes," a key question is how this genetic heterogeneity can be instantiated into common forms of this disorder (Jones and Klim 2013).

Currently, Nadia can no longer be regarded as an exceptionally talented for drawing as a child with ASD so that she was referred to as savant. Certainly, she was extremely talented. However there are number of children as well as of adults with ASD who are artistically talented no less than Nadia was, and most of them, despite IQs equivalent to typically developing (TD) children and to neurotypical adults, have been found to share similar characteristics for drawing with Nadia. If one attempts to search artistic pieces produced by individuals with ASD with by internet (e.g., Google), using "drawing" and "autism" as key words, one will find hundreds of thousands of images immediately, each of which is created by so variable individuals from children to those aged. On the other hand, one will realize the fact that most of the images drawn are those concerning landscape that frequently include animals, buildings robots, military planes, and railway trains, with striking realism whereas there are few drawings of human figures. Otherwise, some geometric patterns are favorable as shown in Fig. 9.4 (left). That was drawn by a 9-year-old boy who had been born in Japan, N.H., under my study (2017a). Based on direct had been diagnosed as ASD by an independent child psychiatrist according to ICD-10 (World Health Organization 1994) as well as DSM-IV (American Psychiatric Association 1994). When drawing spontaneously with no experience of any form of training or exposure to ay art, N.H. used to never do other drawing than geometric patterns. Notice that similar pieces were also often found in the Lascoux cave, each "square" being colored by different colors occasionally. When N.H. was asked to draw a face of anybody, he could do it only poorly as shown Fig. 9.4 (right).

In this regard, the fact should be noticeable that while a lot of the attention concerning ASD focuses on negative behaviors such as those described above in the social context, a number of positive attributes associated with ASD have also been revealed in the non-social context. Individuals with the disorder are likely to be particularly skilled at perceiving details as opposed to whole gestalts. Children with ASD perform better than TD children on the block design test of the Wechsler Intelligence Scale for Children (WISC-IV), which requires taking blocks that are all white, all red, or a mixture of red and white, and putting them together to match a preexisting pattern (Baron-Cohen 2003). They also perform better than TD children on the tests of detection of patterns embedded in more complex patterns and often exhibit superior drawing techniques to TD children (Kanoiyi et al., 2003), though such findings have, so far, led some researchers to suggest that individuals with ASD experience what has been termed "weak central coherence," namely, they fail to grasp the whole of a situation and perceive mainly the constituent parts (Armstrong 2012), a deficit-oriented view of the disorder,.

A collection of drawings and paintings produced by professional adult artists with ASD has been published (Mullin 2009), including artistic pieces by up to 21 such artists who can be referred to as so-called outsider artists. Most of them are characterized by their realism in drawing, but it is exclusively concerned with landscape as the object of drawing but not with human figures, as produced by the first artist documented in the book, Glenn Russ.

## 9.4 Uniqueness of Realism Exhibited by the Ice Age Art as Prehistoric One

Meanwhile, our knowledge about prehistoric arts after the European Ice Age has also accumulated. It reveals the fact that animal images drawn in the Ice Age were exceptionally realistic, as compared to other artistic pieces created in the subsequent period, and no animals depicted in the prehistoric art later than the Ice Age have become less realistic and more crudely drawn than before. Interestingly, moreover, human figures came to be discovered much more abundantly as much more realistic images, conversely. This could be obvious when one compare human figures and animal images depicted in Figs. 9.2 and 9.5. The human figure illustrated in Fig. 9.5 is that of rock art produced by San people in Nambia, the hunter-gatherers, that is well known as "The White Lady in Namibia." Historically, the pioneering researcher of prehistoric art, Henri Breuil, was known to discover it as the first ancient rock art (Bahn 1998). The image is actually that of an adult male (as revealed by the presence of a penis) with much more masterful description that that with which the human

**Fig. 9.5** Representative images of animals and a human in Rock Art

figure was drawn in Lascoux. With respect to animal images, however, the opposite is the case.

In fact, there is a discrepancy in the development of the artistic techniques between "artists of the Ice Age caves" who were precocious when depicting animas and others in the subsequent period. Figure 9.6 presents images of reindeer dated to 300 to 400 B.C. that were discovered in the vicinity of the Chauvet cave. While the images are produced under the developed art of design, foreshortening, three-quarter views, perspective, and shading that were characteristic to the Ice Age art are still missing. This is the case extensively found in ethnic art pieces (Egyptian, South Sea island, Kwakiutl Indian art, etc.) as well as in the Western art before the end of the Middle Age. One has to wait for Renaissance until one can see the realism equivalent to that observed in the Ice Age caves.

In fact, Humphrey (1998) has correctly referred to the fact that no human figures were recorded in the Chauvet cave and merely a few were included in Nadia's collection (Selfe 1977), which were only crudely drawn. Nevertheless, one cannot dismiss that "incomplete human figures" have been discovered occasionally in the Chauvet cave and others, that in all of them, the top of their body was that of an animal (such as a bison for the case of the Chauvet cave, see Plate 83 in Chauvet et al. 1996) and the bottom of the body was that of a human with two legs, and that in such cases, the human part of the body was drawn with as realistic as when animals were drawn. As for human hand shapes, moreover, its precise depiction has been found abundantly.

Concerning Nadia's recordings, too, she was likely to draw legs with shoes as fragmentary human body part besides its head, and when she did that, their images were as realistic as those of animals (Fig. 9.7). Taken together, landscape, including animals, can be depicted with exceptional realism, but the entire human figure, including the head, as a social being could be only crudely drawn by any in the

**Fig. 9.6** Images of reindeer drawn in the vicinity of the Chauvet cave that dated to 300–400 B.C

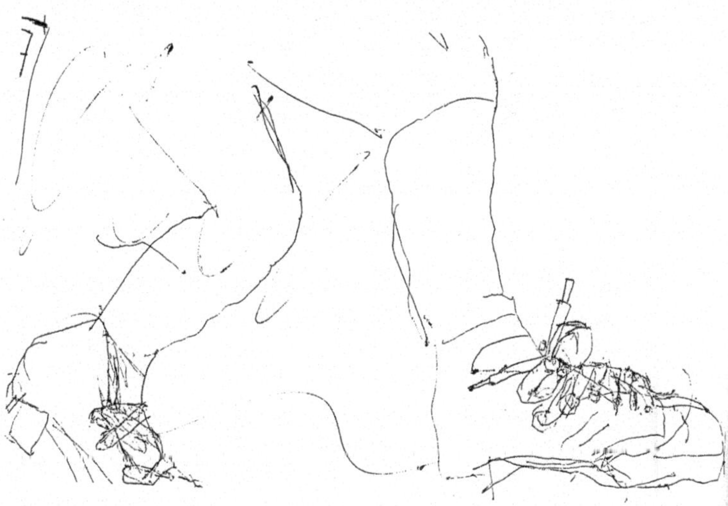

**Fig. 9.7** Drawing her own legs with shoes produced by a 5-year-old girl with ASD, Nadia

Ice Age art and by any one with ASD including Nadia and contemporary outsider professional artists. Temple Grandin (Grandin 1996; Grandin and Cook 2004) has called such unique manner of perception of the landscape, or, if using her expression, "brain's representation of the world" "thinking in pictures." Being with ASD, she describes her own acute visual processes and her particular ability to employ visualization as a means of concretizing events and concepts as outstanding characteristics of many individuals with ASD and states that "one of the most profound mysteries of autism has been the remarkable ability of most autistic people to excel at visual spatial skills." When she was a child, she thought "everyone thought in pictures." Her description of her thought processes is instructive in the light of enormous realism when landscape is depicted by individuals with ASD as "when I do an equipment simulation in my imagination or work on an engineering problem, it is like seeing a videotape in my mind." In order to create new images, she takes "many little parts of images" she has "in the video library" of her imagination and piece them together. She describes that she has video memories of every item she has ever worked with. Most significantly, Grandin (1996) reports that personal relationships themselves "made absolutely no sense to me until I developed visual symbols of doors and windows" with which to visualize the give and take of social interaction. Her acute visual memory and ability to "think in pictures" should what links Grandin to the contemporary and ancient creators of realistic drawings and paintings of landscape including animals who should understand the object in images, forms and visual relationship exclusively.

## 9.5 Visual Realism When Depicting Animals and Intellectual Realism When Depicting Humans

The findings of their "poor" ability to depict humans are, on the other hand, in line with our accumulating knowledge about the development of drawing of human figures in developmental psychology. This reveals the fact that children's early attempts to represent them have been described as "tadpoles" because these representations consist of a circle with arms and legs emanating from it, as shown in Fig. 9.4. These figures appear to have heads but no trunks, which are typically drawn around 3 years of age. Thereafter, between the ages of 3 and 5, as children move from the tadpole to a conventional figure with a body differentiated from the head, they draw transitional figures in which the arms are attached with the legs with body features, such as buttons or stomach, placed with the legs (Winner 2006). Such developmental patterns, as demonstrated in Fig. 9.8, are assumed to be a universal trajectory, according which human figures drawn in the Ice Age cave and by the children with ASD correspond to the level of 5-year-olds.

As explanations for the trajectory, Piaget and Inhelder (1956) are well known to claim that children as young as 5–6 years old do not draw what they see (visual realism) but instead what they know (intellectual realism). They show that children

(3) (4) (5)

**Fig. 9.8** Comparison of drawings of a human figure by a TD children included in the study of Masataka (2017a) when he was three (3), four (4), and five years old (5)

younger than 7 years old drew an apple with a pin stuck in it so that the pin (which the children knew was inside the apple) was visible inside the apple, which thus appeared transparent. The same phenomenon was demonstrated when asking the children to draw a stick shown either from a side view or a foreshortened end view. In fact, there is other ample evidence for intellectual realism in young children. For example, 5-years-olds who did not typically show occlusion were able to do so when they were asked to draw a toy policeman "hiding" behind a wall (Cox 1981). When asked to draw a toy behind the glass, children drew the toy and the glass side-by-side; while asked to draw a toy inside the glass, they drew realistically (Freeman and Jenikoun 1972).

The shift to visual realism has been reasoned to occur because children recognize their drawings as failed in terms of realism and are therefore motivated to invent a better method. Such reasoning will lead us to conclude that the creators in the Ice Age caves like the children with ASD were good observers of landscape including animals, indeed exceptionally good observers as compared with artists in the subsequent period until Renaissance, and depicted it on the basis of visual realism. However, they experienced difficulty of drawing of human figures, and when obliged to do it, they had to rely upon their own knowledge about the figures (intellectual realism).

Such reasoning should be supported by the current knowledge about individuals with ASD who are believed to have remarkable difficulties in social interactions, for instance, making eye contact, engaging in reciprocal interactions, and responding to the emotional cues of others (Dawson et al. 2005). Basic impairments, such as lack of attention to others, often appear within the first year of life (Werner et al. 2000). Infants later diagnosed with ASD exhibit a mean decline in the duration of visual fixation from 2 to 6 months of age (Dawson et al. 1998). By age 2–3 years, broader impairments become evident in social orientation, joint attention, imitation, response to others' emotional expressions, and face recognition (Sigman et al. 1992; Dawson et al. 2002, 2004). Many of the early social impairments in ASD involve the ability to attend to and process information from faces. This evidence has led to the reasoning that, in ASD, impairments in face and emotion processing might play a fundamental role in the dysfunction of the neurological mechanism underlying impairments in social cognition (Dawson et al. 2005).

## 9.6 Co-occurrence of Emergence of Bone Flutes and that of Cave Art

Living a precarious and difficult existence at hunter-gatherers in an environment of bitter glacial winds, dangerous terrain, and unforgiving circumstances, these early modern inhabitants of Europe brought with them new technologies as well as new manners of living unlike those employed by the earlier Neanderthals who first populated the rugged landscape. New objects, unlike the small unchanging number of tools employed by the Neanderthals for thousands of years, indicate the new inhabitants' more elaborated way of living. Early portable objects utilizing a variety of materials, including antler, bone, clay, stone, and likely wood and other materials, were used for spear thrower handles, needles for sewing tailored clothing, beads and items of personal adornment, plaques and batons of unknown function, scrapers, spear points, burins, and a large number of other tools. One of the earliest objects to be found was a bone flute, which indicates that music must have been important for their living.

Given the fact that for early evidence of music in the archaeological record, considerable debate surrounds and that it continues up to the present, the implication of the findings is of great importance. Researchers universally accept the existence of complex musical instruments as an indication of fully modern behavior and advanced symbolic communication. Owing to the scarcity of finds, however, the archeological record of the evolution and spread of music remains incomplete. Although arguments have been made for Neanderthal musical traditions and the presence of musical instruments in Middle Palaeolithic assemblages, concrete evidence for these claims has been wanting.

Once, report has been made that the discovery of bone and ivory flutes were made from the early Aurignancian period of southwestern Germany (Conard et al. 2009), which claimed to have demonstrated the presence of a well-established musical tradition at the time when modern humans colonized Europe, more than 35,000 calendar years ago. However, a more recent study presents negative evidence against the notion (Diendrich 2015), which claims that these are not instruments, nor human made, but products of the most important cave bear scavenger of Europe, hyenas. They occupied mainly cave entrances as dens but went deeper for scavenging into cave bear dens or used, in a few cases, branches/diagonal shafts.

Hyenas left bones in repeating similar tooth mark and crush damage stages, demonstrating a butchering/bone cracking strategy. The femora of subadult cave bears are intermediate in damaging patterns, compared to the adult ones, which were fully crushed to pieces. Hyenas produced round-oval puncture marks in cub femora by the bone-crushing premolar teeth of both upper and lower jaws.

The first Neanderthal cave bone flute from the Middle Palaeolithic was believed to have been discovered in 1920s in Slovenia (Brodar and Brodar 1983). This was a larger cave bear den, and Late Palaeolithic Aurignacian (not Neanderthal) used rock shelter camp site at the entrance. Other cave bear cub femora with holes were then reported from Hungary (Dobosi 1985). This was a smaller cave bear and Ice Age spotted hyena (*Crocuta crocuta*) carnivore den which overlap another Aurignancian camp site. Further proof of the oldest music instruments has also been reported from another cave in Slovenia.

In all, almost all prehistoric bone flutes, known to be the oldest known deliberately made musical instruments, come from a time in prehistory associated with post-Neanderthal activity. The cave in Germany known as Hohle Fels also revealed some of the most interesting examples of the Ice Age cave art in the world. The shape of the cave when seen without the present coverage of trees is reminiscent of the mouth in the "face" of the mountain, perhaps explaining why it was inhabited for over 10,000 years.

The discovery of a bone flute and two fragments of ivory flutes, made there in a level dates ca. 31–33,000 years ago, are said to represent the earliest known flowering of music-making in Stone Age culture. The bone flute with five finger holes, found at Hohe Fels Cave in the hills west of Ulm, was by far the most complete musical instruments recovered from the caves in a region where pieces of other flutes have been turned up in recent years. A three-hole flute carved from mammoth ivory was uncovered a few years ago at another cave, as well as two flutes made from the wing bones of a mute swan. In the same cave, beautiful carvings of animals were also found.

The most significant of the new artifacts was a flute made from a hollow bone from a griffon vulture. The preserved portion is about 8.5 inches long and includes the end of the instrument into which the musician blew. The maker carved two deep V-shaped notches there and four fine lines near the finger holes. The other end appears to have been broken off. Judging by the typical length of these bird bones, 2 or 3 inches are missing. Other flute fragments found in the site nearby have been dated to around 35,000 years ago. Another flute excavated in Austria is believed to be 19,000 years old, and a group of 22 flutes found in the French Pyrenees mountains have been dated up to 30,000 years ago. In addition, several prehistoric bone flutes have also been found in China.

Friedrich Seeberger, a German specialist in ancient music, reproduced the ivory flute in wood. Experimenting with the replica, he found that the ancient flute produced a range of notes comparable in many ways to modern flutes and that the tones were quite harmonic. In addition to these bone flutes, the discovery of several Palaeolithic figurines have been reported in Hohle Fels, which included a horse's head (or possible a bear), water bird of some sort possible in flight, and a "Lowenmensch," a half-lion/half-human figurine (Seeberger 2003).

## 9.7 The Ice Age Caves Where Musical Activity, Using Bone Flutes, Occurred

The study of Palaeo-acoustics has revealed that several ancient structures were built so as to incorporate acoustic phenomena in their design, and remarkably, such effects are considered to have been utilized at several cave systems in Palaeolithic Europe.

The question whether the painted caves of western Europe were once resound to the music of Palaeolithic chants was put forward (Reznikoff and Dauvois 1988). They have studied three caves in the Ariege department at the foot of the French Pyrenees. Their results suggest that the acoustics of the caves played a significant part in determining where the paintings were located, and this observation leads directly to the supposition that music or chants were important elements in cave ceremonies around 20,000 years ago.

The authors of the study rely on the fact that in certain places referred to as "points of resonance," the caves resonate in response to particular notes. They proceeded slowly through the cave using their voices to produce a series of notes spanning almost three octaves, from C to G. They extended the range of notes for a further two octaves by harmonics and whistling. Where there was a resonance response, they recorded the location and the particular note eliciting the response. They used these observation to draw up a resonance map of the cave.

The resonance of the caves is not in itself surprising, but the significance of the study becomes apparent when the authors compare their points of resonance with the location of cave paintings. They draw three main conclusions. First, most of the cave paintings are at or within 1 m of points of resonance. The Grand Salle at Portal, for example, which gave no resonance response, also has a relatively few paintings. Next, most of the points of resonance correspond to locations with cave paintings. Indeed, the best points of resonance are always marked in this way. Finally, the authors claim that the location of some of the paintings can be explained only by the resonance of that particular location. A good example is number 23 at Portal, where a particular effective point of resonance is marked by red painted dots, as there is not enough room for a full painted figure.

The authors remark from their own experience on the impressive effect of cave resonance, which would have been all the more striking in the flickering half-light of the simple lamps or tapers used by the original artists. Drums, flutes, and whistles may have been used in cave rituals. Given the above-mentioned results that bone flutes have been discovered at several Palaeothic sites in Europe of roughly the same age as the paintings, the study is of particular value in drawing attention to the likely importance of music in the rituals of our ancestors.

When standing well back from the painted walls, one claps or creates percussion sounds and records the echo's bouncing back, it turns out that rock art seems to be placed intentionally where echoes are not only unusually loud but are also related to the pictured subject matter. Where hoofed animals are depicted, one easily evokes the echoes of a running herd. If a person is dawn, the echoes of voices seem to emanate from the picture itself.

Walter (1993) has visited rock art sites in Europe, North America, and Australia. At open air sites with paintings, it was found that echoes reverberate on average at a level 8 decibels above the level of the background. At sites without art the average was 3 decibels. In deep caves such as Lascaux and Font-de-Gaume in France, echoes in painted chambers produce thee sound level between 23 and 31 decibels. Deep cave walls painted with cats produce sounds from approximately 1 to 7 decibels. In contrast, surfaces without paint are totally flat.

## 9.8 How Was the Mind of the Ice Age Musicians Understood?

Taken together, it appears sensible to understand that instrumental music emerged in the history as cave art emerged, both being closely entangled with one another. In this regard, an extremely intriguing fact has been presented by the study of children with ASD (Masataka 2017b) that reported the results of the investigation of the capability of aesthetic perceptual judgment of music in male children diagnosed with ASD when compared to age-matched TD male children. Nineteen boys between 4 and 7 years of age with ASD were compared to 28 TD boys while listening to musical stimuli of different aesthetic levels. The results from two musical experiments using the above participants are described here. In the first study, responses to a Mozart minuet and a dissonant altered version of the same Mozart minuet (Trainor and Heinmiller 1998; Masataka 2006) were compared. In this first study, the results indicated that both ASD and TD males preferred listening to the original consonant version of the minuet over the altered dissonant version. With the same participants, the second experiment included musical stimuli from four renowned composers: Mozart and Bach's musical works, both considered consonant in their harmonic structure (Cooper 2001), were compared with music from Schoenberg and Albinoni, two composers who wrote musical works considered exceedingly harmonically dissonant (Boulez 1971; Thompson et al. 2001). In the second study, when the stimuli included consonant or dissonant musical stimuli from different composers, the children with ASD showed greater preference for the aesthetic quality of the highly dissonant music compared to the TD children. While children in both the groups listened to the consonant stimuli of Mozart and Bach music for the same amount of time, the children with ASD listened to the dissonant music of Schoenberg and Albinoni longer than the TD children. As preferring dissonant music is more aesthetically demanding perceptually, these results suggest that ASD male children demonstrate an enhanced capability of aesthetic judgment of music. Subsidiary data collected after the completion of the experiment revealed that absolute pitch ability (Masataka 2011) was prevalent only in the children with ASD, some of whom also possessed extraordinary musical memory.

Apparently, similar characteristics underlie the mind between the "musicians" and the "artists" in the Ice Age caves, being likely those of ASD. Along with the conceptualization of the theory of multiple intelligences (Gardner 2011), the cognitive characteristics documented above in children with ASD with respect to drawing and painting and music can be explained as greater than typical strength of artistic intelligence in themselves although the children experience difficulty with interpersonal communication (Lord et al. 1994; World Health Organization 1994). While the pattern of operation of an individual's mind can be categorized according to the domain toward which that individual is more oriented, individuals with ASD, overall, do not rely upon their social relationships but rather are predisposed to process perceived non-social objects in more depth (Masataka 2017b). For instance, these individuals exhibit enhanced discrimination between auditory stimuli, more accurate local target detection of auditory stimuli, and diminished global interference with auditory processing (Applebaum et al. 1979; Gebauer et al. 2014). These abilities, referred to as "naturalist intelligence" by Gardner (2011), are highly adaptive for living in nature and, in particular, would seem to have conferred an evolutionary advantage upon individuals during prehistoric times, as well as upon individuals in later times who would prefer to live outside any community (Armstrong 2017).

As far as the dawn of the history concerned, the documentation of such individuals appeared as early as the tenth century, in southern Europe, the region between the Alps and the Pyrenees, with them being referred to as "Spielmann (plural: Spielleute)" in German and as "wandering minstrels" in English. Spielleute had not settled abode and instead used to roam about from place to place over a broad range of regions across a number of communities (Hartung 2003). In the thirteenth century, such individuals came to be distributed over all the regions of Europe. While all other individuals who were integrated into feudal society were occupied with predetermined, inherited work, Spielleute were the first who chose this occupation of their own will (Stuart 1999). Born in urban regions in the society, most of them felt difficulty in staying there and decided to live as outcasts. As a result, they were treated as being dishonorable by society members. They were talented in a variety of rudimentary forms of contemporary art performance, performing arts including music and dancing without experiencing any form of professional training. In particular, they were well known to perform mimicking of a variety of sounds heard in the landscape (such as bird songs) while playing stringed and/or woodwind instruments simultaneously (Hartung 1982, 2003). When one of their performances was successful, the melody was memorized no matter how long it was without recording it using musical notation (actually, they were unable to read music) and it became part of their repertoire of routine performance. While such a seemingly mysterious endowment seems also to have been possessed by some famous modern artists who contributed to the subsequent development of classical music, such as Mozart, Bach, and Beethoven (Deutsch 2006), and also by the children with ASD (Heaton et al., 2009; Heaton 2009) in the current study, the findings of abundant bone flutes in the Ice Age caves would indicate the possibility that their existence could also be traced back to such ancient prehistoric period.

## 9.9 Possible Role of Ancient Musicians and Artists in Prehistoric Living

Concerning bone flutes found in the Ice Age caves, their function is highly debatable, but the large numbers discovered reveal that it had a genetic importance. Suggestions include ritual, ceremonial, shamanic, and simple pleasure. Even with respect to evolutionary reason of music, most existing theories contradict each other and cannot provide any plausible explanation. Reviews about the presented evidence for the evolutionary origins of music emphasize innateness, domain specificity for music, and uniqueness to humans in an unambiguous identification of genetic evolution as a source of music origins and invariably conclude our predispositional drive to make and enjoy music (Justus and Hustler 2005; McDemott and Hauser 2003). Indeed, there is accumulated suggestive evidence that for a biological predisposition for music.

A strong argument for the evolutionary origins of music also lies in its universality; music exists in all scientifically documented societies all over the world (Darwin 1871; Masataka 2003). The fact would lead us to reason that music possesses common attributes across cultures, that music exploits the human capacity to entrain to social stimuli, and that music is necessary for the very development of culture. Cultural evolution is based on the ability to create and perceive the sociointentional aspect of meaning. This should be unique to humans and has been crated in music, combining biologically generic, humanly specific, and culturally enactive dimensions. Apparently the evolution of music was based on biological and genetic mechanisms that had already been provided in other animals.

The capacity for culture requires not only the transmission of information but also the context of communication (Masataka 2008). The notion could lean one to reason that music and language constitute complementary components of the human communication toolkit while the power of language lies in its ability to present semantically decomposable propositions. As claimed by Perlovsky (2017), music is directed at increasing a sense of shared intentionality. Its major role is social. It serves as an honest signal and reveals qualities of a signaler to a receiver with nonspecific goals. This property, as the indeterminacy of meaning or floating intentionality, allows for individual interactions while maintaining different goals and meaning that may conflict. Therefore music promotes the alignment of participants' sense of goals.

Such argument could lead one to reason that as humans came to be able to live a successful living in societies (Spikins et al. 2016), the evolution of such communication system would be promoted. This appears particularly of importance for living as a hunter-gatherer. Archaeologically they had come to use such caves where present cave art is found as a shelter, first, to avoid danger of animals and bad weather, and then as living space. However, even staying there was sufficient to secure against possible attacks by some animals such as cave lions and cave bears that were provided with developed teeth and claws. No doubt when individuals were obliged to stay outside together as a group during hunting and foraging, the magnitude of danger should increase dramatically. They must have experienced a sort of psychological dilemma; without exposing to dangers to be scared by animals, one cannot subsist.

## 9.10 Cognitive Function of Music

From the cognitive perspective, living as a hunter-gatherer can be expressed as that used to be accompanied with serious cognitive dissonance (CD), an uncomfortable feeling caused by holding conflicting ideas simultaneously and that people are provided with a motivational drive to reduce dissonance (Festinger 1957). If a person is induced to cease performing a desired action through threat of punishment, dissonance will be experienced. The cognition that the person is not performing the action is dissonant with the cognition that the action is desirable. An effective way of reducing dissonance is by devaluing the action. The greater the threat of punishment, the less the dissonance, because a severe threat is consonant with ceasing to perform the action. Therefore, the milder the threat, the greater will be a person's tendency to devalue the action.

Recently, the findings that such CD can be temporally mitigated if being exposed to music have been reported (Masataka and Perlovasky 2012). In the study, CD was experimentally created in 4-year-old children, using a well-established method (i.e., the induced compliance paradigm), to evaluate several toys (Aronson and Carlsmith 1963), issuing either a mild or a severe threat of punishment for playing alone with one specific toy, and asking the children to re-evaluate the toys at the end of the experimental session. Through this technique, it was expected to be able to compare the effect of a mild threat with that of a severe threat on the attractiveness of playing with the forbidden toy. In addition, in another group who experienced a mild threat, the participants were exposed to music (i.e., one of Mozart's sonatas) while playing alone, whereas no music was played for any other participants under the same circumstances, so that one was enabled to evaluate the impact of the music on the effect of the mild threat. It was hypothesized that even if relatively greater CD was experienced by the children who experienced a mild threat, the degree of the devaluation of the forbidden action would be smaller when the participants were to be exposed to music than when they were not exposed to music if the music served to mitigate CD.

In the experiment, all the children who participated were to be left in a room with a variety of toys. One group of the participants were told that there would be a severe punishment if they played with specific toys and were thereafter asked to rate those toys. The remaining children were told that there would be a mild punishment if they played with such toys. In addition, those participants were sub-classified into two groups: one in which each child was exposed to music (i.e., one of Mozart's sonatas) while playing with the toys, and another in which no music was played so that the efficacy of the music for reducing CD could be evaluated.

The results of the experiment were as expected: the attractiveness of a toy for the children tended to be enhanced if it was merely withdrawn temporarily from them. This tendency was observed in all three of the groups to which the present participants were randomly assigned and is consistent with previously reported findings. When forbidden to play with the toy with no exposure to music, moreover, the children in the group that had experienced a mild threat were more likely to devalue that toy than the children of the group that had experienced a severe threat. These findings are also in accordance with the following notion proposed by the classical theory of CD: when a child experienced a severe threat, his/her cognition that he/she did not play with an attractive toy was consonant with his/her cognition that he/she would have been severely punished if he had played with the toy. On the other hand, when a child refrained from playing with a toy in the absence of a severe threat, he/she experienced dissonance. His/her cognition that he/she did not play with the toy was dissonant with his/her cognition that it was attractive. To reduce this dissonance, he/she devalued the toy. Under the same circumstances, however, the children in the group who were exposed to Mozart's sonata were less likely to devalue the toy. The experience of being exposed to that music appears to have exerted an influence that acted to reconcile such CD.

## 9.11 Neurodiversity Hypothesis

As noted elsewhere (Masataka and Perlovasky 2013), CD is a discomfort caused by holding conflicting cognitions simultaneously (Cooper 2007). It usually leads to devaluation and discarding of conflicting knowledge (Festinger 1957). This theory is among the most influential and extensively studied theories in psychology. It is intimately connected to the entirety of human evolution. At the dawn of human evolution, the emergence of language led to the proliferation of CD (Perlovsky 2010, 2012). If they had not been overcome, language and knowledge would have been discarded and further human evolution would have been stopped in its tracks. This is why the ability of music to mitigate CD could be fundamental for musical cognitive function and music evolution (Perlovsky 2017). In this sense, maintenance of subsistence as a hunter-gather was only possible by executing musical activity: before going hunting or gathering, experiencing to be visually exposed to realistically drawn animals that might be encountered soon, simultaneously listening to music that might mimic landscape sounds of the animals, including their vocalizations, would serve to enhance motivation of the hunters or gatherers while suppressing their fear evoked by the possible danger of the activity itself. This effect would also serve to strengthen the psychological bond among individuals who are going to participate in the coming activity. Such reasoning could be referred to as a neurodiversity hypothesis of the Ice Age cave art.

Neurodiversity refers to the notion that seemingly "impaired" cognitive as well as emotional properties characteristic of developmental disorders such as ASD a neurodevelopmental disorder with impaired linguistic capability and unusual sensory processing, are not necessarily deficits, but fall into normal behavioral variations exhibited by humans. Stated more formally, this notion was recently described as "a concept where neurological differences are to be recognized and respected as any other human variation" (Armstrong 2012). The term was first coined in the late 1990s by New York journalist Harvey Blume and Australian autism activist Judy Singer and has become an important component of the civil rights movement for those with neurologically based disabilities.

Developmentally, the earliest non-cry sounds produced by human infants do not transmit meanings, unlike words uttered by older individuals, but rather reveal fitness or express states. We make no assumption that very young infants intend to communicate with such sounds. On the other hand, adults who receive the sounds are provided with broad reception/interpretation capabilities that allow them to provide differentiated feedback to signals if the adults are neurotypical (Masataka 2003). Such reasoning could lead us to notice the importance of considering how selection pressures might engender an increase in the tendency of an infant to produce spontaneous non-cry vocalization for arguing language evolution, and that in fact, evolution of increased infant spontaneous vocalizations began with developmental steps in individual infants, if they are neurotypical, under the selective pressure of their own neurotypical caregivers. Indeed, during modern human development, infant vocal capabilities emerge at least partly in response to social interaction, where caregivers react to vocal capabilities of infants in accordance with a scaffolding principle requiring parental discernment and intuitive parenting to reinforce vocal exploration and learning. Both endogenous inclination of infants to explore the vocal space and interactive feedback from caregivers thus foster growth in vocal capability. If this assumption about selection pressures for spontaneous vocalizations are valid, it follows that hominin neurotpical parents would also have been selected to become aware of the fitness reflected in neurotypical infant vocalizations and capable of responding to those indicators with selective care and reinforcement of vocalizations.

Taken together, the fact that proliferation of CD caused by the emergence of the language that was prevalent among the members of hunter-gatherer groups can be mitigated by artistic performance of neurodivergent (ASD) individuals, who themselves had failed to acquire languages unlike the majority of neurotypical group members, can be explained by the notion of neurodiversity. Their failure of language learning due to their impairment in the social communication overall can be regarded as a pattern of human variation. The Ice Age art and music could have been developed at the expense of the development of languages in individuals such as those with ASD as selective advantage of enhanced capabilities characteristic of this disorder, i.e., neurodiversity, that may represent a balance toward "folk physics" at the expense of "folk psychology."

**Acknowledgments** The author is greatly indebted to Takeshi Nishimura for his assistance when preparing the earlier version of the manuscript. The author is also grateful to Elizabeth Nakajima for proofreading the English of the manuscript.

**Conflict of Interest Statement** The author declares that the research was conducted in the absence of any commercial or financial relationships that could be construed as a potential conflict of interest.

## References

American Psychiatric Association (1994) Diagnostic and statistical manual of mental disorders, 4th edn. American Psychiatric Association, Washington, DC, pp 1–609
American Psychiatric Association (2013) Diagnostic and statistical manual of mental disorders, 5th edn. American Psychiatric Association, Washington, DC, pp 1–903
Applebaum E, Egel AL, Koegel RL, Imboff B (1979) Measuring musical abilities of autistic children. J Autism Dev Disord 9:279–285
Armstrong T (2012) Neurodiversity in the classroom: strength-based strategies to help students with special needs succeed in school and life. ASCD, Alexandria, pp 1–183
Armstrong T (2017) The healing balm of nature: understanding and supporting the naturalist intelligence in individuals diagnosed with ASD. Phys Life Rev 20:109–111
Aronson E, Carlsmith JM (1963) Effect of the severity of threat on the devaluation of forbidden behavior. J Abnor Soc Psychol 66:584–588
Bahn P (1988) Images of Ice Age art. Facts-on-File, New York, pp 1–364
Bahn PG (1998) *Prehistoric art*. Cambridge University Press, Cambridge, pp 1–302
Baron-Cohen S (2003) The essential difference: the truth about male and female. Basic Books, New York, pp 1–206
Boulez P (1971) Boulez on music today. Faber and Faber, London, pp 1–156
Brodar S, Brodar M (1983) Potocka zijalka: Visokoalpska postaja aurignacienskih lovcev. Razleda Slov Akad Znan Umetn 24:1–213
Chauvet J-M, Deschamps EB, Hillaire C (1996) Dawn of art: the Chauvet Cave. Abrams, New York, pp 1–135
Conard NJ, Malina M, Munzel S (2009) New flutes document the earliest musical tradition in southwestern Germany. Nature 460:737–740
Constantino JN, Hilton C, Todorov C, Law P, Zhang Y, Molloy E et al (2013) Autism recurrence in half siblings: strong support for genetic mechanisms of transmission in ASD. Mol Psychiatry 18:137–138
Cooper JS (2001) The Mozart effect. J R Soc Med 94:170–172
Cooper J (2007) Cognitive dissonance: 50 years of a classic theory. Sage, New York, pp 1–298
Cox MV (1981) One thing behind another: problems of representation in children's drawing. Educ Psychol 1:275–287
Darwin CR (1871) The descent of man, and selection in relation to sex. John Murray, London, pp 1–265
Dawson G, Meltzoff AN, Osterling J, Rinaldi J, Brown E (1998) Children with autism fail to orient to naturally occurring social stimuli. J Autism Dev Disord 28:479–485
Dawson G, Carver L, Meltzoff AN, Panagiotides H, McPartland J, Webb SJ (2002) Neural correlates of face and object recognition in young children with autism, spectrum disorder, developmental delay, and typical development. Child Dev 73:700–717
Dawson G, Toth K, Abbot R, Osterling J, Munson J, Estes A et al (2004) Early social attention impairments in autism: social orientation, joint attention, and attention to distress. Dev Psychol 40:271–283
Dawson G, Webb SJ, McPartland J (2005) Understanding the nature of face processing impairment in autism: insights from behavioral and electrophysiological studies. Dev Neuropsychol 27:403–424
Deutsch D (2006) The enigma of absolute pitch. Acoust Today 2:11–19

Dobosi VT (1985) Jewelry, musical instruments and exotic objects from the Hungarian Palaeolithic. Folia Archaeol 36:7–29
Festinger L (1957) A theory of cognitive dissonance. Stanford University Press, Stanford, pp 1–157
Freeman NH, Jenikoun R (1972) Intellectual realism in children's drawings of a familiar object with distinctive features. Child Dev 43:1116–1121
Gardner H (2011) Frame of mind: the theory of multiple intelligences. Basic Books, New York, pp 1–387
Gebauer L, Skewes J, Westphael G, Heaton P, Vuust P (2014) Intact brain processing of musical emotions in autism spectrum disorder, but more cognitive load and arousal in happy vs sad music. Front Neursci 8:192
Grandin T (1996) Thinking in pictures. Vintage, New York, pp 1–177
Grandin T, Cook K (2004) Developing talents: careers for individuals with asperger syndrome and high-functioning autism. Autism Asperger Publishing, Lenexa, pp 1–185
Hartung W (1982) Die Spielleute: Eine Randgruppe in der Gesellschaft des Mittelalters. Franz Steiner Verlag, Wiesbaden, pp 1–108
Hartung W (2003) Die Spielleute im Mittelalter: Gaukler, Dichter, *Musikanten*. Artemis & Winkler, Dusseldorf, pp 1–405
Heaton P (2009) Assessing musical skills in autistic children who are not savants. Philos Trans R Soc B 364:1443–1447
Humphrey N (1998) Cave art, autism, and the evolution of human mind. Cam Archaeol J 2:165–191
Jones W, Klim A (2013) Attention to eyes is present but in decline in 2-6-month-old infants later diagnosed with autism. Nature 504:427–431
Justus T, Hustler JJ (2005) Fundamental issues in the evolutionary psychology of music: assessing innateness and domain specificity. Music Percept 23:1–27
Kanner L (1943) Autistic disturbances in affective contact. Nerv Child 2:217–250
Leroi-Gourham A, Allain J, Balout L, Bassier C, Bouchez R, Bouchud J et al (1979) Lascoux Inconnu. Centre National de la Recherché Scientifique, Paris, pp 1–381
Lord C, Rutter M, Le Couteur A (1994) Autism diagnostic interview-revised: a revised version of a diagnostic interview for caregivers of individuals with possible pervasive developmental disorders. J Autism Dev Disord 24:659–685
Lord C, Kim SH, Dimartino A (2011) Autism spectrum disorders: general overview. In: Howlin PA, Charman T, Ghaziuddin M (eds) SAGE handbook of developmental disorders. Sage, New York, pp 287–305
Marshack A (1995) Images of the Ice Age. Archaeology 48:228–237
Masataka N (2003) The onset of language. Cambridge University Press, Cambridge, pp 1–281
Masataka N (2006) Preference for consonance over dissonance by hearing newborns of deaf parents and of hearing parents. Dev Sci 9:46–50
Masataka N (2008) The origins of language: unraveling evolutionary forces. Springer, New York, pp 1–155
Masataka N (2011) Enhancement of speech-relevant auditory acuity in absolute pitch possessors. Front Psychol 2:101
Masataka N (2017a) Implications of the idea of neurodiversity for understanding the origins of developmental disorders. Phys Life Rev 20:85–108
Masataka N (2017b) Neurodiversity, giftedness and aesthetic perceptual judgment of music in children with autism. Front Psychol 8:1595
Masataka N, Perlovasky L (2012) The efficacy of musical emotions evoked by Mozart's music for the reconciliation of cognitive dissonance. Sci Rep 2:307
Masataka N, Perlovasky L (2013) Cogntive interefernce can be mitigated by consonant music and facilitated by disoonant music. Sci Rep 3:2028
McDemott J, Hauser M (2003) The origins of music: innateness, uniqueness, and evolution. Music Percept 23:29–59

Mithen S (2007) The singing Neanderthals: the origins of music, language, mind, and body. Cambridge University Press, Cambridge, pp 1–227
Mullin J (2009) Drawing Autism. Akashic, New York, pp 1–160
Perlovsky LI (2010) Musical emotions: functions, origin, evolution. Phys Life Rev 7:2–27
Perlovsky LI (2012) Cognitive function of music, part I. Interdisci Sci Rev 37:129–142
Perlovsky L (2017) Music, passion, and cognitive function. Academic Press, New York, pp 1–186
Piaget J, Inhelder B (1956) The child's conception of space. Routlege, London, pp 1–342
Reznikoff L, Dauvois M (1988) La dimension sonore des grottes ornees. Bull Soc Prehist Franc 85:238–246
Seeberger F (2003) Steinzeit selbst erleben! Theiss, Darmstadt, pp 1–186
Selfe L (1977) Nadia: a case of extraordinary creative ability in an autistic child. Academic Press, New York, pp 1–137
Sigman MD, Kasari C, Kwon J-H, Yirmiya N (1992) Responses to the negative emotions of others by autistic, mentally retarded, and normal children. Child Dev 63:796–807
Spikins P, Wright B, Hodgson D (2016) Are there alternative adaptive strategies to human pro-sociality? The role of collaborative morality in the emergence of personality variation and autistic traits. Time Mind 9:289–313
State MW, Sestan N (2012) The emerging biology of autism spectrum disorders. Science 377:1301–1303
Stuart K (1999) Defiled trades social outcasts: honor and ritual pollution in early modern Germany. Cambridge University Press, Cambridge, pp 1–286
Thompson WF, Schellenberg EG, Husain G (2001) Arousal mood and the Mozart effect. Psychol Sci 12:248–251
Trainor L, Heinmiller BM (1998) The development of evaluative response to music: infants prefer to listen to consonance over dissonance. Infant Behav Dev 21:77–88
Walter SJ (1993) Sound and rock art. Nature 363:6429
Werner M, Dawson D, Osterling J, Dinno N (2000) Recognition of autism spectrum disorder before one year of age: a retrospective study based on home videotapes. J Autism Dev Disord 30:157–162
Winner E (2006) Development in the arts: drawing and music. In: Damon R (ed) Handbook of child psychology, vol 2. Wiley, New York, pp 859–904
World Health Organization (1994) The composite international diagnostic interview, version 1.1. World Health Organization, Geneva, pp 1–632

The manufacturer's authorised representative in the EU is Springer Nature Customer Service Centre GmbH, Europaplatz 3, 69115 Heidelberg, Germany. If you have any concerns regarding our products, please contact ProductSafety@springernature.com

Printed and bound by CPI Group (UK) Ltd, Croydon, CR0 4YY
30/09/2026
02225299-0004